CORROSION ENGINEERING HANDBOOK
SECOND EDITION

CORROSION of
POLYMERS and ELASTOMERS

CORROSION ENGINEERING HANDBOOK
SECOND EDITION

Fundamentals of Metallic Corrosion:
Atmospheric and Media Corrosion of Metals
–ISBN 978-0-8493-8243-7

Corrosion of Polymers and Elastomers
–ISBN 978-0-8493-8245-1

Corrosion of Linings and Coatings:
Cathodic and Inhibitor Protection and Corrosion Monitoring
–ISBN 978-0-8493-8247-5

CORROSION ENGINEERING HANDBOOK
SECOND EDITION

CORROSION of POLYMERS and ELASTOMERS

Philip A. Schweitzer

CRC Press
Taylor & Francis Group
Boca Raton London New York

CRC Press is an imprint of the
Taylor & Francis Group, an **informa** business

CRC Press
Taylor & Francis Group
6000 Broken Sound Parkway NW, Suite 300
Boca Raton, FL 33487-2742

First issued in paperback 2019

© 2007 by Taylor & Francis Group, LLC
CRC Press is an imprint of Taylor & Francis Group, an Informa business

No claim to original U.S. Government works

ISBN-13: 978-0-8493-8245-1 (hbk)
ISBN-13: 978-0-367-38959-8 (pbk)

Library of Congress Cataloging-in-Publication Data

Schweitzer, Philip A.
     Corrosion of polymers and elastomers / author, Philip A. Schweitzer.
          p. cm.
     Includes bibliographical references and index.
     ISBN-13: 978-0-8493-8245-1 (alk. paper)
     ISBN-10: 0-8493-8245-9 (alk. paper)
     1. Plastics--Corrosion. 2. Elastomers--Corrosion. I. Title.

TA455.P5S365 2006
620.1'9204223--dc22   2006014885

Visit the Taylor & Francis Web site at
http://www.taylorandfrancis.com

and the CRC Press Web site at
http://www.crcpress.com

Preface

Corrosion is both costly and dangerous. Billions of dollars are spent annually for the replacement of corroded structures, machinery, and components, including metal roofing, condenser tubes, pipelines, and many other items. In addition to replacement costs are those associated with maintenance to prevent corrosion, inspections, and the upkeep of cathodically protected structures and pipelines. Indirect costs of corrosion result from shutdown, loss of efficiency, and product contamination or loss.

Although the actual replacement cost of an item may not be high, the loss of production resulting from the need to shut down an operation to permit the replacement may amount to hundreds of dollars per hour. When a tank or pipeline develops a leak, product is lost. If the leak goes undetected for a period of time, the value of the lost product could be considerable. In addition, contamination can result from the leaking material, requiring cleanup, and this can be quite expensive. When corrosion takes place, corrosion products build up, resulting in reduced flow in pipelines and reduced efficiency of heat transfer in heat exchangers. Both conditions increase operating costs. Corrosion products may also be detrimental to the quality of the product being handled, making it necessary to discard valuable materials.

Premature failure of bridges or structures because of corrosion can also result in human injury or even loss of life. Failures of operating equipment resulting from corrosion can have the same disastrous results.

When all of these factors are considered, it becomes obvious why the potential problem of corrosion should be considered during the early design stages of any project, and why it is necessary to constantly monitor the integrity of structures, bridges, machinery, and equipment to prevent premature failures.

To cope with the potential problems of corrosion, it is necessary to understand

1. Mechanisms of corrosion
2. Corrosion resistant properties of various materials
3. Proper fabrication and installation techniques
4. Methods to prevent or control corrosion
5. Corrosion testing techniques
6. Corrosion monitoring techniques

Corrosion is not only limited to metallic materials but also to all materials of construction. Consequently, this handbook covers not only metallic materials but also all materials of construction.

Chapter 1 through Chapter 4 covers polymeric (plastic) materials, both thermoplastic and thermoset. An explanation is presented as to the type of corrosive effects of each polymer, its ability to withstand sun, weather, and ozone, along with compatibility tables.

Chapter 5 and Chapter 6 cover elastomeric materials in the same manner that polymers are covered in Chapter 1 through Chapter 4.

It is the intention of this book that regardless of what is being built, whether it be a bridge, tower, pipeline, storage tank, or processing vessel, information for the designer/engineer/maintenance personnel/or whoever is responsible for the selection of material of construction will be found in this book to enable them to avoid unnecessary loss of material through corrosion.

Philip A. Schweitzer

Author

Philip A. Schweitzer is a consultant in corrosion prevention, materials of construction, and chemical engineering based in York, Pennsylvania. A former contract manager and material specialist for Chem-Pro Corporation, Fairfield, New Jersey, he is the editor of the *Corrosion Engineering Handbook* and the *Corrosion and Corrosion Protection Handbook, Second Edition*; and the author of *Corrosion Resistance Tables, Fifth Edition*; *Encyclopedia of Corrosion Technology, Second Edition*; *Metallic Materials*; *Corrosion Resistant Linings and Coatings*; *Atmospheric Degradation and Corrosion Control*; *What Every Engineer Should Know About Corrosion*; *Corrosion Resistance of Elastomers*; *Corrosion Resistant Piping Systems*; *Mechanical and Corrosion Resistant Properties of Plastics and Elastomers* (all titles Marcel Dekker, Inc.), and *Paint and Coatings, Applications and Corrosion Resistance* (Taylor & Francis). Schweitzer received the BChE degree (1950) from Polytechnic University (formerly Polytechnic Institute of Brooklyn), Brooklyn, New York.

Contents

1

Introduction to Polymers

Plastics are an important group of raw materials for a wide array of manufacturing operations. Applications range from small food containers to large chemical storage tanks, from domestic water piping systems to industrial piping systems that handle highly corrosive chemicals, from toys to boat hulls, from plastic wrap to incubators, and a multitude of other products. When properly designed and applied, plastic provides light weight, sturdy/economic/resistant, and corrosion products.

Plastics are polymers. The term *plastic* is defined as "capable of being easily molded," such as putty or wet clay. The term *plastics* was originally adopted to describe the early polymeric materials because they could be easily molded. Unfortunately, many current polymers are quite brittle, and once they are formed they cannot be molded. In view of this, the term *polymer* will be used throughout the book.

There are three general categories of polymers: thermoplastic polymers called *thermoplasts*, thermosetting polymers called *thermosets*, and elastomers called *rubbers*. Thermoplasts are long-chain linear molecules that can be easily formed by heat and pressures at temperatures above a critical temperature referred to as the *glass temperature*. This term was originally applied to glass and was the temperature where glass became plastic and formed. The glass temperatures for many polymers are above room temperature; therefore, these polymers are brittle at room temperature. However, they can be reheated and reformed into new shapes and can be recycled.

Thermosets are polymers that assume a permanent shape or set when heated; although, some will set at room temperature. The thermosets begin as liquids or powders that are reacted with a second material or that through catalyzed polymerization result in a new product whose properties differ from those of either starting material. Examples of a thermoset that will set at room temperatures are epoxies that result from combining an epoxy polymer with a curing agent or catalyst at room temperature. Rather than a long-chain molecule, thermosets consist of a three dimensional network of atoms. Because they decompose on heating, they cannot be reformed or recycled. Thermosets are amorphous polymers.

Elastomers are polymeric materials whose dimensions can be drastically changed by applying a relatively modest force, but they return to their original values when the force is released. The molecules are extensively linked so that when a force is applied, they unlink or uncoil and can be extended in length by approximately 100% with a minimum force and return to their original shape when the force is released. Because their glass temperature is below room temperature, they must be cooled below room temperature to become brittle.

Polymers are the building blocks of plastics. The term is derived from the Greek meaning "many parts." They are large molecules composed of many repeat units that have been chemically bonded into long chains. Wool, silk, and cotton are examples of natural polymers.

The monomeric building blocks are chemically bonded by a process known as polymerization that can take place by one of several methods. In condensation polymerization, the reaction between monomer units or chain endgroups release a small molecule, usually water. This is an equilibrium reaction that will halt unless the by-product is removed. Polymers produced by this process will degrade when exposed to water and high temperatures.

In addition polymerization, a chain reaction appends new monomer units to the growing molecule one at a time. Each new unit creates an active site for the next attachment. The polymerization of ethylene gas (C_2H_4) is a typical example. The process begins with a monomer of ethylene gas in which the carbon atoms are joined by covalent bonds as below:

$$
\begin{array}{cc}
H & H \\
| & | \\
C & = C \\
| & | \\
H & H
\end{array}
$$

Each bond has two electrons, which satisfies the need for the s and p levels to be filled. Through the use of heat, pressure, and a catalyst, the double bonds, believed to be unsaturated, are broken to form single bonds as below:

$$
\begin{array}{cc}
& H \quad H \\
& | \quad | \\
- & C - C - \\
& | \quad | \\
& H \quad H
\end{array}
$$

This resultant structure, called a *mer*, is now free to react with other mers, forming the long-chain molecule shown below:

$$
\begin{array}{cccccccc}
H & H & H & H & H & H & H & H \\
| & | & | & | & | & | & | & | \\
-C-&C-&C-&C-&C-&C-&C-&C- \\
| & | & | & | & | & | & | & | \\
H & H & H & H & H & H & H & H
\end{array}
$$

Most addition polymerization reactions follow a method of chain growth where each chain, once initiated, grows at an extremely rapid rate until terminated. Once terminated, it cannot grow any more except by side reactions.

The year 1868 marked the beginning of the polymer industry with the production of celluloid that was produced by mixing cellulose nitrate with camphor. This produced a molded plastic material that became very hard when dried. Synthetic polymers appeared in the early twentieth century when Leo Bakeland invented Bakelite by combining the two monomers, phenol and formaldehyde. An important paper published by Staudinger in 1920 proposed chain formulas for polystyrene and poloxmethylene. In 1953, he was awarded the Nobel prize for this work in establishing polymer science. In 1934, W.H. Carothers demonstrated that chain polymers could be formed by condensation reactions that resulted in the invention of nylon through polymerization of hexamethylenediamine and adipic acid. Commercial nylon was placed on the market in 1938 by the DuPont Company. By the late 1930s, polystyrene, polyvinyl chloride (PVC), and polymethyl methacrylate (Plexiglass) were in commercial production.

Further development of linear condensation polymers resulted from the recognition that natural fibers such as rubber, sugars, and cellulose were giant molecules of high molecular weight. These are natural condensation polymers, and understanding their structure paved the way for the development of the synthetic condensation polymers such as polyesters, polyamides, polyimides, and polycarbonates. The chronological order of the development of polymers is shown in Table 1.1.

A relatively recent term, *engineering polymers*, has come into play. It has been used interchangeably with the terms *high-performance polymers* and *engineering plastics*. According to the ASM Handbook, engineering plastics are defined as "Synthetic polymers of a resin-based material that have load-bearing characteristics and high-performance properties which permit them to be used in the same manner as metals and ceramics." Others have limited the term to thermoplastics only. Many engineering polymers are reinforced and/or alloy polymers (a blend of polymers). Polyethylene, polypropylene, PVC, and polystyrene, the major products of the polymer industry, are not considered engineering polymers.

Reinforced polymers are those to which fibers have been added that increase the physical properties—especially impact resistance and heat deflection temperatures. Glass fibers are the most common additions, but carbon, graphite, aramid, and boron fibers are also used. In a reinforced polymer, the resin matrix is the continuous phase, and the fiber reinforcement is the discontinuous phase. The function of the resin is to bond the fibers together to provide shape and form and to transfer stresses in the structure from the resin to the fiber. Only high-strength fibers with high modulus are used. Because of the increased stiffness resulting from the fiber

TABLE 1.1

Chronological Development of Polymers

Year	Material
1868	Celluloid
1869	Cellulose nitrate, cellulose propionate, ethyl cellulose
1907	PF resin (Bakelite)
1912	Cellulose acetate, vinyl plastics
1919	Glass-bonded mica
1926	Alkyl polyester
1928	Polyvinyl acetate
1931	Acrylic plastics
1933	Polystyrene plastics, ABS plastics
1937	Polyester-reinforced urethane
1938	Polyamide plastics (nylon)
1940	Polyolefin plastics, polyvinyl aldehyde, PVC, PVC plastisols
1942	Unsaturated polyester
1943	Fluorocarbon resins (Teflon), silicones, polyurethanes
1947	Epoxy resins
1948	Copolymers of butadiene and styrene (ABS)
1950	Polyester fibers, polyvinylidene chloride
1954	Polypropylene plastic
1955	Urethane
1956	POM (acetals)
1957	PC (polycarbonate)
1961	Polyvinylidene fluoride
1962	Phenoxy plastics, polyallomers
1964	Polyimides, polyphenylene oxide (PPO)
1965	Polysulfones, methyl pentene polymers
1970	Polybutylene terephthalate (PBT)
1971	Polyphenylene sulfide
1978	Polyarylate (Ardel)
1979	PET–PC blends (Xenoy)
1981	Polyether block amides (Pebax)
1982	Polyetherether ketone (PEEK)
1983	Polyetheramide (Ultem)
1984	Liquid crystal polymers (Xydar)
1985	Liquid crystal polymers (Vectra)
1988	PVC–SMA blend

reinforcement, these polymers that are noted for their flexibility are not normally reinforced.

Virtually all thermosetting polymers can be reinforced with fibers. Polyester resins are particularly useful on reinforced polymers. They are used extensively in manufacturing very large components such as swimming pools, large tankage, boat hulls, shower enclosures, and building components. Reinforced molding materials such as phenolics, alkyls, or epoxies are extensively used in the electronics industry.

Many thermoplastic polymers are reinforced with fibers. Reinforcement is used to improve physical properties—specifically heat deflection temperature. Glass fibers are the most commonly used reinforcing material. The wear resistance and abrasion resistance of the thermoplastics polymers are improved by the use of aramid reinforcing. Although fibers can be used with any thermoplastics polymer, the following are the most important:

1. Polyamide polymers use glass fiber to control brittleness. Tensile strengths are increased by a factor of 3 and heat deflection temperature increases from 150 to 500°F (66 to 260°C).

2. Polycarbonate compounds using 10, 20, 30, and 40% glass–fiber loading have their physical properties greatly improved.

3. Other polymers benefiting from the addition of glass fibers include polyphenylene sulfide, polypropylene, and polyether sulfone.

Polymers chosen for structural application are usually selected as a replacement for metal. A like replacement of a polymer section for a metallic section will result in a weight savings. In addition, polymers can be easily formed into shapes that are difficult to achieve with metals. By using a polymer, the engineer can design an attractive shape that favors plastic forming and achieve a savings in cost and weight and a cosmetic-improvement. An additional cost savings is realized since the polymer part does not require painting for corrosion protection as would the comparable metal part. Selection of the specific polymer will be based on the mechanical requirements, the temperature, and the chemical enduse environment.

1.1 Additives

Various properties of polymers may be improved by the use of additives. In some instances, the use of additives to improve a specific property may have an adverse effect on certain other properties. Corrosion resistance is a property most often affected by the use of additives. In many cases a polymer's corrosion resistance is reduced as a result of additives being used. Common additives used to improve the performance of thermoplastic polymers are listed here:

Antioxidants	Protect against atmospheric oxidation
Colorants	Dyes and pigments
Coupling agents	Used to improve adhesive bonds
Fillers or extenders	Minerals, metallic powders, and organic compounds used to improve specific properties or to reduce costs

Flame retardants	Change the chemistry/physics of combustion
Foaming agents	Generate cells or gas pockets
Impact modifiers	Materials usually containing an elastomeric compound to reduce brittleness
Lubricants	Substances that reduce friction, heat, and wear between surfaces
Optical brighteners	Organic substances that absorb UV radiation below 3000 Å and emit radiation below 5500 Å
Plasticizers	Increase workability
Reinforcing fibers	Increase strength, modulus, and impact strength
Processing aids	Improve hot processing characteristics
Stabilizers	Control for adjustment of deteriorative physico-chemical reactions during processing and subsequent life

A list of specific fillers and the properties they improve is given in Table 1.2. Many thermoplastic polymers have useful properties without the need for additives. However, other thermoplasts require additives to be useful. For example, PVC benefits from all additives and is practically useless in its pure form. Examples of the effects of additives on specific polymers will be illustrated.

Impact resistance is improved in polybutylene terephathalates, polypropylene, polycarbonate, PVC, acetals (POM), and certain polymer blends by the use of additives. Figure 1.1 shows the increase in impact strength of nylon, polycarbonate, polypropylene, and polystyrene by the addition of 30 wt% of glass fibers.

Glass fibers also increase the strength and moduli of thermoplastic polymers. Figure 1.2 and Figure 1.3 illustrate the effect on the tensile stress and flexural moduli of nylon, polycarbonate, polypropylene, and polystyrene when 30 wt% glass fiber additions have been made.

The addition of 20 wt% of glass fibers also increases the heat distortion temperature. Table 1.3 shows the increase in the HDT when glass fibers have been added to polymers.

1.2   Permeation

All materials are somewhat permeable to chemical molecules, but plastic materials tend to be an order of magnitude greater in their permeability than metals. Gases, vapors, or liquids will permeate polymers. Permeation is molecular migration through microvoids either in the polymer (if the polymer is more or less porous) or between polymer molecules. In neither case is there an attack on the polymer. This action is strictly a physical phenomenon. However, permeation can be detrimental when a polymer is

TABLE 1.2

Fillers and Their Property Contribution to Polymers

Filler	Chemical Resistance	Heat Resistance	Electrical Insulation	Impact Strength	Tensile Strength	Dimensional Stability	Stiffness	Hardness	Electrical Conductivity	Thermal Conductivity	Moisture Resistance	Hardenability
Alumina powder										X	X	
Alumina tetrahydrate			X								X	X
Bronze									X	X		X
Calcium carbonate		X				X	X	X				X
Calcium silicate		X					X	X	X			
Carbon black		X					X	X	X	X		X
Carbon fiber							X		X	X		
Cellulose				X	X	X	X	X				
Alpha cellulose			X		X							
Coal, powdered	X										X	
Cotton, chopped fibers	X		X	X	X		X	X				
Fibrous glass	X	X	X	X	X	X	X	X			X	
Graphite	X				X	X	X	X	X	X		
Jute				X			X					
Kaolin	X	X				X	X	X			X	X
Mica	X	X	X			X	X	X			X	
Molybdenium disulfide	X					X	X	X			X	X
Nylon, chopped fibers	X	X	X	X	X	X	X	X				X
Orlon	X	X	X	X	X	X	X	X		X	X	X
Rayon			X	X	X							
Silica, amorphous			X			X	X	X			X	X
TFE						X	X				X	X
Talc	X	X	X			X		X			X	X
Wood flour	X		X		X	X						X

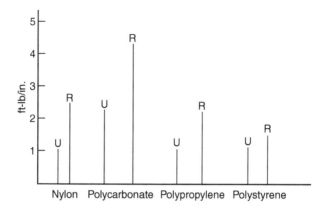

FIGURE 1.1
Izod impact change with glass reinforcement of thermoplastic polymers. U, unreinforced;
R, reinforced.

used to line piping or equipment. In lined equipment, permeation can
result in:

1. Failure of the substrate from corrosive attack.
2. Bond failure and blistering, resulting from the accumulation of
   fluids at the bond when the substrate is less permeable than the
   liner or from corrosion/reaction products if the substrate is
   attacked by the permeant.
3. Loss of contents through substrate and liner as a result of the
   eventual failure of the substrate. In unbonded linings, it is

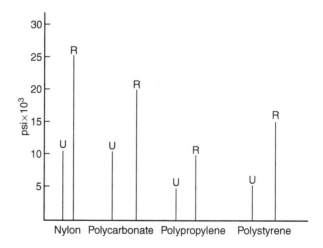

FIGURE 1.2
Increase in tensile strength with glass reinforcement of thermoplastic polymers. U, unreinforced;
R, reinforced.

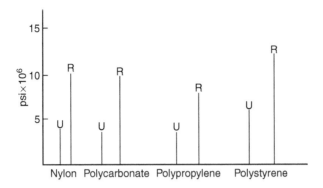

FIGURE 1.3
Increase in flexural modulus with reinforcement of thermoplastic polymers. U, unreinforced; R, reinforced.

important that the space between the liner and support member be vented to the atmosphere, not only to allow minute quantities of permeant vapors to escape, but also to prevent expansion of entrapped air from collapsing the liner.

Permeation is a function of two variables: one relating to diffusion between molecular chains and the other to the solubility of the permeant in the polymer. The driving force of diffusion is the partial pressure of gases and the concentration gradient of liquids. Solubility is a function of the affinity of the permeant for the polymer.

All polymers do not have the same rate of permeation. In fact, some polymers are not affected by permeation. The fluoro-polymers are

TABLE 1.3

Increase of HDT with 20 wt% Glass Fiber Addition to the Polymer

Polymer	HDT at 264 psi 20% Glass (°F/°C)	Increase over Base Polymer (°F/°C)
Acetal copolymer	325/163	95/52
Polypropylene	250/121	110/61
Linear polyethylene	260/127	140/77
Thermoplastic polyester	400/207	230/139
Nylon 6[a]	425/219	305/168
Nylon 6/6[a]	490/254	330/183
ABS	215/102	25/14
Styrene–acrylonitrile	215/102	20/12
Polystyrene	220/104	20/12
Polycarbonate	290/143	20/12
Polysulfone	365/185	20/12

[a] 30 wt% glass fibers.

TABLE 1.4

Vapor Permeation into PTFE[a]

Gases	Permeation g/100 in.^{2}/24 h/mll	
	73°F/23°C	86°F/30°C
Carbon dioxide		0.66
Helium		0.22
Hydrogen chloride, anhydrous		<0.01
Nitrogen		0.11
Acetophenone	0.56	
Benzene	0.36	0.80
Carbon tetrachloride	0.06	
Ethyl alcohol	0.13	
Hydrochloric acid, 20%	<0.01	
Piperdine	0.07	
Sodium hydroxide, 50%	5×10^{-5}	
Sulfuric acid, 98%	1.8×10^{-5}	

[a] Based on PTFE having a specific gravity of >2.2.

particularly affected. Vapor permeation of PTFE is shown in Table 1.4 while Table 1.5 shows the vapor permeation of FEP. Table 1.6 provides permeation data of various gases into PFA and Table 1.7 gives the relative gas permeation into fluoropolymers.

TABLE 1.5

Vapor Permeation into FEP

	Permeation (g/100 in.^{2}/24 h/mil) at		
	73°F/23°C	93°F/35°C	122°F/50°C
Gases			
Nitrogen	0.18		
Oxygen	0.39		
Vapors			
Acetic acid		0.42	
Acetone	0.13	0.95	3.29
Acetophenone	0.47		
Benzene	0.15	0.64	
n-Butyl ether	0.08		
Carbon tetrachloride	0.11	0.31	
Decane	0.72		1.03
Ethyl acetate	0.06	0.77	2.9
Ethyl alcohol	0.11	0.69	
Hexane		0.57	
Hydrochloric acid 20%	<0.01		
Methanol			5.61
Sodium hydroxide	4×10^{-5}		
Sulfuric acid 98%	8×10^{-6}		
Toluene	0.37		2.93

TABLE 1.6

Permeation of Gases into PFA

Gas	Permeation at 77°F/25°C (cc/mil thickness/100 in.^{2}/24 h/atm)
Carbon dioxide	2260
Nitrogen	291
Oxygen	881

There is no relationship between permeation and passage of materials through cracks and voids; although, in both cases, migrating chemicals travel through the polymer from one side to the other.

Some control can be exercised over permeation that is affected by:

1. Temperature and pressure
2. The permeant concentration
3. The thickness of the polymer

Increasing the temperature will increase the permeation rate because the solubility of the permeant in the polymer will increase, and as the temperature rises, polymer chain movement is stimulated, permitting more permeant to diffuse among the chains more easily.

The permeation rates of many gases increase linearly with the partial-pressure gradient, and the same effect is experienced with the concentration of gradients of liquids. If the permeant is highly soluble in the polymer, the permeability increase may be nonlinear. The thickness will generally decrease permeation by the square of the thickness.

The density of the polymer as well as the thickness will have an effect on the permeation rate. The greater the density of the polymer, the fewer voids through which permeation can take place. A comparison of the density of sheets produced from different polymers does not provide an indication of the relative permeation rates. However, a comparison of the sheets' density

TABLE 1.7

Relative Gas Permeation into Fluoropolymers[a]

Gas	PVDF	PTFE	FEP	PFA
Air	27	2000	600	1150
Oxygen	20	1500	2900	—
Nitrogen	30	500	1200	—
Helium	600	35,000	18,000	17,000
Carbon dioxide	100	15,000	4700	7000

[a] Permeation through a 100 μm film at 73°F/23°C. Units=cm^3/m^2 deg bar.

produced from the same polymer will provide an indication of the relative permeation rates. The denser the sheet, the lower the permeation rate.

The thickness of the liner is a factor affecting permeation. For general corrosion resistance, thicknesses of 0.010–0.020 in. are usually satisfactory, depending on the combination of lining material and the specific corrodent. When mechanical factors such as thinning to cold flow, mechanical abuse, and permeation rates are a consideration, thicker linings may be required.

Increasing a lining thickness will normally decrease permeation by the square of the thickness. Although this would appear to be the approach to follow to control permeation, there are some disadvantages. First, as thickness increases, the thermal stresses on the boundary increase that can result in bond failure. Temperature changes and large differences in coefficients of thermal expansion are the most common causes of bond failure. The plastic's thickness and modulus of elasticity are two of the factors that influence these stresses. Second, as the thickness of the lining increases, installation becomes more difficult with a resulting increase in labor costs.

The rate of permeation is also affected by the temperature and the temperature gradient in the lining. Lowering these will reduce the rate of permeation. Lined vessels, such as storage tanks, that are used under ambient conditions provide the best service.

Other factors affecting permeation consist of these chemical and physiochemical properties:

1. Ease of condensation of the permeant. Chemicals that readily condense will permeate at higher rates.

2. The higher the intermolecular chain forces (e.g., van der Waals hydrogen bonding) of the polymer, the lower the permeation rate.

3. The higher the level of crystallinity in the polymer, the lower the permeation rate.

4. The greater the degree of cross-linking within the polymer, the lower the permeation rate.

5. Chemical similarity between the polymer and permeant when the polymer and permeant both have similar functional groups, the permeation rate will increase.

6. The smaller the molecule of the permeant, the greater the permeation rate.

1.3 Absorption

Polymers have the potential to absorb varying amounts of corrodents they come into contact with, particularly organic liquids. This can result in swelling, cracking, and penetration to the substrate of a lined component.

Swelling can cause softening of the polymer, introduce high stresses, and cause failure of the bond on lined components. If the polymer has a high absorption rate, permeation will probably take place. An approximation of the expected permeation and/or absorption of a polymer can be based on the absorption of water. This data is usually available. Table 1.8 provides the water absorption rates for the more common polymers. Table 1.9 gives the absorption rates of various liquids by FEP, and Table 1.10 provides the absorption rates of representative liquids by PFA.

The failures because of absorption can best be understood by considering the "steam cycle" test described in ASTM standards for lined pipe. A section of lined pipe is subjected to thermal and pressure fluctuations. This is repeated for 100 cycles. The steam creates a temperature and pressure gradient through the liner, causing absorption of a small quantity of steam that condenses to water within the inner wall. Upon pressure release or on the reintroduction of steam, the entrapped water can expand to vapor, causing an original micropore. The repeated pressure and thermal cycling enlarge the micropores, ultimately producing visible water-filled blisters within the liner.

In an actual process, the polymer may absorb process fluids, and repeated temperature or pressure cycling can cause blisters. Eventually, the corrodent may find its way to the substrate.

Related effects occur when process chemicals are absorbed that may later react, decompose, or solidify within the structure of the polymer. Prolonged retention of the chemicals may lead to their decomposition within the polymer. Although it is unusual, it is possible for absorbed monomers to polymerize.

Several steps can be taken to reduce absorption. Thermal insulation of the substrate will reduce the temperature gradient across the vessel, thereby

TABLE 1.8

Water Absorption Rates of Polymers

Polymer	Water Absorption 24h at 73°F/23°C (%)
PVC	0.05
CPVC	0.03
PP (Homo)	0.02
PP (Co)	0.05
EHMW PE	<0.01
ECTFE	<0.01
PVDF	<0.04
PVCP (Saran)	Nil
PFA	<0.03
ETFE	0.029
PTFE	<0.01
FEP	<0.01

TABLE 1.9

Absorption of Selected Liquids by FEP[a]

Chemical	Temp. (°F/°C)	Range of Weight Gains (%)
Aniline	365/185	0.3–0.4
Acetophenone	394/201	0.6–0.8
Benzaldehyde	354/179	0.4–0.5
Benzyl alcohol	400/204	0.3–0.4
n-Butylamine	172/78	0.3–0.4
Carbon tetrachloride	172/78	2.3–2.4
Dimethyl sulfide	372/190	0.1–0.2
Nitrobenzene	410/210	0.7–0.9
Perchlorethylene	250/121	2.0–2.3
Sulfuryl chloride	154/68	1.7–2.7
Toluene	230/110	0.7–0.8
Tributyl phosphate	392/200[b]	1.8–2.0

[a] Exposure for 168 hours at their boiling points.
[b] Not boiling.

TABLE 1.10

Absorption of Representative Liquids in PFA

Liquid[a]	Temp. (°F/°C)	Range of Weight Gains (%)
Aniline	365/185	0.3–0.4
Acetophenone	394/201	0.6–0.8
Benzaldehyde	354/179	0.4–0.5
Benzyl alcohol	400/204	0.3–0.4
n-Butylamine	172/78	0.3–0.4
Carbon tetrachloride	172/78	2.3–2.4
Dimethylsulfoxide	372/190	0.1–0.2
Freon 113	117/47	1.2
Isooctane	210/99	0.7–0.8
Nitrobenzene	410/210	0.7–0.9
Perchlorethylene	250/121	2.0–2.3
Sulfuryl chloride	154/68	1.7–2.7
Toluene	230/110	0.7–0.8
Tributyl phosphate	392/200[b]	1.8–2.0
Bromine, anh.	−5/−22	0.5
Chlorine, anh.	248/120	0.5–0.6
Chlorosulfonic acid	302/150	0.7–0.8
Chromic acid 50%	248/120	0.00–0.01
Ferric chloride	212/100	0.0–0.01
Hydrochloric acid 37%	248/120	0.0–0.03
Phosphoric acid, conc.	212/100	0.00–0.01
Zinc chloride	212/100	0.00–0.03

[a] Liquids were exposed for 168 h at the boiling point of the solvents. The acidic reagents were exposed for 168 h.
[b] Not boiling.

TABLE 1.11

Wavelength Regions of the UV

Region	Wavelength (nm)	Characteristics
UV-A	400–315	Causes polymer damage
UV-B	315–200	Includes the shortest wavelengths found at the earth's surface
		Causes severe polymer damage
		Absorbed by window glass
UV-C	280–100	Filtered out by the earth's atmosphere
		Found only in outer space

preventing condensation and subsequent expansion of the absorbed fluids. This also reduces the rate and magnitude of temperature changes, keeping blisters to a minimum. The use of operating procedures or devices that limit the ratio of process pressure reductions or temperature increases will provide added protection.

1.4 Painting of Polymers

Polymers are painted because this is frequently a less expensive process than precolored resins or molded-in coloring. They are also painted when necessary to provide UV protection. However, they are difficult to paint, and proper consideration must be given to the following:

1. Heat distortion point and heat resistance. This determines whether a bake-type paint can be used, and if so, the maximum baking temperature the polymer can tolerate.

2. Solvent resistance. Because different polymers are subject to attack by different solvents, this will dictate the choice of the paint system. Some softening of the surface is desirable to improve adhesion, but a solvent that aggressively attacks the surface and results in cracking or crazing must be avoided.

3. Residual stress. Molded parts may have localized areas of stress. A coating applied in these areas may swell the polymer and cause crazing. Annealing of the part prior to coating will minimize or eliminate the problem.

4. Mold-release residues. If excessive amounts of mold-release compounds remain on the part, adhesion problems are likely. To prevent such a problem, the polymer should be thoroughly rinsed or otherwise cleaned.

5. Plasticizers and other additives. Most polymers are formulated with plasticizers and chemical additives. These materials have a tendency to migrate to the surface and may even soften the coating and destroy adhesion. The specific polymer should be checked to determine whether the coating will cause short- or long-term softening or adhesion problems.

6. Other factors. The long-term adhesion of the coating is affected by the properties of the polymer such as stiffness or rigidity, dimensional stability, and coefficient of expansion. The physical properties of the paint film must accommodate to those of the polymer.

1.5 Corrosion of Polymers

Corrosion of metallic materials takes place via an electro-chemical reaction at a specific corrosion rate. Consequently, the life of a metallic material in a particular corrosive environment can be accurately predicted. This is not the case with polymeric materials.

Polymeric materials do not experience specific corrosion rates. They are usually completely resistant to a specific corrodent (within specific temperature ranges) or they deteriorate rapidly. Polymers are attacked either by chemical reaction or solvation. Solvation is the penetration of the polymer by a corrodent that causes swelling, softening, and ultimate failure. Corrosion of plastics can be classified in the following ways as to attack mechanism:

1. Disintegration or degradation of a physical nature because of absorption, permeation, solvent action, or other factors

2. Oxidation, where chemical bonds are attacked

3. Hydrolysis, where ester linkages are attacked

4. Radiation

5. Thermal degradation involving depolymerization and possibly repolymerization

6. Dehydration (rather uncommon)

7. Any combination of the above

Results of such attacks will appear in the form of softening, charring, crazing, delamination, embrittlement, discoloration, dissolving, or swelling.

The corrosion of polymer matrix composites is also affected by two other factors: the nature of the laminate, and in the case of thermoset resins, the cure. Improper or insufficient cure will adversely affect the corrosion resistance, whereas proper cure time and procedures will generally improve corrosion resistance.

Polymeric materials in outdoor applications are exposed to weather extremes that can be extremely deleterious to the material. The most harmful weather component, exposure to ultraviolet (UV) radiation, can cause embrittlement fading, surface cracking, and chalking. After exposure to direct sunlight for a period of years, most polymers exhibit reduced impact resistance, lower overall mechanical performance, and a change in appearance.

The electromagnetic energy from the sunlight is normally divided into ultraviolet light, visible light, and infrared energy. Infrared energy consists of wavelengths longer than visible red wavelengths and starts above 760 nm. Visible light is defined as radiation between 400 and 760 nm. Ultraviolet light consists of radiation below 400 nm. The UV portion of the spectrum is further subdivided into UV-A, UV-B, and UV-C. The effects of the various wavelengths are shown in Table 1.11.

Because UV is easily filtered by air masses, cloud cover, pollution, and other factors, the amount and spectrum of natural UV exposure is extremely variable. Because the sun is lower in the sky during the winter months, it is filtered through a greater air mass. This creates two important differences between summer and winter sunlight: changes in the intensity of the light and in the spectrum. During winter months, much of the damaging short-wavelength UV light is filtered out. For example, the intensity of UV at 320 nm changes about 8 to 1 from summer to winter. In addition, the short-wavelength solar cutoff shifts from approximately 295 nm in summer to approximately 310 nm in winter. As a result, materials sensitive to UV below 320 nm would degrade only slightly, if at all, during the winter months.

Photochemical degradation is caused by photons or light breaking chemical bonds. For each type of chemical bond, there is a critical threshold wavelength of light with enough energy to cause a reaction. Light of any wavelength shorter than the threshold can break a bond, but longer wavelengths cannot break it. Therefore, the short wavelength cutoff of a light source is of critical importance. If a particular polymer is only sensitive to UV light below 290 nm (the solar cutoff point), it will never experience photochemical deterioration outdoors.

The ability to withstand weathering varies with the polymer type and within grades of a particular resin. Many resin grades are available with UV-absorbing additives to improve weather-ability. However, the higher molecular weigh grades of a resin generally exhibit better weatherability than the lower molecular weight grades with comparable additives. In addition, some colors tend to weather better than others.

Many of the physical property and chemical resistance differences of polymers stem directly from the type and arrangement of atoms in the polymer chains. In the periodic table, the basic elements of nature are placed into classes with similar properties, i.e., elements and compounds that exhibit similar behavior. These classes are alkali metals, alkaline earth metals, transition metals, rare earth series, other metals, nonmetals, and noble (inert) gases.

Of particular importance and interest for thermoplasts is the category known as halogens that are found in the nonmetal category. The elements included in this category are fluorine, chlorine, bromine, and iodine. Since these are the most electro-negative elements in the periodic table, they are the most likely to attract an electron from another element and become part of a stable structure. Of all the halogens, fluorine is the most electronegative, permitting it to strongly bond with carbon and hydrogen atoms, but not well with itself. The carbon–fluorine bond, the predominant bond in PVDF and PTFE that gives it such important properties, is among the strongest known organic compounds. The fluorine acts as a protective shield for other bonds of lesser strength within the main chain of the polymer. The carbon–hydrogen bond, that such plastics as PPE and PP are composed, is considerably weaker. This class of polymers is known as polyolefins. The carbon–chlorine bond, a key bond of PVC, is weaker yet.

The arrangement of elements in the molecule, the symmetry of the structure, and the polymer chains' degree of branching are as important as the specific elements contained in the molecule. Polymers containing the carbon–hydrogen bonds such as polypropylene and polyethylene, and the carbon–chlorine bonds such as PVC and ethylene chlorotrifluoroethylene are different in the important property of chemical resistance from a fully fluorinated polymer such as polytetrafluoroethylene. The latter has a much wider range of corrosion resistance.

The fluoroplastic materials are divided into two groups: fully fluorinated fluorocarbon polymers such as PTFE, FEP, and PPA called perfluoropolymers, and the partially fluorinated polymers such as ETFE, PVDF, and ECTFE that are called fluoropolymers. The polymeric characteristics within each group are similar, but there are important differences between the groups as will be seen later.

2

Thermoplastic Polymers

Thermoplastic materials can be repeatedly re-formed by the application of heat, similar to metallic materials. They are long-chain linear molecules that are easily formed by the application of heat and pressure at temperatures above a critical temperature referred to as the _glass temperature_. Because of this ability to be re-formed by heat, these materials can be recycled. However, thermal aging that results from repeated exposure to the high temperatures required for melting causes eventual degradation of the polymer and limits the number of reheat cycles.

Polymers are formed as the result of a polymerization reaction of a monomer that is a single molecule or substance consisting of single molecules. Copolymers are long-chain molecules formed by the addition reaction of two or more monomers. In essence, they are chains where one mer has been substituted with another mer. When the chain of a polymer is made up of a single repeating section, it is referred to as a homopolymer in contrast to a copolymer. Thermoplastic polymers can be either homopolymers or copolymers. Alloy polymers are blends of different polymers.

In general, thermoplastic materials tend to be tougher and less brittle than thermoset polymers, so they can be used without the need for incorporating fillers. However, all thermoplasts do not fall into this category. Some tend to craze or crack easily, so each case must be considered on its individual merits. By virtue of their basic structure, thermoplastics have been less dimensionally and thermally stable than thermosetting polymers. Therefore, thermosets have offered a performance advantage; although, the lower processing costs for thermoplastics have given the latter a cost advantage. Because of three major developments, thermosets and thermoplastics are now considered on the basis of their performance. First, stability of thermoplastics has been greatly improved by the use of fiber reinforcement. Second, has been the development of the so-called engineering or high-stability, higher-performance polymers that can be reinforced with fiber filler to increase their stability even further. Third, to offset the gains in thermoplastics, lower-cost processing of thermoset polymers, has been developed specifically the screw-injection-molding technology.

The two most common materials used as reinforcement are glass fiber and carbon. When a reinforcing material is used in a thermoplast, the thermoplast is known as a composite. Compatibility of the reinforcing material with the corrodent must be checked as well as the compatibility of the thermoplast. Table 2.1 provides the compatibility of glass fibers with selected corrodents, and Table 2.2 provides the compatibility of carbon fiber

TABLE 2.1

Compatability of Borosilicate Glass with Selected Corrodents

	Maximum Temperature	
Chemical	°F	°C
Acetaldehyde	450	232
Acetamide	270	132
Acetic acid, 10%	400	204
Acetic acid, 50%	400	204
Acetic acid, 80%	400	204
Acetic acid, glacial	400	204
Acetic anhydride	250	121
Acetone	250	121
Adipic acid	210	99
Allyl alcohol	120	49
Allyl chloride	250	121
Alum	250	121
Aluminum chloride, aqueous	250	121
Aluminum chloride, dry	180	82
Aluminum fluoride	X	X
Aluminum hydroxide	250	121
Aluminum nitrate	100	38
Aluminum oxychloride	190	88
Aluminum sulfate	250	121
Ammonium bifluoride	X	X
Ammonium carbonate	250	121
Ammonium chloride, 10%	250	121
Ammonium chloride, 50%	250	121
Ammonium chloride, sat.	250	121
Ammonium fluoride, 10%	X	X
Ammonium fluoride, 25%	X	X
Ammonium hydroxide, 25%	250	121
Ammonium hydroxide, sat.	250	121
Ammonium nitrate	200	93
Ammonium persulfate	200	93
Ammonium phosphate	90	32
Ammonium sulfate, 10–40%	200	93
Amyl acetate	200	93

(continued)

TABLE 2.1 *Continued*

Chemical	Maximum Temperature	
	°F	°C
Amyl alcohol	250	121
Amyl chloride	250	121
Aniline	200	93
Antimony trichloride	250	121
Aqua regia 3:1	200	93
Barium carbonate	250	121
Barium chloride	250	121
Barium hydroxide	250	121
Barium sulfate	250	121
Barium sulfide	250	121
Benzaldehyde	200	93
Benzene	200	93
Benzene sulfonic acid, 10%	200	93
Benzoic acid	200	93
Benzyl alcohol	200	93
Benzyl chloride	200	93
Borax	250	121
Boric acid	300	149
Bromine gas, moist	250	121
Bromine liquid	90	32
Butadiene	90	32
Butyl acetate	250	121
Butyl alcohol	200	93
Butyric acid	200	93
Calcium bisulfite	250	121
Calcium carbonate	250	121
Calcium chlorate	200	93
Calcium chloride	200	93
Calcium hydroxide, 10%	250	121
Calcium hydroxide, sat.	X	X
Calcium hypochlorite	200	93
Calcium nitrate	100	38
Carbon bisulfide	250	121
Carbon dioxide, dry	160	71
Carbon dioxide, wet	160	71
Carbon disulfide	250	121
Carbon monoxide	450	232
Carbon tetrachloride	200	93
Carbonic acid	200	93
Cellosolve	160	71
Chloracetic acid, 50% water	250	121
Chloracetic acid	250	121
Chlorine gas, dry	450	232
Chlorine gas, wet	400	204
Chlorine, liquid	140	60
Chlorobenzene	200	93
Chloroform	200	93
Chlorosulfonic acid	200	93

(continued)

TABLE 2.1 *Continued*

Chemical	Maximum Temperature	
	°F	°C
Chromic acid, 10%	200	93
Chromic acid, 50%	200	93
Citric acid, 15%	200	93
Citric acid, conc.	200	93
Copper chloride	250	121
Copper sulfate	200	93
Cresol	200	93
Cupric chloride, 5%	160	71
Cupric chloride, 50%	160	71
Cyclohexane	200	93
Cyclohexanol		
Dichloroacetic acid	310	154
Dichloroethane (ethylene dichloride)	250	121
Ethylene glycol	210	99
Ferric chloride	290	143
Ferric chloride, 50% in water	280	138
Ferric nitrate, 10–50%	180	82
Ferrous chloride	200	93
Fluorine gas, dry	300	149
Fluorine gas, moist	X	X
Hydrobromic acid, dil.	200	93
Hydrobromic acid, 20%	200	93
Hydrobromic acid, 50%	200	93
Hydrochloric acid, 20%	200	93
Hydrochloric acid, 38%	200	93
Hydrocyanic acid, 10%	200	93
Hydrofluoric acid, 30%	X	X
Hydrofluoric acid, 70%	X	X
Hydrofluoric acid, 100%	X	X
Hypochlorous acid	190	88
Iodine solution, 10%	200	93
Ketones, general	200	93
Lactic acid, 25%	200	93
Lactic acid, conc.	200	93
Magnesium chloride	250	121
Malic acid	160	72
Methyl chloride	200	93
Methyl ethyl ketone	200	93
Methyl isobutyl ketone	200	93
Nitric acid, 5%	400	204
Nitric acid, 20%	400	204
Nitric acid, 70%	400	204
Nitric acid, anhydrous	250	121
Oleum	400	204
Perchloric acid, 10%	200	93
Perchloric acid, 70%	200	93
Phenol	200	93

(continued)

TABLE 2.1 *Continued*

| Chemical | Maximum Temperature | |
	°F	°C
Phosphoric acid, 50–80%	300	149
Picric acid	200	93
Potassium bromide, 30%	250	121
Silver bromide, 10%		
Sodium carbonate	250	121
Sodium chloride	250	121
Sodium hydroxide, 10%	X	X
Sodium hydroxide, 50%	X	X
Sodium hydroxide, conc.	X	X
Sodium hypochloride, 20%	150	66
Sodium hypochloride, conc.	150	66
Sodium sulfide, to 50%	X	X
Stannic chloride	210	99
Stannous chloride	210	99
Sulfuric acid, 10%	400	204
Sulfuric acid, 50%	400	204
Sulfuric acid, 70%	400	204
Sulfuric acid, 90%	400	204
Sulfuric acid, 98%	400	204
Sulfuric acid, 100%	400	204
Sulfurous acid	210	99
Thionyl chloride	210	99
Toluene	250	121
Trichloroacetic acid	210	99
White liquor	210	99
Zinc chloride	210	99

The chemicals listed are in the pure state or in a saturated solution unless otherwise indicated. Compatibility is shown to the maximum allowable temperature for which data is available. Incompatibility is shown by an X. A blank space indicates that data is unavailable.

Source: From P.A. Schweitzer. 2004. *Corrosion Resistance Tables*, Vols. 1–4, 5th ed., New York: Marcel Dekker.

with selected corrodents. Table 2.3 provides the compatibility of impervious graphite with selected corrodents.

The engineering plastics are synthetic polymers of resin-based materials that have load-bearing characteristics and high-performance properties that permit them to be used in the same manner as metals. The major products of the polymer industry that include polyethylene, polypropylene, polyvinyl chloride, and polystyrene are not considered engineering polymers because of their low strength. Many of the engineering plastics are copolymers or alloy polymers. Table 2.4 lists the abbreviations used for the more common thermoplasts.

TABLE 2.2

Compatibility of Carbon Fiber with Selected Corrodents

Chemical	Maximum Temperature	
	°F	°C
Acetaldehyde	340	171
Acetamide	340	171
Acetic acid, 10%	340	171
Acetic acid, 50%	340	171
Acetic acid, 80%	340	171
Acetic acid, glacial	340	171
Acetic anhydride	340	171
Acetone	340	171
Acetyl chloride	340	171
Acrylonitrile	340	171
Adipic acid	340	171
Allyl alcohol	340	171
Allyl chloride	100	38
Alum	340	171
Aluminum chloride, aqueous	340	171
Aluminum chloride, dry	340	171
Aluminum fluoride	340	171
Aluminum hydroxide	340	171
Aluminum nitrate	340	171
Ammonia gas	340	171
Ammonium bifluoride	390	199
Ammonium carbonate	340	171
Ammonium chloride, 10%	340	171
Ammonium chloride, 50%	340	171
Ammonium chloride, sat.	340	171
Ammonium fluoride, 10%	330	166
Ammonium fluoride, 25%	340	171
Ammonium hydroxide, 25%	200	93
Ammonium hydroxide, sat.	220	104
Ammonium nitrate	340	171
Ammonium persulfate	340	171
Ammonium phosphate	340	171
Ammonium sulfate, 10–40%	340	171
Ammonium sulfide	340	171
Amyl acetate	340	171
Amyl alcohol	200	93
Amyl chloride	210	99
Aniline	340	171
Barium carbonate	250	121
Barium chloride	250	121
Barium hydroxide	250	121
Barium sulfate	250	121
Barium sulfide	250	121
Benzaldehyde	340	171
Benzene	200	93
Benzene sulfonic acid, 10%	340	171

(continued)

TABLE 2.2 *Continued*

Chemical	Maximum Temperature	
	°F	°C
Benzoic acid	350	177
Borax	250	121
Boric acid	210	99
Bromine gas, dry	X	X
Bromine gas, moist	X	X
Bromine liquid	X	X
Butadiene	340	171
Butyl acetate	340	171
Butyl alcohol	210	99
n-Butylamine	100	38
Butyl phthalate	90	32
Butyric acid	340	171
Calcium bisulfide	340	171
Calcium bisulfite	340	171
Calcium carbonate	340	171
Calcium chlorate, 10%	140	60
Calcium chloride	340	171
Calcium hydroxide, 10%	200	93
Calcium hydroxide, sat.	250	121
Calcium hypochlorite	170	77
Calcium nitrate	340	171
Calcium oxide	340	171
Calcium sulfate	340	171
Caprylic acid	340	171
Carbon bisulfide	340	171
Carbon dioxide, dry	340	171
Carbon dioxide, wet	340	171
Carbon disulfide	340	171
Carbon monoxide	340	171
Carbon tetrachloride	250	121
Carbonic acid	340	171
Cellosolve	200	93
Chloracetic acid, 50% water	340	171
Chloracetic acid	340	171
Chlorine gas, dry	180	82
Chlorine gas, wet	80	27
Chlorobenzene	340	171
Chloroform	340	171
Chlorosulfonic acid	340	171
Chronic acid, 10%	X	X
Chlorine acid, 50%	X	X
Citric acid, 15%	340	171
Citric acid, conc.	340	171
Copper carbonate	340	171
Copper chloride	340	171
Copper cyanide	340	171
Copper sulfate	340	171
Cresol	400	238

(continued)

TABLE 2.2 *Continued*

Chemical	Maximum Temperature	
	°F	°C
Cupric chloride, 5%	340	171
Cupric chloride, 50%	340	171
Cyclohexane	340	171
Ethylene glycol	340	171
Ferric chloride	340	171
Ferric chloride, 50% in water	340	171
Ferrous nitrate, 10–50%	340	171
Ferrous chloride	340	171
Ferrous nitrate	340	171
Fluorine gas, dry	X	X
Hydrobromic acid, dil.	340	171
Hydrobromic acid, 20%	340	171
Hydrobromic acid, 50%	340	171
Hydrochloric acid, 20%	340	171
Hydrochloric acid, 38%	340	171
Hydrocyanic acid, 10%	340	171
Hydrofluoric acid, 30%	340	171
Hydrofluoric acid, 70%	X	X
Hydrofluoric acid, 100%	X	X
Hypochlorous acid	100	38
Ketones, general	340	171
Lactic acid, 25%	340	171
Lactic acid, conc.	340	171
Magnesium chloride	170	77
Malic acid	100	38
Manganese chloride	400	227
Methyl chloride	340	171
Methyl ethyl ketone	340	171
Methyl isobutyl ketone	340	171
Muriatic acid	340	171
Nitric acid, 5%	180	82
Nitric acid, 20%	140	60
Nitric acid, 70%	X	X
Nitric acid, anhydrous	X	X
Nitrous acid, conc.	X	X
Perchloric acid, 10%	340	171
Perchloric acid, 70%	340	171
Phenol	340	171
Phosphoric acid, 50–80%	200	93
Picric acid	100	38
Potassium bromide, 30%	340	171
Salicyclic acid	340	171
Sodium bromide	340	171
Sodium carbonate	340	171
Sodium chloride	340	171
Sodium hydroxide, 10%	240	116
Sodium hydroxide, 50%	270	132
Sodium hydroxide, conc.	260	127

(continued)

TABLE 2.2 *Continued*

Chemical	Maximum Temperature °F	°C
Sodium hypochlorite, 20%	X	X
Sodium hypochlorite, conc.	X	X
Sodium sulfide, to 50%	120	49
Stannic chloride	340	171
Sulfuric acid, 10%	340	171
Sulfuric acid, 50%	340	171
Sulfuric acid, 70%	340	171
Sulfuric acid, 90%	180	82
Sulfuric acid, 98%	X	X
Sulfuric acid, 100%	X	X
Sulfurous acid	340	171
Toluene	340	171
Trichloroacetic acid	340	171
White liquor	100	38
Zinc chloride	340	171

The chemicals listed are in the pure state or in a saturated solution unless otherwise indicated. Compatibility is shown to the maximum allowable temperature for which data is available. Incompatibility is shown by an X.

Source: From P.A. Schweitzer. 2004. *Corrosion Resistance Tables*, Vols. 1–4, 5th ed., New York: Marcel Dekker.

TABLE 2.3

Compatibility of Impervious Graphite with Selected Corrodent

Chemical	Resin	Maximum Temperature °F	°C
Acetaldehyde	phenolic	460	238
Acetamide	phenolic	460	238
Acetic acid, 10%	furan	400	204
Acetic acid, 50%	furan	400	204
Acetic acid, 80%	furan	400	204
Acetic acid, glacial	furan	400	204
Acetic anhydride	furan	400	204
Acetone	furan	400	204
Acetyl chloride	furan	460	238
Acrylic acid			
Acrylonitrile	furan	400	204
Adipic acid	furan	460	238
Allyl alcohol		X	X
Aluminum chloride, aqueous	phenolic	120	49
Aluminum chloride, dry	furan	460	238
Aluminum fluoride	furan	460	238

(continued)

TABLE 2.3 *Continued*

Chemical	Resin	Maximum Temperature	
		°F	°C
Aluminum hydroxide	furan	250	121
Aluminum nitrate	furan	460	238
Aluminum sulfate	phenolic	460	238
Ammonium bifluoride	phenolic	390	199
Ammonium carbonate	phenolic	460	238
Ammonium chloride, 10%	phenolic	400	204
Ammonium hydroxide, 25%	phenolic	400	204
Ammonium hydroxide, sat.	phenolic	400	204
Ammonium nitrate	furan	460	238
Ammonium persulfate	furan	250	121
Ammonium phosphate	furan	210	99
Ammonium sulfate, 10–40%	phenolic	400	204
Amyl acetate	furan	460	238
Amyl alcohol	furan	400	204
Amyl chloride	phenolic	210	99
Aniline	furan	400	204
Aqua regia 3:1		X	X
Barium chlotide	furan	250	121
Barium hydroxide	phenolic	250	121
Barium sulfate	phenolic	250	121
Barium sulfide	phenolic	250	121
Benzaldehyde	furan	460	238
Benzene	furan	400	204
Benzene sulfonic acid, 10%	phenolic	460	238
Borax	furan	460	238
Boric acid	phenolic	460	238
Bromine gas, dry		X	X
Bromine gas, moist		X	X
Butadiene	furan	460	238
Butyl acetate	furan	460	238
Butyl alcohol	furan	400	204
n-Butylamine	phenolic	210	99
Butyric acid	furan	460	238
Calcium bisulfite	furan	460	238
Calcium carbonate	phenolic	460	238
Calcium chloride	furan	460	238
Calcium hydroxide, 10%	furan	250	121
Calcium hydroxide, sat.	furan	250	121
Calcium hypochlorite	furan	170	77
Calcium nitrate	furan	460	238
Calcium oxide			
Calcium sulfate	furan	460	238
Carbon bisulfide	furan	400	204
Carbon dioxide, dry	phenolic	460	238
Carbon dioxide, wet	phenolic	460	238
Carbon disulfide	furan	400	204
Carbon monoxide	phenolic	460	238
Carbon tetrachloride	furan	400	204

(continued)

TABLE 2.3 *Continued*

Chemical	Resin	Maximum Temperature	
		°F	°C
Carbonic acid	phenolic	400	204
Cellosolve	furan	460	238
Chloracetic acid, 50% water	furan	400	204
Chloracetic acid	furan	400	204
Chlorine gas, dry	furan	400	204
Chlorine, liquid	phenolic	130	54
Chlorobenzene	phenolic	400	204
Chloroform	furan	400	204
Chlorosulfanic acid		X	X
Chromic acid, 10%		X	X
Chromic acid, 50%		X	X
Citric acid, 15%	furan	400	204
Citric acid, conc.	furan	400	204
Copper chloride	furan	400	204
Copper cyanide	furan	460	238
Copper sulfate	phenolic	400	204
Cresol	furan	400	204
Cupric chloride, 5%	furan	400	204
Cupric chloride, 50%	furan	400	204
Cyclohexane	furan	460	238
Ethylene glycol	furan	330	166
Ferric chloride, 60%	phenolic	210	99
Ferric chloride, 50% in water	furan	260	127
Ferric nitrate, 10–50%	furan	210	99
Ferrous chloride	furan	400	204
Fluorine gas, dry	phenolic	300	149
Fluorine gas, moist		X	X
Hydrobromic acid, dil.	furan	120	49
Hydrobromic acid, 20%	furan	250	121
Hydrobromic acid, 50%	furan	120	49
Hydrochloric acid, 20%	phenolic	400	204
Hydrochloric acid, 38%	phenolic	400	204
Hydrocyanic acid, 10%	furan	460	238
Hydrofluoric acid, 30%	phenolic	460	238
Hydrofluoric acid, 70%		X	X
Hydrofluoric acid, 100%		X	X
Iodine Solution, 10%	Phenolic	120	49
Ketones, general	furan	400	204
Lactic acid, 25%	furan	400	204
Lactic acid, conc.	furan	400	204
Magnesium chloride	furan	170	77
Manganese chloride	furan	460	238
Methyl chloride	phenolic	460	238
Methyl ethyl Ketone	furan	460	238
Methyl isobutyl ketone	furan	460	238
Muriatic acid	phenolic	400	204
Nitric acid, 5%	phenolic	220	104
Nitric acid, 20%	phenolic	220	104

(continued)

TABLE 2.3 *Continued*

Chemical	Resin	Maximum Temperature	
		°F	°C
Nitric acid, 70%		X	X
Nitric acid anhydrous		X	X
Nitrous acid conc.		X	X
Oleum		X	X
Phenol	furan	400	204
Phosphoric acid, 50–80%	phenolic	400	204
Potassium bromide, 30%	furan	460	238
Salicylic acid	furan	340	171
Sodium carbonate	furan	400	204
Sodium chloride	furan	400	204
Sodium hydroxide, 10%	furan	400	204
Sodium hydroxide, 50%	furan	400	204
Sodium hydroxide, conc.		X	X
Sodium hypochlorite, 20%		X	X
Sodium hypochlorite conc.		X	X
Stannic chloride	furan	400	204
Sulfuric acid, 10%	phenolic	400	204
Sulfuric acid, 50%	phenolic	400	204
Sulfuric acid, 70%	phenolic	400	204
Sulfuric acid, 90%	phenolic	400	204
Sulfuric acid, 98%		X	X
Sulfuric acid, 100%		X	X
Sulfuric acid, fuming		X	X
Sulfuric acid	phenolic	400	204
Thionyl chloride	phenolic	320	160
Toluene	furan	400	204
Trichloroacetic acid	furan	340	171
Zinc chloride	phenolic	400	204

The chemicals listed are in the pure state or in a saturated solution unless otherwise indicated. Compatibility is shown to the maximum allowable temperature for which data is available. Incompatibility is shown by an X. A blank space indicates that data is unavailable.

Source: From P.A. Schweitzer. 2004. *Corrosion Resistance Tables*, Vols. 1–4, 5th ed., New York: Marcel Dekker.

One of the applications for plastic materials is to resist atmospheric corrosion. In addition to being able to resist attack by specific corrodents in plant operations, they are also able to resist corrosive fumes that may be present in the atmosphere. Table 2.5 provides the resistance of the more common thermoplasts to various atmospheric pollutants, and Table 2.6 gives the allowable temperature range for thermoplastic polymers.

2.1 Joining of Thermoplastics

Thermoplastic materials are joined by either solvent cementing, thermal fusion, or by means of adhesives. Solvent cementing is the easiest and most

TABLE 2.4

Abbreviations Used for Thermoplasts

ABS	Acrylonitrile-butadiene-styrene
CPE	Chlorinated polyether
CPVC	Chlorinated polyvinyl chloride
CTFE	Chlorotrifluoroethylene
ECTFE	Ethylenechlorotrifluoroethylene
ETFE	Ethylenetetrafluoroethylene
FEP	Flouroethylene-propylene copolymer
HDPE	High-density polyethylene
LDPE	Low-density polyethylene
LLDPE	Linear low-density polyethylene
PA	Polyamide (Nylon)
PAI	Polyamide-imide
PAN	Polyacrylonitrile
PAS	Polyarylsulfone
PB	Polybutylene
PBT	Polybutylene terephthalate
PC	Polycarbonate
PE	Polyethylene
PEEK	Polyetheretherketone
PEI	Polyether-imide
PEK	Polyetherketone
PEKK	Polyetherketoneketone
PES	Polyethersulfone
PET	Polyethylene terephthalate
PFA	Perfluoralkoxy
PI	Polyimide
PP	Polypropylene
PPE	Polyphenylene ether
PPO	Polyphenylene oxide
PPS	Polyphenylene sulfide
PPSS	Polyphenylene sulfide sulfone
PSF	Polysulfone
PTFE	Polytetrafluoroethylene (Teflon) also TFE
PUR	Polyurethane
PVC	Polyvinyl chloride
PVDC	Polyvinylidene chloride
PVDF	Polyvinylidene fluoride
PVF	Polyvinyl fluoride
SAN	Styrene-acrylonitrile
UHMWPE	Ultra high molecular weight polyethylene

economical method for joining thermoplasts. Solvent-cemented joints are less sensitive to thermal cycling than joints bonded with adhesives, and they have the same corrosion resistance as the base polymer as do joints made by a thermal-fusion process.

The major disadvantages of solvent cementing are the possibility of the part's stress cracking and the possible hazards of using low vapor point solvents. Adhesive bonding is generally recommended when two dissimilar polymers are joined because of solvent and polymer compatibility problems.

TABLE 2.5

Atmospheric Resistance of Thermoplastic Polymers

Polymer	UV Degradation[a,b]	Moisture[c] Absorption	Weathering	Ozone	SO_2	NO_x	H_2S
ABS	R	0.30	R	X^d	R^e		R^e
CPVC	R	0.03	R	R	R		R^e
ECTFE	R	<0.1	R	R	R	R	R
ETFE	R	<0.029	R	R	R	R	R
FEP	R	<0.01	R	R	R	R	R
HDPE	RS		R	R	R		R
PA	R	0.6–1.2	R	X^d	R^f		X^d
PC	RS	0.15	L^g	R			
PCTFE	R		R	R	R	R	R
PEEK	R	0.50	L^g	R	R		R
PEI	R	0.25	R	R			
PES					X^d		R
PF					R		R
PFA	R	<0.03	R	R	R	R	R
PI	R		R	R			
PP	RS	0.02	L^g	L^g	R	R	R
PPS	R	0.01	R		R		R
PSF	R	0.30	R	R			
PTFE	R	<0.01	R	R	R	R	R
PVC	R	0.05	R	R	R^h	X^d	R
PVDC	R		R	R	R	R	R
PVDF	R	<0.04	R	R	R	R	R
UHMWPE	RS	<0.01	R	R	R		R

[a] R, Resistant.
[b] RS, Resistant only if stabilized with a UV protector.
[c] Water absorption at 73°F/23°C (%).
[d] Not resistant.
[e] Dry only.
[f] Wet only.
[g] L, Limited resistance.
[h] Type 1 only.

A universal solvent should never be used. Solvent cements that have approximately the same solubility parameters as the polymer to be bonded should be selected. Table 2.7 lists the typical solvents used to bond the major polymers.

2.1.1 Use of Adhesive

The physical and chemical properties of both the solidified adhesive and the polymer affect the quality of the bonded joint. Major elements of concern are the corrosion resistance of the adhesive, its thermal expansion coefficient, and the glass transition temperature of the polymer relative to the adhesive.

Large differences in the thermal expansion coefficient between the polymer and adhesive can result in stress at the polymer's joint interface.

TABLE 2.6

Allowable Temperature Range for Thermoplastic Polymers

| Polymer | Allowable Temperature (°F/°C) | |
	Minimum	Maximum
ABS	−40/−40	140/60
CPVC	0/−18	180/82
ECTFE	−105/−76	340/171
ETFE	−370/−223	300/149
FEP	−50/−45	400/205
HDPE	−60/−51	180/82
PA	−60/−51	300/149
PAI	−300/−190	500/260
PB		200/93
PC	−200/−129	250–275/120–135
PCTFE	−105/−76	380/190
PEEK	−85/−65	480/250
PEI	−310/−190	500/260
PES		340/170
PFA	−310/−190	500/260
PI	−310/−190	500–600/260–315
PP	32/0	215/102
PPS		450/230
PSF	−150/−101	300/149
PTFE	−20/−29	500/260
PVC	0/−18	140/60
PVDC	0/−18	175/80
PVDF	−50/−45	320/160
UHWMPE	40/4	200/93

Temperature limits may have to be modified depending upon the corrodent if in direct contact.

TABLE 2.7

Typical Solvents for Solvent Cementing

Polymer	Solvent
PVC	Cyclohexane, tetrahydrofuran, dichlorobenzene
ABS	Methyl ethyl ketone, methyl isobutyl ketone, tetrahydrofuran, methylene chloride
Acetate	Methylene chloride, acetone, chloroform, methyl ethyl ketone, ethyl acetate
Acrylic	Methylene chloride, ethylene dichloride
PA (Nylon)	Aqueous phenol, solutions of resorcinol in alcohol, solutions of Calcium chloride in alcohol
PPO	Trichloroethylene, ethylene dichloride, chloroform, methylene chloride
PC	Methylene chloride, ethylene dichloride
Polystyrene	Methylene chloride, ethylene dichloride, ethylene ketone, trichloroethylene, toluene, xylene
PSF	Methylene chloride

These stresses are compounded by thermal cycling and low-temperature service requirements.

A structural adhesive must have a glass transition temperature higher than the operating temperature to prevent a cohesively weak bond and possible creep problems. Engineering polymers such as polyamide or polyphenylene sulfide have very high glass transition temperatures whereas most adhesives have very low glass transition temperatures, meaning that the weakest link in the joint may be the adhesive.

All polymers cannot be joined by all of the methods cited. Table 2.8 lists the methods by which individual polymers may be joined.

Selection of the proper adhesives is important. Which adhesive is selected will depend upon the specific operating conditions such as chemical environment and temperature.

There is not a best adhesive for universal environments. For example, an adhesive providing maximum resistance to acids, in all probability, will provide poor resistance to bases. It is difficult to select an adhesive that will not degrade in two widely differing chemical environments. In general, the adhesives that are most resistant to high temperatures usually exhibit the best resistance to chemicals and solvents. Table 2.9 provides the relative compatibilities of synthetic adhesives in selected environments.

TABLE 2.8

Methods of Joining Thermoplasts

Polymer	Adhesives	Mechanical Fastening	Solvent Welding	Thermal Welding
ABS	X	X	X	X
Acetals	X	X	X	X
Acrylics	X	X	X	
CPE	X	X		
Ethylene copolymers				X
Fluoroplastics	X			
PA	X	X	X	
PPO	X	X	X	X
Polyesters	X	X	X	
PAI	X	X		
PAS	X	X		
PC	X	X	X	X
PA/ABS	X	X	X	X
PE	X	X		X
PI	X	X		
PPS	X	X		
PP	X	X		X
Polystyrene	X	X	X	X
PSF	X	X		
PVC/Acrylic alloy	X	X		
PVC/ABS alloy	X			X
PVC	X	X	X	X
CPVC	X	X	X	

TABLE 2.9

Chemical Resistance of Adhesives

Adhesive	Type[c]	A	B	C	D	E	F	G	H	J	K	L	M	N
Cyanoacrylate	TS	X		X	X	X	X	3	3	X	X	X	4	4
Polyester+isocyanate	TS	3	2	1	3	3	2	2	2	3	2	2	X	2
Polyester+monomer	TS	X	3	3	X	3	X	2	2	2	X	X	X	X
Urea formaldehyde	TS	3	3	2	X	2	2	2	2	2	2	2	2	2
Melamine formaldehyde	TS	2	2	2	X	2	2	2	2	2	2	2	2	2
Resorcinol formaldehyde	TS	2	2	2	2	2	2	2	2	2	2	2	2	2
Epoxy+polyamine	TS	3	X	2	2	3	X	2	3	1	X	X	1	
Epoxy+polyamide	TS	X	2	2	X	3	2	2	2	1	X	X	3	
Polyimide	TS	1	1	2	4	2	2	2	2	2	2	2	2	2
Acrylic	TS	X	3	1	3	1	2		2	2	2	2	2	2
Cellulose acetate	TP	2	3	1	X	3	X		2	4	X	X	X	X
Cellulose nitrate	TP	3	3	3	3	3	2	2	2	X	X	X	X	X
Polyvinyl acetate	TP	X		3	X	3	3	2	2	X	X	X	X	X
Polyvinyl alcohol	TP	3		X	X	X	X	2	1	3	1	1	1	1
Polyamide	TP	X		X	X	X	2	2	2	X	2	2	2	X
Acrylic	TP	4	3	3	3	3		2			4	4		4
Phenoxy	TP	4	3	4	4	3	2	3	X	X	4	4	X	4

[a] Environment: A, heat; B, cold; C, water; D, hot water; E, acid; F, alkali; G, oil, grease; H, fuels; J, alcohols; K, ketones; L, esters; M, aromatics; N, chlorinated solvents.

[b] Resistance: 1, excellent; 2, good; 3, fair; 4, poor; X, not recommended.

[c] Type: TS, thermosetting adhesive; TP, thermoplastic adhesive.

Epoxy adhesives are generally limited to continuous applications below 300°F (149°C). However, there are epoxy formulations that can withstand short terms at 500°F (260°C) and long-term service at 300–350°F (149–177°C). A combination epoxy–phenolic resin has been developed that will provide an adhesive capability at 700°F (371°C) for short-term operation and continuous operation at 350°F (177°C).

Nitrile–phenolic adhesives have high shear strength up to 250–350°F (121–177°C), and the strength retention on aging at these temperatures is very good. Silicone adhesives are primarily used in nonstructural applications. They have very good thermal stability but low strength.

Polyamide-adhesives have thermal endurance at temperatures greater than 500°F (260°C) that is unmatched by any commercially available adhesive. Their short-term exposure at 1000°F (538°C) is slightly better than the epoxy–phenolic alloy.

The successful application of an adhesive at low temperatures is dependent upon the difference in the coefficient of thermal expansion between adhesive and polymer, the elastic modulus, and the thermal conductivity of the adhesive. Epoxy–polyamide adhesives can be made serviceable at very low temperatures by the addition of appropriate fillers to control thermal expansion.

Epoxy–phenolic adhesives exhibit good adhesive properties at both elevated and low temperatures. Polyurethane and epoxy–nylon systems exhibit outstanding cryogenic properties.

Other factors affecting the life of an adhesive bond are humidity, water immersion, and outdoor weathering. Moisture can affect adhesive strength in two ways. Some polymeric materials, notably ester-based polyurethanes, will revert, i.e., lose hardness, strength, and in the worst case, turn to fluid during exposure to warm humid air. Water can also permeate the adhesive and displace the adhesive at the bond interface. Structural adhesives not susceptible to the reversion phenomenon are also likely to lose adhesive strength when exposed to moisture.

Adhesives exposed outdoors are affected primarily by heat and humidity. Thermal cycling, ultraviolet radiation, and cold are relatively minor factors. Structural adhesives, when exposed to weather, rapidly lose strength during the first six months to a year. After two or three years, the rate of decline usually levels off, depending upon the climate zone, polymer, adhesive, and stress level. The following are important considerations when designing an adhesive joint for outdoor service:

1. The most severe locations are those with high humidity and warm temperatures.

2. Stressed panels deteriorate more rapidly than unstressed panels.

3. Heat-cured adhesive systems are generally more resistant than room-temperature cured systems.

4. With the better adhesives, unstressed bonds are relatively resistant to severe outdoor weathering, although all joints will eventually show some strength loss.

2.2 Acrylonitrile–Butadiene–Styrene (ABS)

ABS polymers are derived from acrylonitrile, butadiene, and styrene and have the following general chemical structure:

Acrylonitrile   Butadiene   Styrene

The properties of ABS polymers can be altered by the relative amounts of acrylonitrile, butadiene, and styrene present. Higher strength, better toughness, greater dimensional stability and other properties can be obtained at the expense of other characteristics.

Pure ABS polymers will be attacked by oxidizing agents and strong acids, and they will stress crack in the presence of certain organic compounds. It is resistant to aliphatic hydrocarbons, but not to aromatic or chlorinated hydrocarbons.

ABS will be degraded by UV light unless protective additives are incorporated into the formulation.

When an ABS alloy or a reinforced ABS is used, all of the alloying ingredients and/or reinforcing materials must be checked for chemical compatibility. The manufacturer should also be checked. Table 2.10 provides the compatibility of ABS with selected corrodents. A more detailed listing may be found in Reference [1].

ABS polymers are used to produce business machine and camera housings, blowers, bearings, gears, pump impellers, chemical tanks, fume hoods, ducts, piping, and electrical conduit.

2.3 Acrylics

Acrylics are based on polymethylmethacrylate and have a chemical structure as follows:

They are sold under the tradenames of Lucite by E.I. duPont and Plexiglass by Rohm and Haas. Because of their chemical structure, acrylic resins are inherently resistant to discoloration and loss of light transmission. Parts molded from acrylic powders in their pure state may be clear and nearly optically perfect. The total light transmission is as high as 92%, and haze measurements average only 1%.

Acrylics exhibit outstanding weatherability. They are attacked by strong solvents, gasoline, acetone, and other similar fluids. Table 2.11 lists the compatibility of acrylics with selected corrodents. Reference [1] provides a more detailed listing.

TABLE 2.10

Compatibility of ABS with Selected Corrodents

Chemical	Maximum Temperature	
	°F	°C
Acetaldehyde	X	X
Acetic acid, 10%	100	38
Acetic acid, 50%	130	54
Acetic acid, 80%	X	X
Acetic acid, glacial	X	X
Acetic anhydride	X	X
Acetone	X	X
Acetyl chloride	X	X
Adipic acid	140	60
Allyl alcohol	X	X
Allyl chloride	X	X
Alum	140	60
Aluminum chloride, aqueous	140	60
Aluminum fluoride	140	60
Aluminum hydroxide	140	60
Aluminum oxychloride	140	60
Aluminum sulfate	140	60
Ammonia gas dry	140	60
Ammonium bifluoride	140	60
Ammonium carbonate	140	60
Ammonium chloride, sat.	140	60
Ammonium fluoride, 10%	X	X
Ammonium fluoride, 25%	X	X
Ammonium hydroxide, 25%	90	32
Ammonium hydroxide, sat.	80	27
Ammonium nitrate	140	60
Ammonium persulfate	140	60
Ammonium phosphate	140	60
Ammonium sulfate, 10–40%	140	60
Ammonium sulfide	140	60
Amyl acetate	X	X

(continued)

TABLE 2.10 *Continued*

Chemical	Maximum Temperature	
	°F	°C
Amyl alcohol	80	27
Amyl chloride	X	X
Aniline	X	X
Antimony trichloride	140	60
Aqua regia 3:1	X	X
Barium carbonate	140	60
Barium chloride	140	60
Barium hydroxide	140	60
Barium sulfate	140	60
Barium sulfide	140	60
Benzaldehyde	X	X
Benzene	X	X
Benzene sulfonic acid, 10%	80	27
Benzoic acid	140	60
Benzyl alcohol	X	X
Benzyl chloride	X	X
Borax	140	60
Boric acid	140	60
Bromine liquid	X	X
Butadiene	X	X
Butyl acetate	X	X
Butyl alcohol	X	X
Butyric acid	X	X
Calcium bisulfite	140	60
Calcium carbonate	100	38
Calcium chlorate	140	60
Calcium chloride	140	60
Calcium hydroxide, sat.	140	60
Calcium hypochlorite	140	60
Calcium nitrate	140	60
Calcium oxide	140	60
Calcium sulfate, 25%	140	60
Carbon bisulfide	X	X
Carbon dioxide, dry	90	32
Carbon dioxide, wet	140	60
Carbon disulfide	X	X
Carbon monoxide	140	60
Carbon tetrachloride	X	X
Carbonic acid	140	60
Cellosolve	X	X
Chloracetic acid	X	X
Chlorine gas, dry	140	60
Chlorine gas, wet	140	60
Chlorine, liquid	X	X
Chlorobenzene	X	X
Chloroform	X	X
Chlorosulfonic acid	X	X
Chromic acid, 10%	90	32

(continued)

TABLE 2.10 *Continued*

Chemical	Maximum Temperature	
	°F	°C
Chromic acid, 50%	X	X
Citric acid, 15%	140	60
Citric acid, 25%	140	60
Copper chloride	140	60
Copper cyanide	140	60
Copper sulfate	140	60
Cresol	X	X
Cyclohexane	80	27
Cyclohexanol	80	27
Dichloroacetic acid	X	X
Dichloroethane (ethylene dichloride)	X	X
Ethylene glycol	140	60
Ferric chloride	140	60
Ferric nitrate, 10–50%	140	60
Ferrous chloride	140	60
Fluorine gas, dry	90	32
Hydrobromic acid, 20%	140	60
Hydrochloric acid, 20%	90	32
Hydrochloric acid, 38%	140	60
Hydrofluoric acid, 30%	X	X
Hydrofluoric acid, 70%	X	X
Hydrofluoric acid, 100%	X	X
Hypochlorous acid	140	60
Ketones, general	X	X
Lactic acid, 25%	140	60
Magnesium chloride	140	60
Malic acid	140	60
Methyl chloride	X	X
Methyl ethyl ketone	X	X
Methyl isobutyl ketone	X	X
Muriatic acid	140	60
Nitric acid, 5%	140	60
Nitric acid, 20%	130	54
Nitric acid, 70%	X	X
Nitric acid, anhydrous	X	X
Oleum	X	X
Perchloric acid, 10%	X	X
Perchloric acid, 70%	X	X
Phenol	X	X
Phosphoric acid, 50–80%	130	54
Picric acid	X	X
Potassium bromide, 30%	140	60
Sodium carbonate	140	60
Sodium chloride	140	60
Sodium hydroxide, 10%	140	60
Sodium hydroxide, 50%	140	60
Sodium hydroxide, conc.	140	60
Sodium hypochlorite, 20%	140	60

(continued)

TABLE 2.10 *Continued*

Chemical	Maximum Temperature	
	°F	°C
Sodium hypochlorite, conc.	140	60
Sodium sulfide, to 50%	140	60
Stannic chloride	140	60
Stannous chloride	100	38
Sulfuric acid, 10%	140	60
Sulfuric acid, 50%	130	54
Sulfuric acid, 70%	X	X
Sulfuric acid, 90%	X	X
Sulfuric acid, 98%	X	X
Sulfuric acid, 100%	X	X
Sulfuric acid, fuming	X	X
Sulfurous acid	140	60
Thionyl chloride	X	X
Toluene	X	X
White liquor	140	60
Zine chloride	140	60

The chemicals listed are in the pure state or in a saturated solution unless otherwise indicated. Compatibility is shown to the maximum allowable temperature for which data is available. Incompatibility is shown by an X. A blank space indicates that data is unavailable.

Source: From P.A. Schweitzer. 2004. *Corrosion Resistance Tables*, Vols. 1–4, 5th ed., New York: Marcel Dekker.

Acrylics are used for lenses, aircraft and building glazing, lighting fixtures, coatings, textile fibers, fluorescent street lights, outdoor signs, and boat windshields.

2.4 Chlotrifluoroethylene (CTFE)

Chlorotrifluorethylene is sold under the tradename of Kel–F. It is a fluorocarbon with the following chemical structure:

$$\left[\begin{array}{c} Cl \;\; F \\ | \;\;\; | \\ C\!-\!C \\ | \;\;\; | \\ F \;\;\; F \end{array} \right]_n$$

Although CTFE has a wide range of corrosion resistance, it has less resistance than PTFE, FEP, and PFA. It is subject to swelling in some

TABLE 2.11

Compatibility of Acrylics with Selected Corrodents

Chemical	
Acetaldehyde	X
Acetic acid, 10%	X
Acetic acid, 50%	X
Acetic acid, 80%	X
Acetic acid, glacial	X
Acetic anhydride	X
Acetone	X
Alum	R
Aluminum chloride, aqueous	R
Aluminum hydroxide	R
Aluminum sulfate	R
Ammonia gas	R
Ammonium carbonate	R
Ammonium chloride, sat.	R
Ammonium hydroxide, 25%	R
Ammonium hydroxide, sat.	R
Ammonium nitrate	R
Ammonium persulfate	R
Ammonium sulfate, 10–40%	R
Amyl acetate	X
Amyl alcohol	R
Aniline	X
Aqua regia 3:1	X
Barium carbonate	R
Barium chloride	R
Barium hydroxide	R
Barium sulfate	R
Benzaldehyde	X
Benzene	X
Benzoic acid	X
Borax	R
Boric acid	R
Bromine, liquid	X
Butyl acetate	X
Butyl alcohol	R
Butyric acid	X
Calcium bisulfite	R
Calcium carbonate	R
Calcium chlorate	R
Calcium chloride	R
Calcium hydroxide, sat.	R
Calcium hypochlorite	R
Calcium sulfate	R
Calcium bisulfide	R
Carbon dioxide, wet	R
Carbon disulfide	R
Carbon tetrachloride	X
Carbonic acid	R

(continued)

TABLE 2.11 *Continued*

Chemical	
Chloroacetic acid, 50% water	X
Chloroacetic acid	X
Chlorobenzene	X
Chloroform	X
Chlorosulfonic acid	X
Chromic acid, 10%	X
Chromic acid, 50%	X
Citric acid, 15%	R
Citric acid, conc.	R
Copper chloride	R
Copper cyanide	R
Copper sulfate	X
Cresol	X
Cupric chloride, 5%	R
Cupric chloride, 50%	R
Dichloroethane	X
Ethylene chloride	R
Ferric chloride	R
Ferric nitrate, 10–50%	R
Ferric chloride	R
Hydrochloric acid, 20%	R
Hydrochloric acid, 38%	R
Hydrofluoric acid, 70%	X
Hydrofluoric acid, 100%	R
Ketones, general	X
Lactic acid, 25%	X
Lactic acid, conc.	X
Magnesium chloride	R
Methyl chloride	X
Methyl ethyl ketone	X
Nitric acid, 5%	R
Nitric acid, 20%	X
Nitric acid, 70%	X
Nitric acid anhydrous	X
Oleum	X
Phenol	R
Phosphoric acid, 50–80%	R
Picric acid	R
Potassium bromide, 30%	R
Sodium carbonate	X
Sodium chloride	R
Sodium hydroxide, 10%	R
Sodium hydroxide, 50%	R
Sodium hydroxide, conc.	R
Sodium hypochlorite, 20%	R
Sodium sulfide, to 50%	R
Stanninc chloride	R
Stannous chloride	R
Sulfuric acid, 10%	R

(continued)

TABLE 2.11 *Continued*

Chemical	
Sulfuric acid, 50%	R
Sulfuric acid, 70%	R
Sulfuric acid, 90%	X
Sulfuric acid, 98%	X
Sulfuric acid, 100%	X
Sulfuric acid fuming	X
Sulfurous acid	R
Toluene	X
Zinc chloride	R

The chemicals listed are in the pure state or in a saturated solution unless otherwise specified. Compatibility at 90°F/32°C is indicated by R. Incompatibility is shown by an X.

Source: From P.A. Schweitzer. 2004. *Corrosion Resistance Tables*, Vols. 1–4, 5th ed., New York: Marcel Dekker.

chlorinated solvents at elevated temperatures and is attacked by the same chemicals that attack PTFE. Refer to Table 2.12 for the compatibility of CTFE with selected corrodents. Reference [1] provides a more detailed listing.

TABLE 2.12

Compatibility of CTFE with Selected Corrodents

Chemical	Maximum Temperature	
	°F	°C
Acetamide	200	93
Acetaldehyde	122	50
Acetic acid, 95–100%	122	50
Acetic acid glacial	122	50
Acetonitrile	100	38
Acetophenone	200	93
Acetyl chloride, dry	100	38
Allyl alcohol	200	93
Aluminum chloride, 25%	200	93
Ammonia anhydrous	200	93
Ammonia, wet	200	93
Ammonium chloride, 25%	200	93
Ammonium hydroxide, conc.	200	93
Amyl acetate	100	38
Aniline	122	50
Aqua regia	200	93
Barium sulfate	300	149

(continued)

TABLE 2.12 *Continued*

Chemical	Maximum Temperature	
	°F	°C
Benzaldehyde	200	93
Benzene	200	93
Benzontrile	200	93
Benzyl alcohol	200	93
Benzyl chloride	100	38
Borax	300	149
Bromine gas, dry	100	38
Bromine gas, moist	212	100
Butadiene	200	93
Butane	200	93
n-Butylamine	X	
Butyl ether	100	38
Calcium chloride	300	149
Calcium sulfate	300	149
Carbon dioxide	300	149
Carbon disulfide	212	100
Cellusolve	200	93
Chlorine, dry	X	
Chloroacetic acid	125	52
Cupric chloride, 5%	300	149
Cyclohexanone	100	38
Dimethyl aniline	200	93
Dimethyl phthalate	200	93
Ethanolamine	X	
Ethylene dichloride	100	38
Ethylene glycol	200	92
Fluorine gas, dry	X	
Fluorine gas, moist	70	23
Fuel oils	200	93
Gasoline	200	93
Glycerine	212	100
Heptane	200	93
Hexane	200	93
Hydrobromic acid, 37%	300	149
Hydrochloric acid, 100%	300	149
Hydrofluoric acid, 50%	200	93
Magnesium chloride	300	149
Magnesium sulfate	212	100
Mineral oil	200	93
Motor oil	200	93
Naphtha	200	93
Naphthalene	200	93
Nitromethane	200	93
Oxalic acid, 50%	212	100
Perchloric acid	200	93
Phosphorus trichloride	200	93
Potassium manganate	200	93
Seawater	250	121

(continued)

TABLE 2.12   *Continued*

Chemical	Maximum Temperature	
	°F	°C
Sodium bicarbonate	300	149
Sodium bisulfate	140	60
Sodium carbonate	300	149
Sodium chromate, 10%	250	121
Sodium hypochlorite	250	121
Sodium sulfate	300	149
Sodium thiosulfate	300	149
Stannous chloride	212	100
Sulfur	260	127
Sulfur dioxide, wet	300	149
Sulfur dioxide, dry	300	149
Sulfuric acid, 0–100%	212	100
Sulfuric acid, fuming	200	93
Turpentine	200	93
Vinegar	200	93
Water, fresh	250	121
Zinc sulfate	212	100

The chemicals listed are in the pure state or in a saturated solution, unless otherwise noted. Compatibility is shown to the maximum allowable temperature for which data is available. Incompatibility is shown by an X.

2.5   Ethylenechlorotrifluoroethylene (ECTFE)

ECTFE is a 1:1 alternating copolymer of ethylene and chlorotrifluoroethylene having a chemical structure as follows:

$$
\begin{array}{cccc}
\text{H} & \text{H} & \text{H} & \text{H} \\
| & | & | & | \\
-\text{C}-\text{C}-\text{C}-\text{C}- \\
| & | & | & | \\
\text{H} & \text{H} & \text{F} & \text{Cl}
\end{array}
$$

This chemical structure provides the polymer with a unique combination of properties. It possesses excellent chemical resistance from cryogenic to 340°F (117°C) with continuous service to 300°F (149°C) and excellent abrasion resistance. It is sold under the tradename of Halar by Ausimont U.S.A.

ECTFE's resistance to permeation by oxygen, carbon dioxide, chlorine gas, or hydrochloric acid is 10–100 times better than that of PTFE or FEP. Water absorption is less than 0.01%.

The chemical resistance of ECTFE is outstanding. It is resistant to most of the common corrosive chemicals encountered in industry. Included in this list of chemicals are strong mineral and oxidizing acids, alkalies, metal etchants, liquid oxygen, and practically all organic solvents except hot amines (aniline, dimethylamine, etc.) No known solvent dissolves or stress cracks ECTFE at temperatures up to 250°F (120°C).

Some halogenated solvents can cause ECTFE to become slightly plasticized when it comes into contact with them. Under normal circumstances, that does not impair the usefulness of the polymer. When the part is removed from contact with the solvent and allowed to dry, its mechanical properties return to their original values, indicating that no chemical attack has taken place. As with other fluoropolymers, ECTFE will be attacked by metallic sodium and potassium.

The useful properties of ECTFE are maintained on exposure to cobalt-60 radiation of 20 Mrads. ECTFE also exhibits excellent resistance to weathering and UV radiation. Table 2.13 shows the compatibility of ECTFE with selected corrodents. Reference [1] provides a more extensive listing.

ECTFE has found application as pipe and vessel liners and in piping systems, chemical process equipment, and high-temperature wire and cable insulation.

TABLE 2.13

Compatibility of ECTFE with Selected Corrodents

Chemical	Maximum Temperature	
	°F	°C
Acetic acid, 10%	250	121
Acetic acid, 50%	250	121
Acetic acid, 80%	150	66
Acetic acid, glacial	200	93
Acetic anhydride	100	38
Acetone	150	66
Acetyl chloride	150	66
Acrylonitrile	150	66
Adipic acid	150	66
Allyl chloride	300	149
Alum	300	149
Aluminum chloride, aqueous	300	149
Aluminum chloride, dry		
Aluminum fluoride	300	149
Aluminum hydroxide	300	149
Aluminum nitrate	300	149
Aluminum oxychloride	150	66
Aluminum sulfate	300	149
Ammonia gas	300	149
Ammonium bifluoride	300	149
Ammonium carbonate	300	149
Ammonium chloride, 10%	290	143
Ammonium chloride, 50%	300	149
Ammonium chloride, sat.	300	149

(continued)

TABLE 2.13   *Continued*

Chemical	Maximum Temperature	
	°F	°C
Ammonium fluoride, 10%	300	149
Ammonium fluoride, 25%	300	149
Ammonium hydroxide, 25%	300	149
Ammonium hydroxide, sat.	300	149
Ammonium nitrate	300	149
Ammonium persulfate	150	66
Ammonium phosphate	300	149
Ammonium sulfate, 10–40%	300	149
Ammonium sulfide	300	149
Amyl acetate	160	71
Amyl alcohol	300	149
Amyl chloride	300	149
Aniline	90	32
Antimony trichloride	100	38
Aqua regia 3:1	250	121
Barium carbonate	300	149
Barium chloride	300	149
Barium hydroxide	300	149
Barium sulfate	300	149
Barium sulfide	300	149
Benzaldehyde	150	66
Benzene	150	66
Benzenesulfonic acid, 10%	150	66
Benzoic acid	250	121
Benzyl alcohol	300	149
Benzyl chloride	300	149
Borax	300	149
Boric acid	300	149
Bromine gas, dry	X	X
Bromine, liquid	150	66
Butadiene	250	121
Butyl acetate	150	66
Butyl alcohol	300	149
Butyric acid	250	121
Calcium bisulfide	300	149
Calcium bisulfite	300	149
Calcium carbonate	300	149
Calcium chlorate	300	149
Calcium chloride	300	149
Calcium hydroxide, 10%	300	149
Calcium hydroxide, sat.	300	149
Calcium hypochlorite	300	149
Calcium nitrate	300	149
Calcium oxide	300	149
Calcium sulfate	300	149
Caprylic acid	220	104
Carbon bisulfide	80	27
Carbon dioxide, dry	300	149

(continued)

TABLE 2.13 *Continued*

Chemical	Maximum Temperature	
	°F	°C
Carbon dioxide, wet	300	149
Carbon disulfide	80	27
Carbon monoxide	150	66
Carbon tetrachloride	300	149
Carbonic acid	300	149
Cellsolve	300	149
Chloroacetic acid, 50% water	250	121
Chloroacetic acid	250	121
Chlorine gas, dry	150	66
Chlorine gas, wet	250	121
Chlorine, liquid	250	121
Chlorobenzene	150	66
Chloroform	250	121
Chlorosulfonic acid	80	27
Chromic acid, 10%	250	121
Chromic acid, 50%	250	121
Citric acid, 15%	300	149
Citric acid, conc.	300	149
Copper carbonate	150	66
Copper chloride	300	149
Copper cyanide	300	149
Copper sulfate	300	149
Cresol	300	149
Cupric chloride, 5%	300	149
Cupric chloride, 50%	300	149
Cyclohexane	300	149
Cyclohexanol	300	149
Ethylene glycol	300	149
Ferric chloride	300	149
Ferric chloride, 50% in water	300	149
Ferric nitrate, 10–50%	300	149
Ferrous chloride	300	149
Ferrous nitrate	300	149
Fluorine gas, dry	X	X
Fluorine gas, moist	80	27
Hydrobromic acid, dil.	300	149
Hydrobromic acid, 20%	300	149
Hydrobromic acid, 50%	300	149
Hydrochloric acid, 20%	300	149
Hydrochloric acid, 38%	300	149
Hydrocyanic acid, 10%	300	149
Hydrofluoric acid, 30%	250	121
Hydrofluoric acid, 70%	240	116
Hydrofluoric acid, 100%	240	116
Hypochlorous acid	300	149
Iodine solution, 10%	250	121
Lactic acid, 25%	150	66
Lactic acid, conc.	150	66

(continued)

TABLE 2.13 *Continued*

Chemical	Maximum Temperature	
	°F	°C
Magnesium chloride	300	149
Malic acid	250	121
Methyl chloride	300	149
Methyl ethyl ketone	150	66
Methyl isobutyl ketone	150	68
Muriatic acid	300	149
Nitric acid, 5%	300	149
Nitric acid, 20%	250	121
Nitric acid, 70%	150	66
Nitric acid, anhydrous	150	66
Nitrous acid, conc.	250	121
Oleum	X	X
Perchloric acid, 10%	150	66
Perchloric acid, 70%	150	66
Phenol	150	66
Phosphoric acid, 50–80%	250	121
Picric acid	80	27
Potassium bromide, 30%	300	149
Salicylic acid	250	121
Sodium carbonate	300	149
Sodium chloride	300	149
Sodium hydroxide, 10%	300	149
Sodium hydroxide, 50%	250	121
Sodium hydroxide, conc.	150	66
Sodium hypochlorite, 20%	300	149
Sodium hypochlorite, conc.	300	149
Sodium sulfide, to 50%	300	149
Stannic chloride	300	149
Stannous chloride	300	149
Sulfuric acid, 10%	250	121
Sulfuric acid, 50%	250	121
Sulfuric acid, 70%	250	121
Sulfuric acid, 90%	150	66
Sulfuric acid, 98%	150	66
Sulfuric acid, 100%	80	27
Sulfuric acid, fuming	300	149
Sulfurous acid	250	121
Thionyl chloride	150	66
Toluene	150	66
Trichloroacetic acid	150	66
White liquor	250	121
Zinc chloride	300	149

The chemicals listed are in the pure state or in a saturated solution unless otherwise indicated. Compatibility is shown to the maximum allowable temperature for which data is available. Incompatibility is shown by an X. A blank space indicate that data is unavailable.

Source: From P.A. Schweitzer. 2004. *Corrosion Resistance Tables*, Vols. 1–4, 5th ed., New York: Marcel Dekker.

2.6 Ethylene Tetrafluoroethylene (ETFE)

ETFE is an alternating copolymer of ethylene and tetrafluoroethylene by duPont under the tradename of Tefzel and has the following structural formula:

```
     H  H  H  H
     |  |  |  |
    -C--C--C--C-
     |  |  |  |
     H  H  F  F
```

ETFE is a high-temperature fluoropolymer with a maximum service temperature of 300°F (149°C).

Tefzel is a rugged thermoplastic with an outstanding balance of properties. It can be reinforced with carbon or glass fibers, being the first fluoroplastic that can be reinforced, not merely filled. Because the resin will bond to the fibers, strength, stiffness, creep resistance, heat distortion temperature, and dimensional stability are enhanced.

ETFE is inert to strong mineral acids, halogens, inorganic bases, and metal salt solutions. Carboxylic acids, aldehydes, aromatic and aliphatic hydrocarbons, alcohols, ketones, esters, chlorocarbons, and classic polymer solvents have little effect on ETFE. Tefzel is also weather resistant and ultraviolet ray resistant.

Very strong oxidizing acids such as nitric, organic bases such as amines, and sulfuric acid at high concentrations and near their boiling points will affect ETFE to various degrees. Refer to Table 2.14 for the compatibility of ETFE with selected corrodents. Reference [1] provides a more extensive listing.

ETFE finds applications in process equipment, piping, chemical ware, wire insulation, tubing, and pump components.

2.7 Fluorinated Ethylene–Propylene (FEP)

Fluorinated ethylene-propylene is a fully fluorinated thermoplastic with some branching, but it mainly consists of linear chains having the following formula:

```
     F   F   F   F   F
     |   |   |   |   |
    -C---C---C---C---C-
     |   |   |   |   |
     F   F   |   F   F
             |
           F-C-F
             |
             F
```

TABLE 2.14

Compatibility of ETFE with Selected Corrodents

Chemical	Maximum Temperature	
	°F	°C
Acetaldehyde	200	93
Acetamide	250	121
Acetic acid, 10%	250	121
Acetic acid, 50%	250	121
Acetic acid, 80%	230	110
Acetic acid, glacial	230	110
Acetic anhydride	300	149
Acetone	150	66
Acetyl chloride	150	66
Acrylonitrile	150	66
Adipic acid	280	138
Allyl alcohol	210	99
Allyl chloride	190	88
Alum	300	149
Aluminum chloride, aqueous	300	149
Aluminum chloride, dry	300	149
Aluminum fluoride	300	149
Aluminum hydroxide	300	149
Aluminum nitrate	300	149
Aluminum oxychloride	300	149
Aluminum sulfate	300	149
Ammonium bifluoride	300	149
Ammonium carbonate	300	149
Ammonium chloride, 10%	300	149
Ammonium chloride, 50%	290	143
Ammonium chloride, sat.	300	149
Ammonium fluoride, 10%	300	149
Ammonium fluoride, 25%	300	149
Ammonium hydroxide, 25%	300	149
Ammonium hydroxide, sat.	300	149
Ammonium nitrate	230	110
Ammonium persulfate	300	149
Ammonium phosphate	300	149
Ammonium sulfate, 10–40%	300	149
Ammonium sulfide	300	149
Amyl acetate	250	121
Amyl alcohol	300	149
Amyl chloride	300	149
Aniline	230	110
Antimony trichloride	210	99
Aqua regia 3:1	210	99
Barium carbonate	300	149
Barium chloride	300	149
Barium hydroxide	300	149
Barium sulfate	300	149
Barium sulfide	300	149

(continued)

TABLE 2.14 *Continued*

Chemical	Maximum Temperature	
	°F	°C
Benzaldehyde	210	99
Benzene	210	99
Benzenesulfonic acid, 10%	210	99
Benzoic acid	270	132
Benzyl alcohol	300	149
Benzyl chloride	300	149
Borax	300	149
Boric acid	300	149
Bromine gas, dry	150	66
Bromine water, 10%	230	110
Butadiene	250	121
Butyl acetate	230	110
Butyl alcohol	300	149
n-Butylamine	120	49
Butyl phthalate	150	66
Butyric acid	250	121
Calcium bisulfide	300	149
Calcium carbonate	300	149
Calcium chlorate	300	149
Calcium chloride	300	149
Calcium hydroxide, 10%	300	149
Calcium hydroxide, sat.	300	149
Calcium hypochlorite	300	149
Calcium nitrate	300	149
Calcium oxide	260	127
Calcium sulfate	300	149
Caprylic acid	210	99
Carbon bisulfide	150	66
Carbon dioxide, dry	300	149
Carbon dioxide, wet	300	149
Carbon disulfide	150	66
Carbon monoxide	300	149
Carbon tetrachloride	270	132
Carbonic acid	300	149
Cellosolve	300	149
Chloroacetic acid, 50% water	230	110
Chloroacetic acid, 50%	230	110
Chlorine gas, dry	210	99
Chlorine gas, wet	250	121
Chlorine, water	100	38
Chlorobenzene	210	99
Chloroform	230	110
Chlorosulfonic acid	80	27
Chromic acid, 10%	150	66
Chromic acid, 50%	150	66
Chromyl chloride	210	99
Citric acid, 15%	120	49
Copper chloride	300	149

(continued)

TABLE 2.14 *Continued*

Chemical	Maximum Temperature	
	°F	°C
Copper cyanide	300	149
Copper sulfate	300	149
Cresol	270	132
Cupric chloride, 5%	300	149
Cyclohexane	300	149
Cyclohexanol	250	121
Dibutyl phthalate	150	66
Dichloroacetic acid	150	66
Ethylene glycol	300	149
Ferric chloride, 50% in water	300	149
Ferric nitrate, 10–50%	300	149
Ferrous chloride	300	149
Ferrous nitrate	300	149
Fluorine gas, dry	100	38
Fluorine gas, moist	100	38
Hydrobromic acid, dil.	300	149
Hydrobromic acid, 20%	300	149
Hydrobromic acid, 50%	300	149
Hydrochloric acid, 20%	300	149
Hydrochloric acid, 38%	300	149
Hydrocyanic acid, 10%	300	149
Hydrofluoric acid, 30%	270	132
Hydrofluoric acid, 70%	250	121
Hydrofluoric acid, 100%	230	110
Hypochlorous acid	300	149
Lactic acid, 25%	250	121
Lactic acid, conc.	250	121
Magnesium chloride	300	149
Malic acid	270	132
Manganese chloride	120	49
Methyl chloride	300	149
Methyl ethyl ketone	230	110
Methyl isobutyl ketone	300	149
Muriatic acid	300	149
Nitric acid, 5%	150	66
Nitric acid, 20%	150	66
Nitric acid, 70%	80	27
Nitrous acid, anhydrous	X	X
Nitrous acid, conc.	210	99
Oleum	150	66
Perchloric acid, 10%	230	110
Perchloric acid, 70%	150	66
Phenol	210	99
Phosphoric acid, 50–80%	270	132
Picric acid	130	54
Potassium bromide, 30%	300	149
Salicylic acid	250	121

(continued)

TABLE 2.14 *Continued*

Chemical	Maximum Temperature	
	°F	°C
Sodium carbonate	300	149
Sodium chloride	300	149
Sodium hydroxide, 10%	230	110
Sodium hydroxide, 50%	230	110
Sodium hypochlorite, 20%	300	149
Sodium hypochlorite, conc.	300	149
Sodium sulfide, to 50%	300	149
Stannic chloride	300	149
Stannous chloride	300	149
Sulfuric acid, 10%	300	149
Sulfuric acid, 50%	300	149
Sulfuric acid, 70%	300	149
Sulfuric acid, 90%	300	149
Sulfuric acid, 98%	300	149
Sulfuric acid, 100%	300	149
Sulfurous acid, fuming	120	49
Sulfurous acid	210	99
Thionyl chloride	210	99
Toluene	250	121
Trichloroacetic acid	210	99
Zinc chloride	300	149

The chemicals listed are in the pure state or in a saturated solution unless otherwise indicated. Compatibility is shown to the maximum allowable temperature for which date is available. Incompatibility is shown by an X. A blank space indicates that data is unavailable.

Source: From P.A. Schweitzer. 2004. *Corrosion Resistance Tables*, Vols. 1–4, 5th ed., New York: Marcel Dekker.

FEP has a maximum operating temperature of 375°F (190°C). After prolonged exposure at 400°F (204°C), it exhibits changes in physical strength. To improve some physical and mechanical properties, the polymer is filled with glass fibers.

FEP basically exhibits the same corrosion resistance as PTFE with few exceptions but at lower operating temperatures. It is resistant to practically all chemicals except for extremely potent oxidizers such as chlorine trifluoride and related compounds. Some chemicals will attack FEP when present in high concentrations at or near the service temperature limit. Refer to Table 2.15 for the compatibility of FEP with selected corrodents. Reference [1] provides a more detailed listing.

FEP is not degraded by UV light, and it has excellent weathering resistance.

FEP finds extensive use as a lining material for process vessels and piping, laboratory ware, and other process equipment.

TABLE 2.15

Compatibility of FEP with Selected Corrodents

Chemical	Maximum Temperature	
	°F	°C
Acetaldehyde	200	93
Acetamide	400	204
Acetic acid, 10%	400	204
Acetic acid, 50%	400	204
Acetic acid, 80%	400	204
Acetic acid, glacial	400	204
Acetic anhydride	400	204
Acetone[a]	400	204
Acetyl chloride	400	204
Acrylic acid	200	93
Acrylonitrile	400	204
Adipic acid	400	204
Allyl alcohol	400	204
Allyl chloride	400	204
Alum	400	204
Aluminum acetate	400	204
Aluminum chloride, aqueous	400	204
Aluminum chloride, dry	300	149
Aluminum fluoride[b]	400	204
Aluminum hydroxide	400	204
Aluminum nitrate	400	204
Aluminum oxychloride	400	204
Aluminum sulfate	400	204
Ammonia gas[b]	400	204
Ammonium bifluoride[a]	400	204
Ammonium carbonate	400	204
Ammonium chloride, 10%	400	204
Ammonium chloride, 50%	400	204
Ammonium chloride, sat.	400	204
Ammonium fluoride, 10%[b]	400	204
Ammonium fluoride, 25%[b]	400	204
Ammonium hydroxide, 25%	400	204
Ammonium hydroxide, sat.	400	204
Ammonium nitrate	400	204
Ammonium persulfate	400	204
Ammonium phosphate	400	204
Ammonium sulfate, 10–40%	400	204
Ammonium sulfide	400	204
Ammonium sulfite	400	204
Amyl acetate	400	204
Amyl alcohol	400	204
Amyl chloride	400	204
Aniline[a]	400	204
Antimony trichloride	250	121
Aqua regia 3:1	400	204
Barium carbonate	400	204

(continued)

TABLE 2.15 *Continued*

Chemical	Maximum Temperature	
	°F	°C
Barium chloride	400	204
Barium hydroxide	400	204
Barium sulfate	400	204
Barium sulfide	400	204
Benzaldehyde[a]	400	204
Benzene[a,b]	400	204
Benzenesulfonic acid, 10%	400	204
Benzoic acid	400	204
Benzyl alcohol	400	204
Benzyl chloride	400	204
Borax	400	204
Boric acid	400	204
Bromine gas, dry[b]	200	93
Bromine gas, moist[b]	200	93
Bromine, liquid[a,b]	400	204
Butadiene[a]	400	204
Butyl acetate	400	204
Butyl alcohol	400	204
n-Butylamine[a]	400	204
Butyric acid	400	204
Calcium bisulfide	400	204
Calcium bisulfite	400	204
Calcium carbonate	400	204
Calcium chlorate	400	204
Calcium chloride	400	204
Calcium hydroxide, 10%	400	204
Calcium hydroxide, sat.	400	204
Calcium hypochlorite	400	204
Calcium nitrate	400	204
Calcium oxide	400	204
Calcium sulfate	400	204
Caprylic acid	400	204
Carbon bisulfide[b]	400	204
Carbon dioxide, dry	400	204
Carbon dioxide, wet	400	204
Carbon disulfide	400	204
Carbon monoxide	400	204
Carbon tetrachloride[a,b,c]	400	204
Carbonic acid	400	204
Cellosolve	400	204
Chloroacetic acid, 50% water	400	204
Chloroacetic acid	400	204
Chlorine gas, dry	X	X
Chlorine gas, wet[b]	400	204
Chlorine, liquid[a]	400	204
Chlorobenzene[b]	400	204
Chloroform[b]	400	204
Chlorosulfonic acid[a]	400	204

(continued)

TABLE 2.15 *Continued*

Chemical	Maximum Temperature	
	°F	°C
Chromic acid, 10%	400	204
Chromic acid, 50%[a]	400	204
Chromyl chloride	400	204
Citric acid, 15%	400	204
Citric acid, conc.	400	204
Copper acetate	400	204
Copper carbonate	400	204
Copper chloride	400	204
Copper cyanide	400	204
Copper sulfate	400	204
Cresol	400	204
Cupric chloride, 5%	400	204
Cupric chloride, 50%	400	204
Cyclohexane	400	204
Cyclohexanol	400	204
Dibutyl phthalate	400	204
Dichloroacetic acid	400	204
Dichloroethane (ethylene dichloride[b])	400	204
Ethylene glycol	400	204
Ferric chloride	400	204
Ferric chloride, 50% in water[a]	260	127
Ferric nitrate, 10–50%	260	127
Ferrous chloride	400	204
Ferrous nitrate	400	204
Fluorine gas, dry	200	93
Fluorine gas, moist	X	X
Hydrobromic acid, dil.	400	204
Hydrobromic acid, 20%[b,c]	400	204
Hydrobromic acid, 50%[b,c]	400	204
Hydrochloric acid, 20%[b,c]	400	204
Hydrochloric acid, 38%[b,c]	400	204
Hydrocyanic acid, 10%	400	204
Hydrofluoric acid, 30%[b]	400	204
Hydrofluoric acid, 70%[b]	400	204
Hydrofluoric acid, 100%[b]	400	204
Hypochlorous acid	400	204
Iodine solution, 10%[b]	400	204
Ketones, general	400	204
Lactic acid, 25%	400	204
Lactic acid, conc.	400	204
Magnesium chloride	400	204
Malic acid	400	204
Manganese chloride	300	149
Methyl chloride[b]	400	204
Methyl ethyl ketone[b]	400	204
Methyl isobutyl ketone[b]	400	204
Muriatic acid[b]	400	204
Nitric acid, 5%[b]	400	204

(continued)

TABLE 2.15 *Continued*

Chemical	Maximum Temperature	
	°F	°C
Nitric acid, 20%[b]	400	204
Nitric acid, 70%[b]	400	204
Nitric acid, anhydrous[b]	400	204
Nitrous acid, conc.	400	204
Oleum	400	204
Perchloric acid, 10%	400	204
Perchloric acid, 70%	400	204
Phenol[b]	400	204
Phosphoric acid, 50–80%	400	204
Picric acid	400	204
Potassium bromide, 30%	400	204
Salicylic acid	400	204
Silver bromide, 10%	400	204
Sodium carbonate	400	204
Sodium chloride	400	204
Sodium hydroxide, 10%[a]	400	204
Sodium hydroxide, 50%	400	204
Sodium hydroxide, conc.	400	204
Sodium hypochlorite, 20%	400	204
Sodium hypochlorite, conc.	400	204
Sodium sulfide, to 50%	400	204
Stannic chloride	400	204
Stannous chloride	400	204
Sulfuric acid, 10%	400	204
Sulfuric acid, 50%	400	204
Sulfuric acid, 70%	400	204
Sulfuric acid, 90%	400	204
Sulfuric acid, 98%	400	204
Sulfuric acid, 100%	400	204
Sulfuric acid, fuming[b]	400	204
Sulfurous acid	400	204
Thionyl chloride[b]	400	204
Toluene[b]	400	204
Trichloroacetic acid	400	204
White liquor	400	204
Zinc chloride[c]	400	204

The chemicals listed are in the pure state or in a saturated solution unless otherwise indicated. Compatibility is shown to the maximum allowable temperature for which date is available. Incompatibility is shown by an X. A blank space indicates that data is unavailable.

[a] Material will be absorbed.
[b] Material will permeate.
[c] Material can cause stress cracking.

Source: From P.A. Schweitzer. 2004. *Corrosion Resistance Tables*, Vols. 1–4, 5th ed., New York: Marcel Dekker.

2.8 Polyamides (PA)

Polyamides are also known as nylons. Polyamide polymers are available in several grades and are identified by the number of carbon atoms in the diamine and dibasic acid used to produce the particular grade. For example, nylon 6/6 is the reaction product of hexamethylenediamine and adipic acid, both of which are compounds containing six carbon atoms. Some of the commonly commercially available nylons are 6, 6/6, 6/10, 11, and 12. Their structural formulas are as follows:

$$\left[-N-(CH_2)_5 - \overset{\overset{\displaystyle O}{\|}}{C} - \right]_n$$

Nylon 6

$$\left[-N-(CH_2)_6 - N - \overset{\overset{\displaystyle O}{\|}}{C} - (CH_2)_4 - \overset{\overset{\displaystyle O}{\|}}{C} - \right]_n$$

Nylon 6/6

$$\left[-N-(CH_2)_6 - N - \overset{\overset{\displaystyle O}{\|}}{C} - (CH_2)_8 - \overset{\overset{\displaystyle O}{\|}}{C} - \right]_n$$

Nylon 6/10

$$\left[-N-(CH_2)_{10} - \overset{\overset{\displaystyle O}{\|}}{C} - \right]_n$$

Nylon 11

$$\left[-N-(CH_2)_{11} - \overset{\overset{\displaystyle O}{\|}}{C} - \right]_n$$

Nylon 12

Grades 6 and 6/6 are the strongest structurally, grades 6/10 and 11 have the lowest moisture absorption, best electrical, properties and best dimensional stability, and grades 6, 6/6, and 6/10 are the most flexible. Nylon 12 has the same advantages as grades 6/10 and 11, but it is available at a lower cost because it is more easily and economically processed.

Polyamide resins retain useful mechanical properties over a temperature range of -60 to $+400°F$ (-51 to $+204°C$.) Both short-term and long-term temperature effects must be considered. Such properties as stiffness and toughness are affected by short-term exposure to elevated temperatures. There is also the possibility of stress relief and its effect on dimensions.

In long term exposure at high temperatures, there is the possibility of gradual oxidative embrittlement. Such applications should make use of a heat-stabilized grade of resin.

The polyamides exhibit excellent resistance to a broad range of chemicals and harsh environments. They have good resistance to most inorganic alkalies, particularly ammonium hydroxide and ammonia, even at elevated temperatures and to sodium and potassium hydroxides at ambient temperatures. They also display good resistance to almost all inorganic salts and to almost all hydrocarbons and petroleum-based fuels.

Polyamides are also resistant to UV degradation, weathering, and ozone. Refer to Table 2.16 for the compatibility of PA with selected corrodents. A more detailed listing will be found in Reference [1].

TABLE 2.16

Compatibility of Polyamides with Selected Corrodents

Chemical	Maximum Temperature	
	°F	°C
Acetaldehyde	X	X
Acetamide	250	121
Acetic acid, 10%	200	93
Acetic acid, 50%	X	X
Acetic acid, 80%	X	X
Acetic acid, glacial	X	X
Acetic anhydride	200	93
Acetone	80	27
Acetyl chloride	X	X
Acrylonitrile	80	27
Allyl alcohol	80	27
Alum	X	X
Aluminum chloride, aqueous	X	X
Aluminum chloride, dry	X	X
Aluminum fluoride	80	27
Aluminum hydroxide	250	121
Aluminum nitrate	80	27
Aluminum sulfate	140	60
Aluminum gas	200	93
Ammonium carbonate	240	160
Ammonium chloride, 10%	200	93
Ammonium chloride, 50%	200	93
Ammonium fluoride, 10%	80	27
Ammonium fluoride, 25%	80	27
Ammonium hydroxide, 25%	250	121

(continued)

TABLE 2.16 *Continued*

Chemical	Maximum Temperature	
	°F	°C
Ammonium hydroxide, sat.	250	121
Ammonium nitrate	190	88
Ammonium persulfate	X	X
Ammonium phosphate	80	27
Ammonium sulfite	80	27
Amyl acetate	150	66
Amyl alcohol	200	93
Amyl chloride	X	X
Aniline	X	X
Antimony trichloride	X	X
Aqua regia 3:1	X	X
Barium carbonate	80	27
Barium chloride	250	121
Barium hydroxide	80	27
Barium sulfate	80	27
Barium sulfide	80	27
Benzaldehyde	150	66
Benzene	250	121
Benzenesulfonic acid, 10%	X	X
Benzoic acid	80	27
Benzyl alcohol	200	93
Benzyl chloride	250	121
Borax	200	93
Boric acid	X	X
Bromine gas, dry	X	X
Bromine gas, moist	X	X
Bromine, liquid	X	X
Butadiene	80	27
Butyl acetate	250	121
Butyl alcohol	200	93
n-Butylamine	200	93
Butyric acid	X	X
Calcium bisulfite	140	60
Calcium carbonate	200	93
Calcium chloride	250	121
Calcium hydroxide, 10%	150	66
Calcium hydroxide, sat.	150	66
Calcium hypochlorite	X	X
Calcium nitrate	X	X
Calcium oxide	80	27
Calcium sulfate	80	27
Caprylic acid	230	110
Carbon bisulfide	80	27
Carbon dioxide, dry	80	27
Carbon disulfide	80	27
Carbon monoxide	80	27
Carbon tetrachloride	250	121
Carbonic acid	100	38

(continued)

TABLE 2.16 *Continued*

Chemical	Maximum Temperature	
	°F	°C
Cellosolve	250	121
Chloroacetic acid, 50% water	X	X
Chloroacetic acid	X	X
Chlorine gas, dry	X	X
Chlorine gas, wet	X	X
Chlorine, liquid	X	X
Chlorobenzene	250	121
Chloroform	130	54
Chlorosulfonic acid	X	X
Chromic acid, 10%	X	X
Chromic acid, 50%	X	X
Citric acid, 15%	200	93
Chloroacetic acid	X	X
Citric acid, conc.	200	93
Copper carbonate	80	27
Copper chloride	X	X
Copper cyanide	80	27
Copper sulfate	140	60
Cresol	100	38
Cupric chloride, 5%	X	X
Cupric chloride, 50%	X	X
Cyclohexane	250	121
Cyclohexanol	280	121
Dibutyl phthalate	80	27
Dichloroethane (ethylene dichloride)	80	27
Ethylene glycol	200	93
Ferric chloride	X	X
Ferric chloride 50% in water	X	X
Ferric nitrate, 10–50%	X	X
Ferrous chloride	X	X
Fluorine gas, dry	X	X
Fluorine gas, moist	X	X
Hydrobromic acid, dil.	X	X
Hydrobromic acid, 20%	X	X
Hydrobromic acid, 50%	X	X
Hydrochloric acid, 20%	X	X
Hydrochloric acid, 38%	X	X
Hydrofluoric acid, 30%	X	X
Hydrofluoric acid, 70%	X	X
Hydrofluoric acid, 100%	X	X
Lactic acid, 25%	200	93
Lactic acid, conc.	200	93
Magnesium chloride, 50%	240	116
Malic acid	X	X
Manganese chloride, 37%	X	X
Methyl chloride	80	27
Methyl ethyl ketone	240	116
Methyl isobutyl ketone	110	42

(continued)

TABLE 2.16 *Continued*

Chemical	Maximum Temperature	
	°F	°C
Muriatic acid	X	X
Nitric acid, 5%	X	X
Nitric acid, 20%	X	X
Nitric acid, 70%	X	X
Nitric acid, anhydrous	X	X
Nitrous acid, conc.	X	X
Oleum	X	X
Phenol	X	X
Phosphoric acid, 50–80%	X	X
Picric acid	X	X
Potassium bromide, 30%	210	99
Salicylic acid	80	27
Sodium carbonate, to 30%	240	116
Sodium chloride	230	110
Sodium hydroxide, 10%	250	121
Sodium hydroxide, 50%	250	121
Sodium hypochlorite, 20%	X	X
Sodium hypochlorite, conc.	X	X
Sodium sulfide, to 50%	230	110
Stannic chloride	80	27
Stannous chloride	X	X
Sulfuric acid, 10%	X	X
Sulfuric acid, 50%	X	X
Sulfuric acid, 70%	X	X
Sulfuric acid, 90%	X	X
Sulfuric acid, 98%	X	X
Sulfuric acid, 100%	X	X
Sulfuric acid, fuming	X	X
Sulfurous acid	90	32
Toluene	200	93
Trichloroacetic acid	X	X
White liquor	80	27
Zinc chloride	X	X

The chemicals listed are in the pure state or in a saturated solution unless otherwise indicated. Compatibility is shown to the maximum allowable temperature for which data is available. Incompatibility is shown by an X. A blank space indicate that data is unavailable.

Source: From P.A. Schweitzer. 2004. *Corrosion Resistance Tables*, Vols. 1–4, 5th ed., New York: Marcel Dekker.

PAs are used to produce gears, cans, bearings, wire insulation, pipe fittings, and hose fittings. PA/ABS blends are used for appliances, lawn and garden equipment, power tools, and sporting goods. In the automotive industry, the blend is used to produce interior functional components, fasteners, housings, and shrouds.

2.9 Polyamide–Imide (PAI)

Polyamide–imides are heterocyclic polymers having an atom of nitrogen in one of the rings of the molecular chain. A typical chemical structure is shown below:

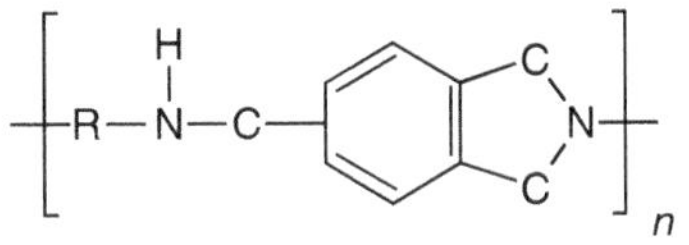

This series of thermoplasts can be used at high and low temperatures, and as such, it finds applications in the extreme environments of space. The temperature range is −310°F to +500°F (−190°C to +260°C). PAI is resistant to acetic acid and phosphoric acid up to 35% and sulfuric acid up to 30%.

PAI is not resistant to sodium hydroxide. PAI has resistance to UV light degradation. Refer to Table 2.17 for the compatibility of PAI with selected corrodents. Reference 1 provides a more detailed listing.

PAI finds applications in underhood applications and as bearings and pistons in compressors.

TABLE 2.17

Compatibility of PAI with Selected Corrodents

| | Maximum Temperature | |
Chemical	°F	°C
Acetic acid, 10%	200	93
Acetic acid, 50%	200	93
Acetic acid, 80%	200	93
Acetic acid glacial	200	93
Acetic anhydride	200	93
Acetone	80	27
Acetyl chloride, dry	120	49
Aluminum sulfate, 10%	220	104
Ammonium chloride, 10%	200	93
Ammonium hydroxide, 25%	200	93
Ammonium hydroxide, sat.	200	93
Ammonium nitrate, 10%	200	93
Ammonium sulfate, 10%	200	93
Amyl acetate	200	93
Aniline	200	93
Barium chloride, 10%	200	93
Benzaldehyde	200	93
Benzene	80	27

(continued)

TABLE 2.17 *Continued*

Chemical	Maximum Temperature	
	°F	°C
Benzenesulfonic acid, 10%	X	
Benzyl chloride	120	49
Bromine gas, moist	120	49
Butyl acetate	200	93
Butyl alcohol	200	93
n-Butylamine	200	93
Calcium chloride	200	93
Calcium hypochlorite	X	
Cellosolve	200	93
Chlorobenzene	200	93
Chloroform	120	49
Chromic acid, 10%	200	93
Cyclohexane	200	93
Cyclohexanol	200	93
Dibutyl phthalate	200	93
Ethylene glycol	200	93
Hydrochloric acid, 20%	200	93
Hydrochloric acid, 38%	200	93
Lactic acid, 25%	200	93
Lactic acid conc.	200	93
Magnesium chloride, dry	200	93
Methyl ethyl ketone	200	93
Oleum	120	49
Sodium carbonate, 10%	200	93
Sodium chloride, 10%	200	93
Sodium hydroxide, 10%	X	
Sodium hydroxide, 50%	X	
Sodium hydroxide conc.	X	
Sodium hypochlorite, 10%	200	93
Sodium hypochlorite conc.	X	
Sodium sulfide, to 50%	X	
Sulfuric acid, 10%	200	93
Sulfuric acid, fuming	120	49
Toluene	200	93

The chemicals listed are in the pure state or in a saturated solution unless otherwise indicated. Compatibility is shown to the maximum allowable temperature for which data is available. Incompatibility is shown by an X.

Source: From P.A. Schweitzer. 2004. *Corrosion Resistance Tables*, Vols. 1–4, 5th ed., New York: Marcel Dekker.

2.10 Polybutylene (PB)

Polybutylene is a semicrystalline polyolefin thermoplasitic based on poly-1-butene and includes homopolymers and a series of copolymers (butene/ethylene). This thermoplast has the following structural formula:

$$-\overset{\overset{\displaystyle H}{|}}{\underset{\underset{\displaystyle H}{|}}{C}}-\overset{\overset{\displaystyle H}{|}}{\underset{\underset{\displaystyle H}{|}}{C}}-\overset{\overset{\displaystyle H}{|}}{\underset{\underset{\underset{\underset{\underset{\displaystyle H}{|}}{\overset{\displaystyle H}{H-C-H}}}{|}}{|}}{C}}-$$

Polybutylene has an upper temperature limit of 200°F (93°C). It possesses a combination of stress cracking resistance, chemical resistance, and abrasion resistance. It is resistant to acids, bases, soaps, and detergents. Polybutylene is partially soluble in aromatic and chlorinated hydrocarbons above 140°F (60°C) and is not completely resistant to aliphatic solvents at room temperature. Chlorinated water will cause pitting attack. Refer to Table 2.18 for the compatibility of polybutylene with selected corrodents. PB is subject to degradation by UV light.

Applications include piping, chemical process equipment, and fly ash bottom ash lines containing abrasive slurries. PB is also used for molded appliance parts.

TABLE 2.18

Compatibility of PB with Selected Corrodents

Acetic acid	R
Acetic anhydride	R
Allyl alcohol	X
Aluminum chloride	X
Ammonium chloride	X
Ammonium hydroxide	R
Amyl alcohol	X
Aniline	R
Benzaldehyde	R
Benzene	R
Benzoic acid	R
Boric acid	R
Butyl alcohol	X
Calcium carbonate	R
Calcium hydroxide	R
Calcium sulfate	R
Carbonic acid	R
Chloracetic acid	R
Chlorobenzene	R
Citric acid	R
Cyclohexane	R
Detergents	R
Lactic acid	R
Malic acid	R
Methyl alcohol	X
Phenol	R

(continued)

TABLE 2.18 *Continued*

Picric acid	R
Propyl alcohol	X
Salicylic acid	R
Soaps	R
Sodium carbonate	R
Sodium hydroxide, 10%	R
Sodium hydroxide, 50%	R
Toluene	R
Trichloroacetic acid	R
Water (chlorine free)	R
Xylene	R

Materials in the pure state or saturated solution unless otherwise specified. R, PB is resistant at 70°F/23°C; X, PB is not resistant.

2.11 Polycarbonate (PC)

This thermoplast is produced by General Electric under the tradename of Lexan. Polycarbonates are classified as engineering thermoplasts because of their high performance in engineering designs. The generalized formula is:

Immersion in water and exposure to high humidity at temperatures up to 212°F (100°C) have little effect on dimensions. Polycarbonates are among the most stable polymers in a wet environment. The operating temperature range for polycarbonates is from −200°F to +250−275°F (−129°C to +120−135°C).

PC is resistant to aliphatic hydrocarbons and weak acids and has limited resistance to weak alkalies. It is resistant to most oils and greases. Polycarbonate will be attacked by strong alkalies and strong acids and is soluble in ketones, esters, and aromatic and chlorinated hydrocarbons. PC is not affected by UV light and has excellent weatherability. Table 2.19 lists the compatibility of PC with selected corrodents. Reference [1] provides a more detailed listing.

Because of its extremely high impact resistance and good clarity, it is widely used for windows in chemical equipment and glazing in chemical plants. It also finds wide use in outdoor energy management devices, network interfaces, electrical wiring blocks, telephone equipment, and lighting diffusers, globes and housings.

TABLE 2.19

Compatibility of PC with Selected Corrodents

Acetic acid, 5%	R
Acetic acid, 10%	R
Acetone	X
Ammonia, 10%	X
Benzene	X
Butyl acetate	X
Cadmium chloride	R
Carbon disulfide	X
Carbon tetrachloride	X
Chlorobenzene	X
Chloroform	X
Citric acid, 10%	R
Diesel oil	X
Dioxane	X
Edible oil	R
Ethanol	X
Ether	X
Ethyl acetate	X
Ethylene chloride	X
Ethylene glycol	R
Formic acid	X
Fruit juice	R
Fuel oil	R
Glycerine	X
Heptane/hexane	R
Hydrochloric acid, 38%	R
Hydrochloric acid, 2%	R
Hydrogen peroxide, 30%	R
Hydrogen peroxide, 0–5%	R
Ink	R
Linseed oil	R
Methanol	X
Methyl ethyl ketone	X
Methylene chloride	X
Milk	R
Nitric acid, 2%	R
Paraffin oil	R
Phosphoric acid, 10%	X
Potassium hydroxide, 50%	X
Potassium dichromate	R
Potassium permanganate, 10%	R
Silicone oil	R
Soap solution	R
Sodium bisulfite	R
Sodium carbonate, 10%	R
Sodium chloride, 10%	R
Sodium hydroxide, 50%	X
Sodium hydroxide, 5%	X
Sulfuric acid, 98%	X
Sulfuric acid, 2%	R

(continued)

TABLE 2.19 *Continued*

Toluene	X
Vaseline	R
Water, cold	R
Water, hot	X
Wax, molten	R
Xylene	X

R, material is resistant at 73°F/20°C; X, material is not resistant.

2.12 Polyetheretherketone (PEEK)

PEEK is a linear polyaromatic thermoplast having the following chemical structure:

It is a proprietary product of ICI and is an engineering polymer suitable for applications that require mechanical strength with the need to resist difficult thermal and chemical environments. PEEK has an operating temperature range of −85 to +480°F (−65 to +250°C).

PEEK is not chemically attacked by water. It has excellent long-term resistance to water at both ambient and elevated temperatures. It also has excellent rain erosion resistance. Because PEEK is not hydrolyzed by water at elevated temperatures in a continuous cycle environment, the material may be steam sterilized using conventional sterilization equipment.

PEEK is insoluble in all common solvents and has excellent resistance to a wide range of organic and inorganic solvents.

It also exhibits excellent resistance to hard (gamma) radiation, absorbing over 1000 Mrads of radiation without suffering significant damage.

Like most polyaromatics, PEEK is subject to degradation by UV light during outdoor weathering. However, testing has shown that over a twelve-month period for both natural and pigmented moldings, the effect is minimal. In more extreme weathering conditions, painting or pigmenting of the polymer will protect it from excessive property degradation.

Refer to Table 2.20 for the compatibility of PEEK with selected corrodents. Reference [1] provides a more detailed listing.

TABLE 2.20

Compatibility of PEEK with Selected Corrodents

Chemical	Maximum Temperature	
	°F	°C
Acetaldehyde	80	27
Acetic acid, 10%	80	27
Acetic acid, 50%	140	60
Acetic acid, 80%	140	60
Acetic acid, glacial	140	60
Acetone	210	99
Acrylic acid	80	27
Acrylonitrile	80	27
Aluminum sulfate	80	27
Ammonia gas	210	99
Ammonium hydroxide, sat.	80	27
Aniline	200	93
Aqua regia 3:1	X	X
Benzaldehyde	80	27
Benzene	80	27
Benzoic acid	170	77
Boric acid	80	27
Bromine gas, dry	X	X
Bromine gas, moist	X	X
Calcium carbonate	80	27
Calcium chloride	80	27
Calcium hydroxide, 10%	80	27
Calcium hydroxide, sat.	100	38
Carbon dioxide, dry	80	27
Carbon tetrachloride	80	27
Carbonic acid	80	27
Chlorine gas, dry	80	27
Chlorine, liquid	X	X
Chlorobenzene	200	93
Chloroform	80	27
Chlorosulfonic acid	80	27
Chromic acid, 10%	80	27
Chromic acid, 50%	200	93
Citric acid, conc.	170	77
Cyclohexane	80	27
Ethylene glycol	160	71
Ferrous chloride	200	93
Fluorine gas, dry	X	X
Fluorine gas, moist	X	X
Hydrobromic acid, dil.	X	X
Hydrobromic acid, 20%	X	X
Hydrobromic acid, 50%	X	X
Hydrochloric acid, 20%	100	38
Hydrochloric acid, 38%	100	38
Hydrofluoric acid, 30%	X	X
Hydrofluoric acid, 70%	X	X

(continued)

TABLE 2.20 *Continued*

Chemical	Maximum Temperature	
	°F	°C
Hydrofluoric acid, 100%	X	X
Hydrogen sulfide, wet	200	93
Ketones, general	80	27
Lactic acid, 25%	80	27
Lactic acid, conc.	100	38
Magnesium hydroxide	100	38
Methyl alcohol	80	27
Methyl ethyl ketone	370	188
Naphtha	200	93
Nitric acid, 5%	200	93
Nitric acid, 20%	200	93
Nitrous acid, 10%	80	27
Oxalic acid, 5%	X	X
Oxalic acid, 10%	X	X
Oxalic acid, sat.	X	X
Phenol	140	60
Phosphoric acid, 50–80%	200	93
Potassium bromide, 30%	140	60
Sodium carbonate	210	99
Sodium hydroxide, 10%	220	104
Sodium hydroxide, 50%	180	82
Sodium hydroxide, conc.	200	93
Sodium hypochlorite, 20%	80	27
Sodium hypochlorite, conc.	80	27
Sulfuric acid, 10%	80	27
Sulfuric acid, 50%	200	93
Sulfuric acid, 70%	X	X
Sulfuric acid, 90%	X	X
Sulfuric acid, 98%	X	X
Sulfuric acid, 100%	X	X
Sulfuric acid, fuming	X	X
Toluene	80	27
Zinc chloride	100	38

The Chemicals listed are in the pure state or in a saturated solution unless otherwise indicated. Compatibility is shown to the maximum allowable temperature for which data is available. Incompatibility is shown by an X. A blank space indicates that data is unavailable.

Source: From P.A. Schweitzer. 2004. *Corrosion Resistance Tables*, Vols. 1–4, 5th ed., New York: Marcel Dekker.

PEEK is used in many bearing applications because of the following advantages:

1. High strength and load carrying capacity
2. Low friction
3. Good dimensional stability
4. Long life
5. A high, continuous service temperature of 500°F (260°C)
6. Excellent wear abrasion and fatigue resistance
7. Outstanding mechanical properties

PEEK has also found application in a number of aircraft exterior applications such as radomes and fairings as a result of its excellent erosion resistance.

2.13 Polyether–Imide (PEI)

PEI is a high-temperature engineering polymer with the structural formula shown below:

Polyether–imides are noncrystalline polymers made up of alternating aromatic ether and imide units. The molecular structure has rigidity, strength, and impact resistance in fabricated parts over a wide range of temperatures. PEI is one of the strongest thermoplasts even without reinforcement.

PEI has a better chemical resistance than most noncrystalline polymers. It is resistant to acetic and hydrochloric acids, weak nitric and sulfuric acids, and alcohol. The unfilled polymer complies with FDA regulations and can be used in food and medical applications. Table 2.21 lists the compatibility of PEI with selected corrodents. Reference [1] provides a more comprehensive listing.

PEI is useful in applications where high heat and flame resistance low NBS smoke evolution, high tensile and flexural strength, stable electrical properties, over a wide range of temperatures, and frequencies, chemical resistance, and superior finishing characteristics are required. One such application is under the hood of automobiles for connecters and MAP sensors.

TABLE 2.21

Compatibility of PEI with Selected Corrodents

Acetic acid, glacial	R
Acetic acid, 80%	R
Acetic acid, 10%	R
Acetone	R
Ammonia, 10%	X
Ammonium hydroxide, 25%	R
Ammonium Hydroxide, sat.	R
Benzoic acid	R
Carbon tetrachloride	R
Chloroform	X
Citric acid, 10%	R
Citric acid, conc.	R
Cyclohexane	R
Diesel oil	R
Ethanol	R
Ether	R
Ethyl acetate	X
Ethylene glycol	R
Formic acid, 10%	R
Fruit juice	R
Fuel oil	R
Gasoline	R
Heptane/hexane	R
Hydrochloric acid, 38%	R
Hydrochloric acid, 2%	R
Hydrochloric acid, 30%	R
Iodine solution, 10%	R
Isopropanol	R
Methanol	R
Methyl ethyl ketone	X
Methyl chloride	X
Motor oil	R
Nitric acid, 20%	R
Nitric acid, 2%	R
Perchloroethylene	R
Phosphoric acid, 50–80%	R
Potassium hydroxide, 10%	R
Potassium permanganate, 10%	R
Sodium chloride, 10%	R
Sodium hydroxide, 10%	R
Sodium hypochlorite, 20%	R
Sodium hypochlorite, conc.	R
Sulfuric acid, 10%	R
Sulfuric acid, 50%	R
Sulfuric acid, fuming	X
Toluene	R
Trichloroethylene	X
Water, cold	R
Water, hot	R

R, material resistant at 73°F/20°C; X, material not resistant.

2.14 Polyether Sulfone (PES)

Polyether sulfone is a high-temperature engineering thermoplastic with the combined characteristics of high thermal stability and mechanical strength. It is a linear polymer with the following structure:

PES also goes under the name *Ultem*.

Small dimensional changes may occur as a result of the *polymer's* absorbing water from the atmosphere. At equilibrium water content for 65% relative humidity, these dimensional changes are of the order of magnitude of 0.15%; whereas, in boiling water, they are approximately 0.3%. The water may be removed by heating at 300–355°F (149–180°C). This predrying will also prevent outgassing at elevated temperatures.

PES has excellent resistance to aliphatic hydrocarbons, some chlorinated hydrocarbons, and aromatics. It is also resistant to most inorganic chemicals. Hydrocarbons and mineral oils, greases, and transmission fluids have no effect on PES.

Polyether sulfone will be attacked by strong oxidizing acids, but glass-fiber reinforced grades are resistant to more dilute acids. PES is soluble is highly polar solvents and is subject to stress cracking in ketones and esters. Polyether sulfone does not have good outdoor weathering properties. Since it is susceptible to degradation by UV light, if used outdoors, it must be stabilized by incorporating carbon black or by painting.

Refer to Table 2.22 for the compatibility of PES with selected corrodents. Reference [1] provides a more comprehensive listing.

Polyether sulfone has found application in

Medical appliances
Chemical plants
Fluid handling
Aircraft and aerospace appliances
Instrumentation housings
Office equipment
Photocopier parts
Electrical components
Automotive (carbureter parts, fuse boxes)

TABLE 2.22

Compatibility of PES with Selected Corrodents

Chemical	Maximum Temperature	
	°F	°C
Acetic acid, 10%	80	27
Acetic acid, 50%	140	60
Acetic acid, 80%	200	93
Acetic acid, glacial	200	93
Acetone	X	X
Ammonia, gas	80	27
Aniline	X	X
Benzene	80	27
Benzenesulfonic acid, 10%	100	38
Benzoic acid	80	27
Carbon tetrachloride	80	27
Chlorobenzene	X	X
Chlorosulfonic acid	X	X
Chromic acid, 10%	X	X
Chromic acid, 50%	X	X
Citric acid, conc.	80	27
Ethylene glycol	100	38
Ferrous chloride	100	38
Hydrochloric acid, 20%	140	60
Hydrochloric acid, 38%	140	60
Hydrogen sulfide, wet	80	27
Methyl ethyl ketone	X	X
Naphtha	80	27
Nitric acid, 5%	80	27
Nitric acid, 20%	X	X
Oxalic acid, 5%	80	27
Oxalic acid, 29%	80	27
Oxalic acid, sat.	80	27
Phenol	X	X
Phosphoric acid, 50–80%	200	93
Potassium bromide, 30%	140	60
Sodium carbonate	80	27
Sodium chloride	80	27
Sodium hydroxide, 20%	80	27
Sodium hydroxide, 50%	80	27
Sodium hypochlorite, 20%	80	27
Sodium hypochlorite, conc.	80	27
Sulfuric acid, 10%	80	27
Sulfuric acid, 50%	X	X
Sulfuric acid, 70%	X	X
Sulfuric acid, 90%	X	X
Sulfuric acid, 98%	X	X
Sulfuric acid, 100%	X	X
Sulfuric acid, fuming	X	X
Toluene	80	27

The chemicals listed are in the pure state or in a saturated solution unless otherwise indicated. Compatibility is shown to the maximum allowable temperature for which data is available. Incompatibility is shown by an X.

Source: From P.A. Schweitzer. 2004. *Corrosion Resistance Tables*, Vols. 1–4, 5th ed., New York: Marcel Dekker.

2.15 Perfluoralkoxy (PFA)

Perfluoralkoxy is a fully fluorinated polymer having the following formula:

$$
\begin{array}{ccccc}
\text{F} & \text{F} & \text{F} & \text{F} & \text{F} \\
| & | & | & | & | \\
-\text{C}-\text{C}-\text{C}-\text{C}-\text{C}- \\
| & | & | & | & | \\
\text{F} & \text{F} & \text{O} & \text{F} & \text{F} \\
& & | \\
& & \text{R}_f
\end{array}
\qquad R_f = C_n F_{2n} + 1
$$

PFA lacks the physical strength of PTFE at elevated temperatures but has somewhat better physical and mechanical properties than FEP above 300°F (149°C) and can be used up to 500°F (260°C). Like PTFE, PFA is subject to permeation by certain gases and will absorb selected chemicals. Refer to Table 2.23 for the absorption of certain liquids by PFA. Perfluoralkoxy also performs well at cryogenic temperatures.

TABLE 2.23

Absorption of Liquids by PFA

Liquid[a]	Temperature (°F/°C)	Range of Weight Gains (%)
Aniline	365/185	0.3–0.4
Acetophenone	394/201	0.6–0.8
Benzaldehyde	354/179	0.4–0.5
Benzyl alcohol	400/204	0.3–0.4
n-Butylamine	172/78	0.3–0.4
Carbon tetrachloride	172/78	2.3–2.4
Dimethyl sulfoxide	372/190	0.1–0.2
Freon 113	117/47	1.2
Isooctane	210/99	0.7–0.8
Nitrobenzene	410/210	0.7–0.9
Perchloroethylene	250/121	2.0–2.3
Sulfuryl chloride	154/68	1.7–2.7
Toluene	230/110	0.7–0.8
Tributyl phosphate[b]	392/200	1.8–2.0
Bromine, anhydrous	−5/−22	0.5
Chlorine, anhydrous	248/120	0.5–0.6
Chlorosulfonic acid	302/150	0.7–0.8
Chromic acid, 50%	248/120	0.00–0.01
Ferric chloride	212/100	0.00–0.01
Hydrochloric acid, 37%	248/120	0.00–0.03
Phosphoric acid, conc.	212/100	0.00–0.01
Zinc chloride	212/100	0.00–0.03

[a] Samples were exposed for 168 hours at the boiling point of the solvent. Exposure of the acidic reagent was for 168 h.
[b] Not boiling.

PFA is inert to strong mineral acids, organic bases, inorganic oxidizers, aromatics, some aliphatic hydrocarbons, alcohols, aldehydes, ketones, ethers, esters, chlorocarbons, fluorocarbons, and mixtures of these.

Perfluoralkoxy will be attacked by certain halogenated complexes containing fluorine. This includes chlorine trifluoride, bromine trifluoride, iodine pentafluoride, and fluorine. It is also subject to attack by such metals as sodium or potassium, particularly in their molten states. Refer to Table 2.24 for the compatibility of PFA with selected corrodents. Reference [1] provides a more detained listing. PFA has excellent weatherability and is not subject to UV degradation.

TABLE 2.24

Compatibility of PFA with Selected Corrodents

	Maximum Temperature	
Chemical	°F	°C
Acetaldehyde	450	232
Acetamide	450	232
Acetic acid, 10%	450	232
Acetic acid, 50%	450	232
Acetic acid, 80%	450	232
Acetic acid, glacial	450	232
Acetic anhydride	450	232
Acetone	450	232
Acetyl chloride	450	232
Acrylonitrile	450	232
Adipic acid	450	232
Allyl alcohol	450	232
Allyl chloride	450	232
Alum	450	232
Aluminum chloride, aqueous	450	232
Aluminum fluoride	450	232
Aluminum hydroxide	450	232
Aluminum nitrate	450	232
Aluminum oxychloride	450	232
Aluminum sulfate	450	232
Ammonia gas[a]	450	232
Ammonium bifluoride[a]	450	232
Ammonium carbonate	450	232
Ammonium chloride, 10%	450	232
Ammonium chloride, 50%	450	232
Ammonium chloride, sat.	450	232
Ammonium fluoride, 10%[a]	450	232
Ammonium fluoride, 25%[a]	450	232
Ammonium hydroxide, 25%	450	232
Ammonium hydroxide, sat.	450	232
Ammonium nitrate	450	232
Ammonium persulfate	450	232
Ammonium phosphate	450	232
Ammonium sulfate, 10–40%	450	232
Ammonium sulfide	450	232
Amyl acetate	450	232
Amyl alcohol	450	232

(continued)

TABLE 2.24 *Continued*

Chemical	Maximum Temperature	
	°F	°C
Amyl chloride	450	232
Aniline[b]	450	232
Antimony trichloride	450	232
Aqua regia 3:1	450	232
Barium carbonate	450	232
Barium chloride	450	232
Barium hydroxide	450	232
Barium sulfate	450	232
Barium sulfide	450	232
Benzaldehyde[b]	450	232
Benzene[a]	450	232
Benzene sulfonic acid, 10%	450	232
Benzoic acid	450	232
Benzyl alcohol[b]	450	232
Benzyl chloride[a]	450	232
Borax	450	232
Boric acid	450	232
Bromine gas, dry[a]	450	232
Bromine, liquid[a,b]	450	232
Butadiene[a]	450	232
Butyl acetate	450	232
Butyl alcohol	450	232
n-Butylamine[b]	450	232
Butyric acid	450	232
Calcium bisulfide	450	232
Calcium bisulfite	450	232
Calcium carbonate	450	232
Calcium chlorate	450	232
Calcium chloride	450	232
Calcium hydroxide, 10%	450	232
Calcium hydroxide, sat.	450	232
Calcium hypochlorite	450	232
Calcium nitrate	450	232
Calcium oxide	450	232
Calcium sulfate	450	232
Caprylic acid	450	232
Carbon bisulfide[a]	450	232
Carbon dioxide, dry	450	232
Carbon dioxide, wet	450	232
Carbon disulfide[a]	450	232
Carbon monoxide	450	232
Carbon tertrachloride[a,b,c]	450	232
Carbonic acid	450	232
Chloroacetic acid, 50% water	450	232
Chloroacetic acid	450	232
Chlorine gas, dry	X	X
Chlorine gas, wet[a]	450	232
Chlorine, liquid[b]	X	X
Chlorobenzene[a]	450	232
Chloroform[a]	450	232
Chlorosulfonic acid[b]	450	232
Chromic acid, 10%	450	232
Chromic acid, 50%[b]	450	232

(continued)

TABLE 2.24 *Continued*

Chemical	Maximum Temperature	
	°F	°C
Chromyl chloride	450	232
Citric acid, 15%	450	232
Citric acid, conc.	450	232
Copper carbonate	450	232
Copper carbonate	450	232
Copper chloride	450	232
Copper cyanide	450	232
Copper sulfate	450	232
Cresol	450	232
Cupric chloride, 5%	450	232
Cupric chloride, 50%	450	232
Cyclohexane	450	232
Cyclohexanol	450	232
Dibutyl phthalate	450	232
Dichloroacetic acid	450	232
Dichloroethane (ethylene di chloride)[a]	450	232
Ethylene glycol	450	232
Ferric chloride	450	232
Ferric chloride, 50% in water[b]	450	232
Ferric nitrate, 10–50%	450	232
Ferrous chloride	450	232
Ferrous nitrate	450	232
Florine gas, dry	X	X
Fluorine gas, moist	X	X
Hydrobromic acid, oil[a,c]	450	232
Hydrobromic acid, 20%[a,c]	450	232
Hydrobromic acid, 50%[a,c]	450	232
Hydrochloric acid, 20%[a,c]	450	232
Hydrochloric acid, 38%[a,c]	450	232
Hydrocyanic acid, 10%	450	232
Hydrofluoric acid, 30%[a]	450	232
Hydrofluoric acid, 70%[a]	450	232
Hydrofluoric acid, 100%[a]	450	232
Hydrochlorus acid	450	232
Iodine solution, 10%[a]	450	232
Ketones, general	450	232
Lactic acid, 25%	450	232
Lactic acid, conc.	450	232
Magnesium chloride	450	232
Malic acid	450	232
Methyl chloride[a]	450	232
Methyl ethyl ketone[a]	450	232
Methyl isobutyl ketone[a]	450	232
Muriatic acid[a]	450	232
Nitric acid, 5%[a]	450	232
Nitric acid, 20%[a]	450	232
Nitric acid, 70%[a]	450	232
Nitric acid, anhydrous[a]	450	232
Nitrous acid, 10%	450	232
Oleum	450	232
Perchloric acid, 10%	450	232
Perchloric acid, 70%	450	232

(continued)

TABLE 2.24 *Continued*

Chemical	Maximum Temperature	
	°F	°C
Phenol[a]	450	232
Phosphoric acid, 50–80%[b]	450	232
Picric acid	450	232
Potassium bromide, 30%	450	232
Salicylic acid	450	232
Sodium carbonate	450	232
Sodium chloride	450	232
Sodium hydroxide, 10%	450	232
Sodium hydroxide, 50%	450	232
Sodium hydroxide, conc.	450	232
Sodium hypochlorite, 20%	450	232
Sodium hypochlorite, conc.	450	232
Sodium sulfide, to 50%	450	232
Stannic chloride	450	232
Stannous chloride	450	232
Sulfuric acid, 10%	450	232
Sulfuric acid, 50%	450	232
Sulfuric acid, 70%	450	232
Sulfuric acid, 90%	450	232
Sulfuric acid, 98%	450	232
Sulfuric acid, 100%	450	232
Sulfuric acid, fuming[a]	450	232
Sulfurous acid	450	232
Thionyl chloride[a]	450	232
Toluene[a]	450	232
Trichloroacetic acid	450	232
White liquor	450	232
Zinc chloride[b]	450	232

The chemicals listed are in the pure state or in a saturated solution unless otherwise indicated. Compatibility is shown to the maximum allowable temperature for which data is available. Incompatibility is shown by an X. A blank space indicates that data is unavailable.

[a] Material will permeate.
[b] Material will be absorbed.
[c] Material will cause stress cracking.

Source: From P.A. Schweitzer. 2004. *Corrosion Resistance Tables*, Vols. 1–4, 5th ed., New York: Marcel Dekker.

PFA finds many applications in the chemical process industry for corrosion resistance. Applications includes lining for pipes and vessels. Reference [2] details the use of PFA in lined process pipe and information on lining of vessels can be found in References [3,4].

2.16 Polytetrafluoroethylene (PTFE)

PTFE is marketed under the tradename of Teflon by DuPont and the tradename Halon by Ausimont USA. It is a fully fluorinated thermoplastic having the following formula:

$$
\begin{array}{c}
\text{F} \quad \text{F} \\
| \quad | \\
-\text{C}-\text{C}- \\
| \quad | \\
\text{F} \quad \text{F}
\end{array}
$$

PTFE has an operating temperature range of from $-20°$F to $+430°$C ($-29°$C to $+212°$C). This temperature range is based on the physical and mechanical properties of PTFE. When handling aggressive chemicals, it may be necessary to reduce the upper temperature limit.

PTFE is unique in its corrosion resistance properties. It is virtually inert in the presence of most materials. There are very few chemicals that will attack PTFE at normal use temperatures. Among materials that will attack PTFE are the most violent oxidizing and reducing agents known. Elemental sodium removes fluorine from the molecule. The other alkali metals (potassium, lithium, etc.) act in a similar manner.

Fluorine and related compounds (e.g., chlorine trifluoride) are absorbed into PTFE resin to such a degree that the mixture becomes sensitive to a source of ignition such as impact. These potent oxidizers should be only handled with great care and a recognition of the potential hazards.

The handling of 80% sodium hydroxide, aluminum chloride, ammonia, and certain amines at high temperatures have the same effect as elemental sodium. Slow oxidation attack can be produced by 70% nitric acid under pressure at 480°F (250°C).

PTFE has excellent weathering properties and is not degraded by UV light. Refer to Table 2.25 for the compatibility of PTFE with selected corrodents. Reference [1] provides a more detailed listing.

Applications for PTFE extend from exotic space-age usages to molded parts and wire and cable insulation to consumer use as a coating for cookware. One of PTFE's largest uses is for corrosion protection, including linings for tanks and piping. Applications in the automotive industry take advantage of the low surface friction and chemical stability using it in seals and rings for transmission and power steering systems and in seals for shafts, compressors and shock absorbers.

Reference [2] will provide details of the usage of PTFE in piping applications.

2.17 Polyvinylidene Fluoride (PVDF)

Polyvinylidene fluoride is a crystalline, high molecular weight polymer containing 50% fluorine. It is similar in chemical structure to PTFE except

TABLE 2.25

Compatibility of PTFE with Selected Corrodents

	Maximum Temperature	
Chemical	°F	°C
Acetaldehyde	450	232
Acetamide	450	232
Acetic acid, 10%	450	232
Acetic acid, 50%	450	232
Acetic acid, 80%	450	232
Acetic acid, glacial	450	232
Acetic anhydride	450	232
Acetone	450	232
Acetyl chloride	450	232
Acrylonitrile	450	232
Adipic acid	450	232
Allyl alcohol	450	232
Allyl chloride	450	232
Alum	450	232
Aluminum chloride, aqueous	450	232
Aluminum fluoride	450	232
Aluminum hydroxide	450	232
Aluminum nitrate	450	232
Aluminum oxychloride	450	232
Aluminum sulfate	450	232
Ammonia gas[a]	450	232
Ammonium bifluoride	450	232
Ammonium carbonate	450	232
Ammonium chloride, 10%	450	232
Ammonium chloride, 50%	450	232
Ammonium chloride, sat.	450	232
Ammonium fluoride, 10%	450	232
Ammonium fluoride, 25%	450	232
Ammonium hydroxide, 25%	450	232
Ammonium hydroxide, sat.	450	232
Ammonium nitrate	450	232
Ammonium persulfate	450	232
Ammonium phosphate	450	232
Ammonium sulfate, 10–40%	450	232
Ammonium sulfide	450	232
Amyl acetate	450	232
Amyl alcohol	450	232
Amyl chloride	450	232
Aniline	450	232
Antimony trichloride	450	232
Aqua regia 3:1	450	232
Barium carbonate	450	232
Barium chloride	450	232
Barium hydroxide	450	232
Barium sulfate	450	232
Barium sulfide	450	232

(continued)

TABLE 2.25 *Continued*

Chemical	Maximum Temperature	
	°F	°C
Benzaldehyde	450	232
Benzene[a]	450	232
Benzene sulfonic acid, 10%	450	232
Benzoic acid	450	232
Benzyl alcohol	450	232
Benzyl chloride	450	232
Borax	450	232
Boric acid	450	232
Bromine gas, dry[a]	450	232
Bromine, liquid[a]	450	232
Butadiene[a]	450	232
Butyl acetate	450	232
Butyl alcohol	450	232
n-Butylamine	450	232
Butyric acid	450	232
Calcium bisulfide	450	232
Calcium bisulfite	450	232
Calcium carbonate	450	232
Calcium chlorate	450	232
Calcium chloride	450	232
Calcium hydroxide, 10%	450	232
Calcium hydroxide, sat.	450	232
Calcium hypochlorite	450	232
Calcium nitrate	450	232
Calcium oxide	450	232
Calcium sulfate	450	232
Caprylic acid	450	232
Carbon bisulfide[a]	450	232
Carbon dioxide, dry	450	232
Carbon dioxide, wet	450	232
Carbon disulfide	450	232
Carbon monoxide	450	232
Carbon tetrachloride[b]	450	232
Carbonic acid	450	232
Chloroacetic acid, 50% water	450	232
Chloroacetic acid	450	232
Chlorine gas, dry	X	X
Chlorine gas, wet[a]	450	232
Chlorine, liquid	X	X
Chlorobenzene[a]	450	232
Chloroform[a]	450	232
Chlorosulfonic acid	450	232
Chromic acid, 10%	450	232
Chromic acid, 50%	450	232
Chromyl chloride	450	232
Citric acid, 15%	450	232
Citric acid, conc.	450	232
Copper carbonate	450	232

(continued)

TABLE 2.25 *Continued*

Chemical	Maximum Temperature	
	°F	°C
Copper chloride	450	232
Copper cyanide, 10%	450	232
Copper sulfate	450	232
Cresol	450	232
Cupric chloride, 5%	450	232
Cupric chloride, 50%	450	232
Cyclohexane	450	232
Cyclohexanol	450	232
Dibutyl phthalate	450	232
Dichloroacetic acid	450	232
Dichloroethane (ethylene dichloride)[a]	450	232
Ethylene glycol	450	232
Ferric chloride	450	232
Ferric chloride, 50% in water	450	232
Ferric nitrate, 10–50%	450	232
Ferrous chloride	450	232
Ferrous nitrate	450	232
Fluorine gas, dry	X	X
Fluorine gas, moist	X	X
Hydrobromic acid, dil.[a,b]	450	232
Hydrobromic acid, 20%[b]	450	232
Hydrobromic acid, 50%[b]	450	232
Hydrochloric acid, 20%[b]	450	232
Hydrochloric acid, 38%[b]	450	232
Hydrocyanic acid, 10%	450	232
Hydrofluoric acid, 30%[a]	450	232
Hydrofluoric acid, 70%[a]	450	232
Hydrofluoric acid, 100%[a]	450	232
Hypochlorous acid	450	232
Iodine solution, 10%[a]	450	232
Ketones, general	450	232
Lactic acid, 25%	450	232
Lactic acid, conc.	450	232
Magnesium chloride	450	232
Malic acid	450	232
Methyl chloride[a]	450	232
Methyl ethyl ketone[a]	450	232
Methyl isobutyl ketone[b]	450	232
Muriatic acid[a]	450	232
Nitric acid, 5%[a]	450	232
Nitric acid, 20%[a]	450	232
Nitric acid, 70%[a]	450	232
Nitric acid, anhydrous[a]	450	232
Nitrous acid, 10%	450	232
Oleum	450	232
Perchloric acid, 10%	450	232
Perchloric acid, 70%	450	232

(continued)

TABLE 2.25   *Continued*

Chemical	Maximum Temperature	
	°F	°C
Phenol[a]	450	232
Phosphoric acid, 50–80%	450	232
Picric acid	450	232
Potassium bromide, 30%	450	232
Salicylic acid	450	232
Sodium carbonate	450	232
Sodium chloride	450	232
Sodium hydroxide, 10%	450	232
Sodium hydroxide, 50%	450	232
Sodium hydroxide, conc.	450	232
Sodium hypochlorite, 20%	450	232
Sodium hypochlorite, conc.	450	232
Sodium sulfide, to 50%	450	232
Stannic chloride	450	232
Stannous chloride	450	232
Sulfuric acid, 10%	450	232
Sulfuric acid, 50%	450	232
Sulfuric acid, 70%	450	232
Sulfuric acid, 90%	450	232
Sulfuric acid, 98%	450	232
Sulfuric acid, 100%	450	232
Sulfuric acid, fuming[a]	450	232
Sulfurous acid	450	232
Thionyl chloride	450	232
Toluene[a]	450	232
Trichloroacetic acid	450	232
White liquor	450	232
Zinc chloride[c]	450	232

The chemicals listed are in the pure state or in a saturated solution unless otherwise indicated. Compatibility is shown to the maximum allowable temperature for which data is available. Incompatibility is shown by an X. A blank space indicates that data is unavailable.

[a] Material will permeate.
[b] Material will cause stress cracking.
[c] Material will be absorbed.

Source: From P.A. Schweitzer. 2004. *Corrosion Resistance Tables*, Vols. 1–4, 5th ed., New York: Marcel Dekker.

that it is not fully fluorinated. The chemical structure is as follows:

$$
\begin{array}{cc}
\text{F} & \text{F} \\
| & | \\
-\text{C}-\text{C}- \\
| & | \\
\text{H} & \text{H}
\end{array}
$$

Much of the strength and chemical resistance of PVDF is maintained through an operating temperature range of -40 to $+320°F$ (-40 to $+160°C$).

Approval has been granted by the Food and Drug Administration for repeated use in contact with food in food handling and processing equipment.

PVDF is chemically resistant to most acids, bases, and organic solvents. It is also resistant to wet or dry chlorine, bromine, and other halogens.

It should not be used with strong alkalies, fuming acids, polar solvents, amines, ketones, and esters. When used with strong alkalies, it stress cracks. Refer to Table 2.26 for the compatibility of PVDF with selected corrodents. Reference [1] provides a more detailed listing.

TABLE 2.26

Compatibility of PVDF with Selected Corrodents

	Maximum Temperature	
Chemical	°F	°C
Acetaldehyde	150	66
Acetamide	90	32
Acetic acid, 10%	300	149
Acetic acid, 50%	300	149
Acetic acid, 80%	190	88
Acetic acid, glacial	190	88
Acetic anhydride	100	38
Acetone	X	X
Acetyl chloride	120	49
Acrylic acid	150	66
Acrylonitrile	130	54
Adipic acid	280	138
Allyl alcohol	200	93
Allyl chloride	200	93
Alum	180	82
Aluminum acetate	250	121
Aluminum chloride, aqueous	300	149
Aluminum chloride, dry	270	132
Aluminum fluoride	300	149
Aluminum hydroxide	260	127
Aluminum nitrate	300	149
Aluminum oxychloride	290	143
Aluminum sulfate	300	149
Ammonia gas	270	132
Ammonium bifluoride	250	121
Ammonium carbonate	280	138
Ammonium chloride, 10%	280	138
Ammonium chloride, 50%	280	138
Ammonium chloride, sat.	280	138
Ammonium fluoride, 10%	280	138
Ammonium fluoride, 25%	280	138
Ammonium hydroxide, 25%	280	138
Ammonium hydroxide, sat.	280	138
Ammonium nitrate	280	138

(continued)

TABLE 2.26 *Continued*

Chemical	Maximum Temperature	
	°F	°C
Ammonium persulfate	280	138
Ammonium phosphate	280	138
Ammonium sulfate, 10–40%	280	138
Ammonium sulfide	280	138
Ammonium sulfite	280	138
Amyl acetate	190	88
Amyl alcohol	280	138
Amyl chloride	280	138
Aniline	200	93
Antimony trichloride	150	66
Aqua regia 3:1	130	54
Barium carbonate	280	138
Barium chloride	280	138
Barium hydroxide	280	138
Barium sulfate	280	138
Barium sulfide	280	138
Benzaldehyde	120	49
Benzene	150	66
Benzene sulfonic acid, 10%	100	38
Benzoic acid	250	121
Benzyl alcohol	280	138
Benzyl chloride	280	138
Borax	280	138
Boric acid	280	138
Bromine gas, dry	210	99
Bromine gas, moist	210	99
Bromine, liquid	140	60
Butadiene	280	138
Butyl acetate	140	60
Butyl alcohol	280	138
n-Butylamine	X	X
Butyric acid	230	110
Calcium bisulfide	280	138
Calcium bisulfite	280	138
Calcium carbonate	280	138
Calcium chlorate	280	138
Calcium chloride	280	138
Calcium hydroxide, 10%	270	132
Calcium hydroxide, sat.	280	138
Calcium hypochlorite	280	138
Calcium nitrate	280	138
Calcium oxide	250	121
Calcium sulfate	280	138
Caprylic acid	220	104
Carbon bisulfide	80	27
Carbon dioxide, dry	280	138
Carbon dioxide, wet	280	138
Carbon disulfide	80	27

(continued)

TABLE 2.26 *Continued*

Chemical	Maximum Temperature	
	°F	°C
Carbon monoxide	280	138
Carbon tetrachloride	280	138
Carbonic acid	280	138
Cellosolve	280	138
Chloroacetic acid, 50% water	210	99
Chloroacetic acid	200	93
Chlorine gas, dry	210	99
Chlorine gas, wet, 10%	210	99
Chlorine, liquid	210	99
Chlorobenzene	220	104
Chloroform	250	121
Chlorosulfonic acid	110	43
Chromic acid, 10%	220	104
Chromic acid, 50%	250	121
Chromyl chloride	110	43
Citric acid, 15%	250	121
Citric acid conc.	250	121
Copper acetate	250	121
Copper carbonate	250	121
Copper chloride	280	138
Copper cyanide	280	138
Copper sulfate	280	138
Cresol	210	99
Cupric chlorides, 5%	270	132
Cupric chloride, 50%	270	132
Cyclohexane	250	121
Cyclohexanol	210	99
Dibutyl phthalate	80	27
Dichloroacetic acid	120	49
Dichloroethane (ethylene dichloride)	280	138
Ethylene glycol	280	138
Ferric chloride	280	138
Ferrous chloride, 50% in water	280	138
Ferrous nitrate, 10–50%	280	138
Ferrous chloride	280	138
Ferrous nitrate	280	138
Fluorine gas, dry	80	27
Fluorine gas, moist	80	27
Hydrobromic acid, dil.	260	127
Hydrobromic acid, 20%	280	138
Hydrobromic acid, 50%	280	138
Hydrochloric acid, 20%	280	138
Hydrochloric acid, 38%	280	138
Hydrocyanic acid, 10%	280	138
Hydrofluoric acid, 30%	260	127
Hydrofluoric acid, 70%	200	93

(continued)

TABLE 2.26 *Continued*

Chemical	Maximum Temperature	
	°F	°C
Hydrofluoric acid, 100%	200	93
Hypochlorous acid	280	138
Iodine solution, 10%	250	121
Ketones, general	110	43
Lactic acid, 25%	130	54
Lactic acid, conc.	110	43
Magnesium chloride	280	138
Malic acid	250	121
Manganese chloride	280	138
Methyl chloride	X	X
Methyl ethyl ketone	X	X
Methyl isobutyl ketone	110	43
Muriatic acid	280	138
Nitric acid, 5%	200	93
Nitric acid, 20%	180	82
Nitric acid, 70%	120	49
Nitric acid, anhydrous	150	6
Nitric acid, conc.	210	99
Oleum	X	X
Perchloric acid, 10%	210	99
Perchloric acid, 70%	120	49
Phenol	200	93
Phosphoric acid, 50–80%	220	104
Picric acid	80	27
Potassium bromide, 30%	280	138
Salicylic acid	220	104
Silver bromide, 10%	250	121
Sodium carbonate	280	138
Sodium chloride	280	138
Sodium hydroxide, 10%	230	110
Sodium hydroxide, 50%	220	104
Sodium hydroxide, conc.	150	66
Sodium hypochlorite, 20%	280	138
Sodium hypochlorite, conc.	280	138
Sodium sulfide, to 50%	280	138
Stannic chloride	280	138
Stannous chloride	280	138
Sulfuric acid, 10%	250	121
Sulfuric acid, 50%	220	104
Sulfuric acid, 70%	220	104
Sulfuric acid, 90%	210	99
Sulfuric acid, 98%	140	60
Sulfuric acid, 100%	X	X
Sulfuric acid, fuming	X	X
Sulfurous acid	220	104
Thionyl chloride	X	X
Toluene	X	X
Trichloroacetic acid	130	54

(continued)

TABLE 2.26 *Continued*

| | Maximum Temperature | |
Chemical	°F	°C
White liquor	80	27
Zinc chloride	260	127

The chemicals listed are in the pure state or in a saturated solution unless otherwise indicated. Compatibility is shown to the maximum allowable temperature for which data is available. Incompatibility is shown by an X. A blank space indicates that the date is unavailable.

Source: From P.A. Schweitzer. 2004. *Corrosion Resistance Tables*, Vols. 1–4, 5th ed., New York: Marcel Dekker.

PVDF also withstands UV light in the "visible" range and gamma radiations up to 100 Mrad.

Polyvinylidene fluoride finds many applications in the corrosion resistance field, being used as a lining material for vessels and piping, as solid piping, column packing, valving, pumps, and other processing equipment. Reference [2] provides details of the usage of PVDF piping.

Polyvinylidene fluoride is manufactured under the tradename of Kynar by Elf Atochem, Solef by Solvay, Hylar by Ausimont USA, and super Pro 230 and ISO by Asahi/America.

2.18 Polyethylene (PE)

Polyethylenes are probably among the best-known thermoplasts. Polyethylene is produced in various grades that differ in molecular structure, crystallinity, molecular weight, and molecular distribution. They are a member of the olefin family. The basic chemical structure is:

$$-\overset{\displaystyle \overset{H}{|}}{\underset{\displaystyle \underset{H}{|}}{C}}-\overset{\displaystyle \overset{H}{|}}{\underset{\displaystyle \underset{H}{|}}{C}}-$$

PE is produced by polymerizing ethylene gas obtained from petroleum hydrocarbons. Changes in the polymerizing conditions are responsible for the various types of PE.

Physical and mechanical properties differ in density and molecular weight. The three main classifications of density are low, medium, and high. These specific gravity ranges are 0.91–0.925, 0.925–0.940, and 0.940–0.965. These grades are sometimes referred to as Type I, II, and III.

All polyethylenes are relatively soft, and hardness increases as density increases. Generally, the higher the density, the better the dimensional stability and physical properties, particularly as a function of temperature. The thermal stability of polyethylenes ranges from 190°F (88°C) for the low-density material up to 250°F/121°C for the high-density material.

Industry practice breaks the molecular weight of polyethylene into four distinct classifications that are:

Medium molecular weight less than 100,000
High molecular weight 110,000 to 250,000
Extra high molecular weight 250,000 to 1,500,000
Ultra high molecular weight 1,500,000 and higher

Usually, the ultrahigh molecular weight material has a molecular weight of at least 3.1 million.

Every polyethylene resin consists of a mixture of large and small molecules, molecules of high and low molecular weight. The molecular weight distribution gives a general picture of the ratio of the large, medium, and small molecules in the resin. If the resin contains molecules close to the average weight, the distribution is called *narrow*. If the resin contains a wide variety of weights, the distribution is called *broad*.

Figure 2.1 depicts this in graph form.

The two varieties of PE generally used for corrosive applications are EHMW and UHMW. Polyethylene exhibits a wide range of corrosion resistance, ranging from potable water to corrosive wastes. It is resistant to most mineral acids, including sulfuric up to 70% concentrations, inorganic salts including chlorides, alkalies, and many organic acids. It is not resistant to bromine, aromatics, or chlorinated hydrocarbons. Refer to Table 2.27 for

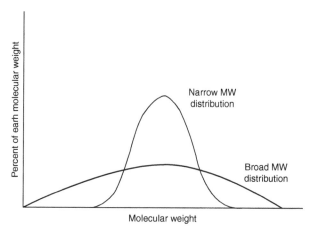

FIGURE 2.1
Schematic illustration of molecular weight distribution.

TABLE 2.27

Compatibility of EHMWPE with Selected Corrodents

Chemical	Maximum Temperature	
	°F	°C
Acetaldehyde, 40%	90	32
Acetamide		
Acetic acid, 10%	140	60
Acetic acid, 50%	140	60
Acetic acid, 80%	80	27
Acetic acid, glacial		
Acetic anhydride	X	X
Acetone	120	49
Acetyl chloride		
Acrylic acid		
Acrylonitrile	150	66
Adipic acid	140	60
Allyl alcohol	140	60
Allyl chloride	80	27
Alum	140	60
Aluminum acetate		
Aluminum chloride, aqueous	140	60
Aluminum chloride, dry	140	60
Aluminum fluoride	140	60
Aluminum hydroxide	140	60
Aluminum nitrate		
Aluminum oxychloride		
Aluminium sulfate	140	60
Ammonia gas	140	60
Ammonium bifluoride		
Ammonium carbonate	140	60
Ammonium chloride, 10%	140	60
Ammonium chloride, 50%	140	60
Ammonium chloride, sat.	140	60
Ammonium fluoride, 10%	140	60
Ammonium fluoride, 25%	140	60
Ammonium hydroxide, 25%	140	60
Ammonium hydroxide, sat.	140	60
Ammonium nitrate	140	60
Ammonium persulfate	140	60
Ammonium phosphate	80	27
Ammonium sulfate, 10–40%	140	60
Ammonium sulfide	140	60
Ammonium sulfite		
Amyl acetate	140	60
Amyl alcohol	140	60
Amyl chloride	X	X
Aniline	130	54
Antimony trichloride	140	60
Aqua regia 3:1	130	54
Barium carbonate	140	60
Barium chloride	140	60

(continued)

TABLE 2.27 *Continued*

Chemical	Maximum Temperature	
	°F	°C
Barium hydroxide	140	60
Barium sulfate	140	60
Barium sulfide	140	60
Benzaldehyde	X	X
Benzene	X	X
Benzenesulfonic acid, 10%	140	60
Benzoic acid	140	60
Benzyl alcohol	170	77
Benzyl chloride		
Borax	140	60
Boric acid	140	60
Bromine gas, dry	X	X
Bromine gas, moist	X	X
Bromine, liquid	X	X
Butadiene	X	X
Butyl acetate	90	32
Butyl alcohol	140	60
n-Butylamine	X	X
Butyric acid	130	54
Calcium bisulfide	140	60
Calcium bisulfite	80	27
Calcium carbonate	140	60
Calcium chlorate	140	60
Calcium chloride	140	60
Calcium hydroxide, 10%	140	60
Calcium hydroxide, sat.	140	60
Calcium hypochlorite	140	60
Calcium nitrate	140	60
Calcium oxide	140	60
Calcium sulfate	140	60
Caprylic acid		
Carbon bisulfide	X	X
Carbon dioxide, dry	140	60
Carbon dioxide, wet	140	60
Carbon disulfide	X	X
Carbon monoxide	140	60
Carbon tetrachloride	X	X
Carbonic acid	140	60
Cellosolve		
Chloroacetic acid, 50% in water	X	X
Chloroacetic acid	X	X
Chlorine gas, dry	80	27
Chlorine gas, wet, 10%	120	49
Chlorine, liquid	X	X
Chlorobenzene	X	X
Chloroform	80	27
Chlorosulfonic acid	X	X

(continued)

TABLE 2.27 *Continued*

Chemical	Maximum Temperature	
	°F	°C
Chromic acid, 10%	140	60
Chromic acid, 50%	90	32
Chromyl chloride		
Citric acid, 15%	140	60
Citric acid, conc.	140	60
Copper acetate		
Copper carbonate		
Copper chloride	140	60
Copper cyanide	140	60
Copper sulfate	140	60
Cresol	80	27
Cupric chloride, 5%	80	27
Cupric chloride, 50%		
Cyclohexane	130	54
Cyclohexanol	170	77
Dibutyl phthalate	80	27
Dichloroacetic acid	73	23
Dichloroethane		
(ethylene dichloride)	X	X
Ethylene glycol	140	60
Ferric chloride	140	60
Ferric chloride, 50% in water	140	60
Ferric nitrate, 10–50%	140	60
Ferrous chloride	140	60
Ferrous nitrate	140	60
Fluorine gas, dry	X	X
Fluorine gas, moist	X	X
Hydrobromic acid, dil.	140	60
Hydrobromic acid, 20%	140	60
Hydrobromic acid, 50%	140	60
Hydrochloric acid, 20%	140	60
Hydrochloric acid, 38%	140	60
Hydrocyanic acid, 10%	140	60
Hydrofluoric acid, 30%	80	27
Hydrofluoric acid, 70%	X	X
Hydrofluoric acid, 100%	X	X
Hypochlorous acid		
Iodine solution, 10%	80	27
Ketones, general	X	X
Lactic acid, 25%	140	60
Lactic acid, conc.	140	60
Magnesium chloride	140	60
Malic acid	100	38
Manganese chloride	80	27
Methyl chloride	X	X
Methyl ethyl ketone	X	X
Methyl isobutyl ketone	80	27
Muriatic acid	140	60

(continued)

TABLE 2.27   *Continued*

Chemical	Maximum Temperature	
	°F	°C
Nitric acid, 5%	140	60
Nitric acid, 20%	140	60
Nitric acid, 70%	X	X
Nitric acid, anhydrous	X	X
Nitrous acid, conc.		
Oleum		
Perchloric acid, 10%	140	60
Perchloric acid, 70%	X	X
Phenol	100	38
Phosphoric acid, 50–80%	100	38
Picric acid	100	38
Potassium bromide, 30%	140	60
Salicylic acid		
Silver bromide, 10%		
Sodium carbonate	140	60
Sodium chloride	140	60
Sodium hydroxide, 10%	170	77
Sodium hydroxide, 50%	170	77
Sodium hydroxide, conc.		
Sodium hypochlorite, 20%	140	60
Sodium hypochlorite, conc.	140	60
Sodium Sulfide, to 50%	140	60
Stannic chloride	140	60
Stannous chloride	140	60
Sulfuric acid, 10%	140	60
Sulfuric acid, 50%	140	60
Sulfuric acid, 70%	80	27
Sulfuric acid, 90%	X	X
Sulfuric acid, 98%	X	X
Sulfuric acid, 100%	X	X
Sulfuric acid, fuming	X	X
Sulfurous acid	140	60
Thionyl chloride	X	X
Toluene	X	X
Trichloroacetic acid	140	60
White liquor		
Zinc chloride	140	60

The chemicals listed are in the pure state or in a saturated solution unless otherwise indicated. Compatibility is shown to the maximum allowable temperature for which data is available. Incompatibility is shown by an X. A blank space indicates that data is unavailable.

Source: From P.A. Schweitzer. 2004. *Corrosion Resistance Tables*, Vols. 1–4, 5th ed., New York: Marcel Dekker.

the compatibility of EHMW PE with selected corrodents and Table 2.28 for HMW PE. Reference [1] provides a more comprehensive listing.

Polyethylene is subject to degradation by UV radiation. If exposed outdoors, carbon black must be added to the formulation for protection against UV degradation.

TABLE 2.28

Compatibility of HMWPE with Selected Corrodents

	Maximum Temperature	
Chemical	°F	°C
Acetaldehyde	X	
Acetamide	140	60
Acetic acid, 10%	140	60
Acetic acid, 50%	140	60
Acetic acid, 80%	80	27
Acetic anhydride	X	
Acetone	80	27
Acetyl chloride	X	
Acrylonitrile	150	66
Adipic acid	140	60
Allyl alcohol	140	60
Allyl chloride	110	43
Alum	140	60
Aluminum chloride, aqueous	140	60
Aluminum chloride, dry	140	60
Aluminum fluoride	140	60
Aluminum hydroxide	140	60
Aluminum nitrate	140	60
Aluminum sulfate	140	60
Ammonium gas	140	60
Ammonium bifluoride	140	60
Ammonium carbonate	140	60
Ammonium chloride, 10%	140	60
Ammonium chloride, 50%	140	60
Ammonium chloride, sat.	140	60
Ammonium fluoride, 10%	140	60
Ammonium fluoride, 25%	140	60
Ammonium hydroxide, 25%	140	60
Ammonium hydroxide, sat.	140	60
Ammonium nitrate	140	60
Ammonium persulfate	150	66
Ammonium phosphate	80	27
Ammonium sulfate to 40%	140	60
Ammonium sulfide	140	60
Ammonium sulfite	140	60
Amyl acetate	140	60
Amyl alcohol	140	60
Amyl chloride	X	
Aniline	130	44
Antimony trichloride	140	60

(continued)

TABLE 2.28 *Continued*

Chemical	Maximum Temperature	
	°F	°C
Aqua regia 3:1	130	44
Barium carbonate	140	60
Barium chloride	140	60
Barium hydroxide	140	60
Barium sulfate	140	60
Barium sulfide	140	60
Benzaldehyde	X	
Benzene	X	
Benzoic acid	140	60
Benzyl alcohol	X	
Borax	140	60
Boric acid	140	60
Bromine gas, dry	X	
Bromine gas, moist	X	
Bromine, liquid	X	
Butadiene	X	
Butyl acetate	90	32
Butyl alcohol	140	60
n-Butylamine	X	
Butyric acid	X	
Calcium bisulfide	140	60
Calcium bisulfite	140	60
Calcium carbonate	140	60
Calcium chlorate	140	60
Calcium chloride	140	60
Calcium hydroxide, 10%	140	60
Calcium hydroxide, sat.	140	60
Calcium hypochlorite	140	60
Calcium nitrate	140	60
Calcium oxide	140	60
Calcium sulfate	140	60
Carbon bisulfide	X	
Carbon dioxide, dry	140	60
Carbon dioxide, wet	140	60
Carbon disulfide	X	
Carbon monoxide	140	60
Carbon tetrachloride	X	
Carbonic acid	140	60
Cellosolve	X	
Chloroacetic acid	X	
Chlorine gas, dry	X	
Chlorine gas, wet	X	
Chlorine, liquid	X	
Chlorobenzene	X	
Chloroform	X	
Chlorosulfonic acid	X	
Chromic acid, 10%	140	60
Chromic acid, 50%	90	32

(continued)

TABLE 2.28 *Continued*

Chemical	Maximum Temperature	
	°F	°C
Citric acid, 15%	140	60
Citric acid, conc.	140	60
Copper chloride	140	60
Copper cyanide	140	60
Copper sulfate	140	60
Cresol	X	
Cupric chloride, 5%	140	60
Cupric chloride, 50%	140	60
Cyclohexane	80	27
Cyclohexonol	80	27
Dibutyl phthalate	80	27
Dichloroethane	80	27
Ethylene glycol	140	60
Ferric chloride	140	60
Ferrous chloride	140	60
Ferrous nitrate	140	60
Fluorine gas, dry	X	
Fluorine gas, moist	X	
Hydrobromic acid, dil.	140	60
Hydrobromic acid, 20%	140	60
Hydrobromic acid, 50%	140	60
Hydrochloric acid, 20%	140	60
Hydrochloric acid, 38%	140	60
Hydrocyanic acid, 10%	140	60
Hydrofluoric acid, 30%	140	60
Hydrofluoric acid, 70%	X	
Hydrochlorous acid	150	66
Iodine solution, 10%	80	27
Ketones, general	80	27
Lactic acid, 25%	150	66
Magnesium chloride	140	60
Malic acid	140	60
Manganese chloride	80	27
Methyl chloride	X	
Methyl ethyl ketone	X	
Methyl isobutyl ketone	80	27
Nitric acid, 5%	140	60
Nitric acid, 20%	140	60
Nitric acid, 70%	X	
Nitric acid, anhydrous	X	
Nitrous acid, conc.	120	49
Perchloric acid, 10%	140	60
Perchloric acid, 70%	X	
Phenol	100	38
Phosphoric acid, 50–80%	100	38
Picric acid	100	38
Potassium bromide, 30%	140	60
Salicylic acid	140	60
Sodium carbonate	140	60

(continued)

TABLE 2.28 *Continued*

Chemical	Maximum Temperature	
	°F	°C
Sodium chloride	140	60
Sodium hydroxide, 10%	150	66
Sodium hydroxide, 50%	150	66
Sodium hypochlorite, 20%	140	60
Sodium hypochlorite, conc.	140	60
Sodium sulfide, to 50%	140	60
Stannic chloride	140	60
Stannous chloride	140	60
Sulfuric acid, 10%	140	60
Sulfuric acid, 50%	140	60
Sulfuric acid, 70%	80	27
Sulfuric acid, 90%	X	
Sulfuric acid, 98%	X	
Sulfuric acid, 100%	X	
Sulfuric acid, fuming	X	
Sulfurous acid	140	60
Thionyl chloride	X	
Toluene	X	
Trichloroacetic acid	80	27
Zinc chloride	140	60

The chemicals listed are in the pure state or in a saturated solution unless otherwise indicated. Compatibility is shown to the maximum allowable temperature for which data is available. Incompatibility is shown by an X.

Source: From P.A. Schweitzer. 2004. *Corrosion Resistance Tables*, Vols. 1–4, 5th ed., New York: Marcel Dekker. With permission.

Applications for polyethylene vary depending upon the grade of resin. EHMW and UHMW polyethylenes find applications where corrosion resistance is required. Among such applications are structural tanks and covers, fabricated parts, and piping. The usage for piping is of major importance. PE pipe is used to transport natural gas, potable water systems, drainage piping, corrosive wastes, and underground fire main water. Reference [2] details usage of polyethylene piping.

2.19   Polyethylene Terephthalate (PET)

Polyethylene terephthalate is a semicrystalline engineering thermoplastic having the following structure:

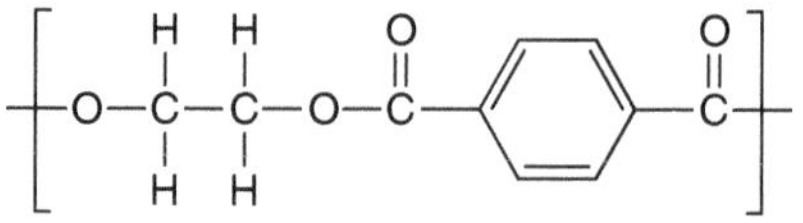

PET is resistant to dilute mineral acids, aliphatic hydrocarbons, aromatic hydrocarbons, ketones, and esters with limited resistance to hot water and washing soda. It is not resistant to alkaline and chlorinated hydrocarbons. PET has good resistance to UV degradation and weatherability. Refer to Table 2.29 for the compatibility of PET with selected corrodents.

PET is used in the automotive industry for housings, racks, minor parts, and latch mechanisms. It can be painted to match metal body panels and has been used as fenders.

PET is also used in water purification, food handling equipment, and pump and valve components.

TABLE 2.29

Compatibility of PET with Selected Corrodents

Acetic acid, 5%	R
Acetic acid, 10%	R
Acetone	R
Ammonia, 10%	X
Benzene	R
Butyl acetate	R
Cadmium chloride	R
Carbon disulfide	R
Carbon tetrachloride	R
Chlorobenzene	X
Chloroform	X
Citric acid, 10%	R
Diesel oil	R
Dioxane	R
Edible oil	R
Ethanol	R
Ether	R
Ethyl acetate	R
Ethylene chloride	X
Ethylene glycol	R
Formic acid	R
Fruit juice	R
Fuel oil	R
Gasoline	R
Glycerine	R
Heptane/hexane	R
Hydrochloric acid, 38%	X
Hydrochloric acid, 2%	R
Hydrogen peroxide, 30%	R
Hydrogen peroxide, 0–5%	R
Ink	R
Linseed oil	R
Methanol	R
Methyl ethyl ketone	R
Methylene chloride	X
Motor oil	R
Nitric acid, 2%	R
Paraffin oil	R

(continued)

TABLE 2.29 *Continued*

Phosphoric acid, 10%	R
Potassium hydroxide, 50%	X
Potassium dichromate, 10%	R
Potassium permanganate, 10%	R
Silicone oil	R
Soap solution	R
Sodium bisulfite	R
Sodium carbonate, 10%	R
Sodium chloride, 10%	R
Sodium hydroxide, 50%	X
Sodium nitrate	R
Sodium thiosulfate	R
Sulfuric acid, 98%	X
Sulfuric acid, 2%	R
Toluene	R
Vaseline	R
Water, cold	R
Water, hot	R
Wax, molten	R
Xylene	R

R, material resistant at 73°F/20°C; X, material not resistant.

2.20 Polyimide (PI)

Polyimides are heterocyclic polymers having an atom of hydrogen in one of the rings in the molecular chain. The atom of hydrogen is in the inside ring as shown below:

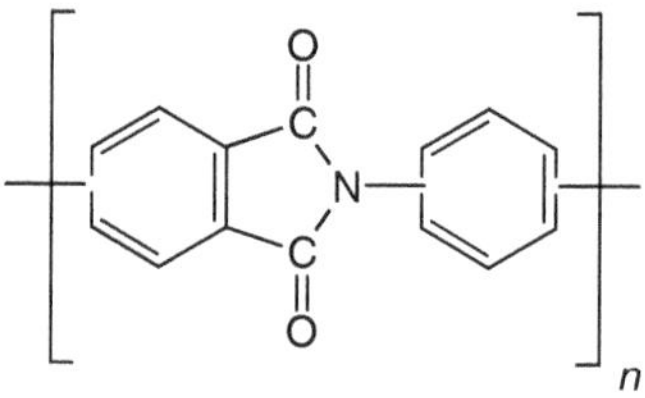

The fused rings provide chain stiffness essential to high-temperature strength retention. The low concentration of hydrogen provides oxidative resistance by preventing thermal degradation fracture of the chain. Polyimides exhibit outstanding properties, resulting from their combination of high-temperature stability up to 500–600°F (260–315°C) continuous service and to 990°F (482°C) for intermediate use.

The polyimides have excellent chemical and radiation inertness and are not subject to UV degradation. High oxidative resistance is another important property.

Polyimides find applications that require high heat resistance and under-the-hood applications in the automotive industry. Other applications include bearings, compressors, valves, and piston rings.

2.21 Polyphenylene Oxide (PPO)

Noryls, patented by G.E. Plastics, are amorphous modified polyphenylene oxide resins. The basic phenylene oxide structure is as follows:

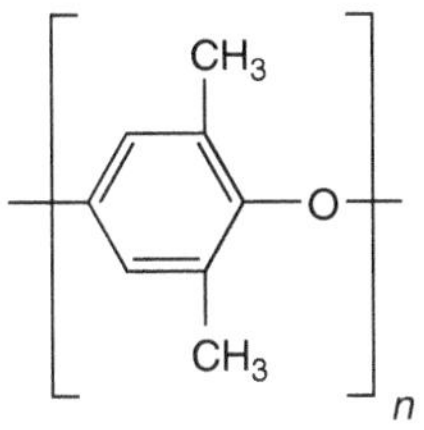

Several grades of the resin are produced to provide a choice of performance characteristics to meet a wide range of engineering application requirements. PPO maintains excellent mechanical properties over a temperature range of from below $-40°F$ ($-40°C$) to above $300°F$ ($149°C$).

Polyphenylene oxide has excellent resistance to aqueous environments, dilute mineral acids, and dilute alkalies. It is not resistant to aliphatic hydrocarbons, aromatic hydrocarbons, ketones, esters, or chlorinated hydrocarbons. Refer to Table 2.30 for the compatibility of PPO with selected corrodents. Reference [1] provides a more extensive listing.

PPO finds application in business equipment, appliances, electronics, and electrical devices.

TABLE 2.30

Compatibility of PPO with Selected Corrodents

Acetic acid, 5%	R
Acetic acid, 10%	R
Acetone	X
Ammonia, 10%	R
Benzene	X
Carbon tetrachloride	X
Chlorobenzene	X
Chloroform	X
Citric acid, 10%	R
Copper sulfate	R
Cyclohexane	X
Cyclohexanone	X
Diesel oil	R
Dioxane	X
Edible oil	R
Ethanol	R
Ethyl acetate	R
Ethylene chloride	X
Ethylene glycol	R
Formaldehyde, 30%	R

(continued)

TABLE 2.30 *Continued*

Formic acid	R
Fruit juice	R
Fuel oil	R
Gasoline	R
Glycerin	R
Hexane/heptane	R
Hydrochloric acid, 38%	R
Hydrochloric acid, 2%	R
Hydrogen peroxide, 30%	R
Hydrogen peroxide, 0–5%	R
Hydrogen sulfide	R
Linseed oil	R
Methanol	R
Methyl ethyl ketone	X
Milk	R
Motor oil	R
Nitric acid, 2%	R
Paraffin oil	R
Phosphoric acid, 10%	R
Potassium hydroxide, 50%	R
Potassium dichromate	R
Potassium permanganate, 10%	R
Silicone oil	R
Soap solution	R
Sodium carbonate, 10%	R
Sodium chloride, 10%	R
Sodium hydroxide, 50%	R
Sodium hydroxide, 5%	R
Styrene	X
Sulfuric acid, 98%	R
Sulfuric acid, 2%	R
Trichloroethylene	X
Urea, aqueous	R
Water, cold	R
Water, hot	R
Wax, molten	R
Xylene	X

R, material resistant at 73°F/20°C; X, material not resistant.

2.22   Polyphenylene Sulfide (PPS)

Polyphenylene sulfide is an engineering polymer capable of use at elevated temperatures. It has the following chemical structure:

PPS has a symmetrical, rigid backbone chain consisting of recurring para-substituted benzene rings and sulfur atoms. It is sold under the tradename of Ryton. Long-term exposure in air at 450°F (230°C) has no effect on the mechanical properties of PPS.

PPS has exceptional chemical resistance. It is resistant to aqueous inorganic salts and bases and many inorganic solvents. Relatively few materials react with PPS at high temperatures. It can also be used under highly oxidizing conditions.

Chlorinated solvents, some halegenated gases, and alkyl amines will attack PPS. It stress cracks in the presence of chlorinated solvents.

Weak and strong alkalies have no effect. Refer to Table 2.31 for the compatibility of PPS with selected corrodents. Reference [1] provides a more detailed listing.

PPS has good resistance to UV light degradation that can be increased by formulating it with carbon black.

In the automotive industry, PPS molded components are used in electrical fuel handling and emission control systems. The chemical process industry makes use of PPS in valve and pump components as well as other industrial applications.

TABLE 2.31

Compatibility of PPS with Selected Corrodents

Chemical	Maximum Temperature	
	°F	°C
Acetaldehyde	230	110
Acetamide	250	121
Acetic acid, 10%	250	121
Acetic acid, 50%	250	121
Acetic acid, 80%	250	121
Acetic acid, glacial	190	88
Acetic anhydride	280	138
Acetone	260	127
Acrylic acid, 25%	100	38
Acrylonitrile	130	54
Adipic acid	300	149
Alum	300	149
Aluminum acetate	210	99
Aluminum chloride, aqueous	300	149
Aluminum chloride, dry	270	132
Aluminum hydroxide	250	121
Aluminum nitrate	250	121
Aluminum oxychloride	460	238
Ammonia gas	250	121
Ammonium carbonate	460	238
Ammonium chloride, 10%	300	149
Ammonium chloride, 50%	300	149

(continued)

TABLE 2.31 *Continued*

Chemical	Maximum Temperature	
	°F	°C
Ammonium chloride, sat.	300	149
Ammonium hydroxide, 25%	250	121
Ammonium hydroxide, sat.	250	121
Ammonium nitrate	250	121
Ammonium phosphate, 65%	300	149
Ammonium sulfate, 10–40%	300	149
Amyl acetate	300	149
Amyl alcohol	210	99
Amyl chloride	200	93
Aniline	300	149
Barium carbonate	200	93
Barium chloride	200	93
Barium hydroxide	200	93
Barium sulfate	220	104
Barium sulfide	200	93
Benzaldehyde	250	121
Benzene	300	149
Benzenesulfonic acid, 10%	250	121
Benzoic acid	230	110
Benzyl alcohol	200	93
Benzyl chloride	300	149
Borax	210	99
Boric acid	210	99
Bromine gas, dry	X	X
Bromine gas, moist	X	X
Bromine, liquid	X	X
Butadiene	100	38
Butyl acetate	250	121
Butyl alcohol	200	93
n-Butylamine	200	93
Butyric acid	240	116
Calcium bisulfite	200	93
Calcium carbonate	300	149
Calcium chloride	300	149
Calcium hydroxide, 10%	300	149
Calcium hydroxide, sat.	300	149
Carbon bisulfide	200	93
Carbon dioxide, dry	200	93
Carbon bisulfide	200	93
Carbon tetrachloride	120	49
Cellosolve	220	104
Chloroacetic acid	190	88
Chlorine gas, dry	X	X
Chlorine gas, wet	X	X
Chlorine, liquid	200	93
Chloroform	150	66
Chlorosulfonic acid	X	X
Chromic acid, 10%	200	93

(continued)

TABLE 2.31 *Continued*

Chemical	Maximum Temperature	
	°F	°C
Chromic acid, 50%	200	93
Citric acid, 15%	250	121
Citric acid, conc.	250	121
Copper acetate	300	149
Copper chloride	220	104
Copper cyanide	210	99
Copper sulfate	250	121
Cresol	200	93
Cupric chloride, 5%	300	149
Cyclohexane	190	88
Cyclohexanol	250	121
Dichloroethane (ethylene dichloride)	210	99
Ethylene glycol	300	149
Ferric chloride	210	99
Ferric chloride, 50% in water	210	99
Ferric nitrate, 10–50%	210	99
Ferrous chloride	210	99
Ferrous nitrate	210	99
Fluorine gas, dry	X	X
Hydrobromic acid, dil.	200	93
Hydrobromic acid, 20%	200	93
Hydrobromic acid, 50%	200	93
Hydrochloric acid, 20%	230	110
Hydrochloric acid, 38%	210	99
Hydrocyanic acid, 10%	250	121
Hydrofluoric acid, 30%	200	93
Lactic acid, 25%	250	121
Lactic acid, conc.	250	121
Magnesium chloride	300	149
Methyl ethyl ketone	200	93
Methyl isobutyl ketone	250	121
Muriatic acid	210	99
Nitric acid, 5%	150	66
Nitric acid, 20%	100	38
Oleum	80	27
Phenol, 88%	300	149
Phosphoric acid, 50–80%	220	104
Potassium bromide, 30%	200	93
Sodium carbonate	300	149
Sodium chloride	300	149
Sodium hydroxide, 10%	210	99
Sodium hydroxide, 50%	210	99
Sodium hypochlorite, 5%	200	93
Sodium hypochlorite, conc.	250	121
Sodium sulfide, to 50%	230	110
Stannic chloride	210	99
Sulfuric acid, 10%	250	121

(continued)

TABLE 2.31 *Continued*

Chemical	Maximum Temperature	
	°F	°C
Sulfuric acid, 50%	250	121
Sulfuric acid, 70%	250	121
Sulfuric acid, 90%	220	104
Sulfuric acid, fuming	80	27
Sulfurous acid, 10%	200	93
Thionyl chloride	X	X
Toluene	300	149
Zinc chloride, 70%	250	121

The chemicals listed are in the pure state or in a saturated solution unless otherwise indicated. Compatibility is shown to the maximum allowable temperature for which data is available. Incompatibility is shown by an X. A blank space indicates that data is unavailable.

Source: From P.A. Schweitzer. 2004. *Corrosion Resistance Tables*, Vols. 1–4, 5th ed., New York: Marcel Dekker.

2.23　Polypropylene (PP)

Polypropylene is one of the most common and versatile thermoplastics. It is closely related to polyethylene, both of which are members of a group known as polyolefins. The polyolefins are composed of only hydrogen and carbon. Within the chemical structure of PP, a distinction is made between isotactic PP and atactic PP. The isotactic form accounts for 97% of the polypropylene produced. This form is highly ordered having the structure shown below:

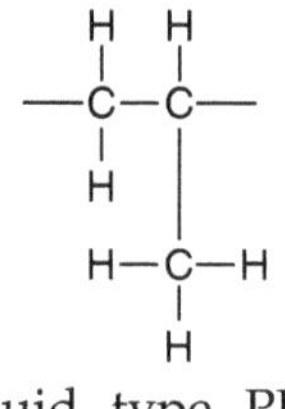

Atactic PP is a viscous liquid type PP having a propylene matrix. Polypropylene can be produced either as a homopolymer or as a copolymer with polyethylene. The copolymer has a structure as follows:

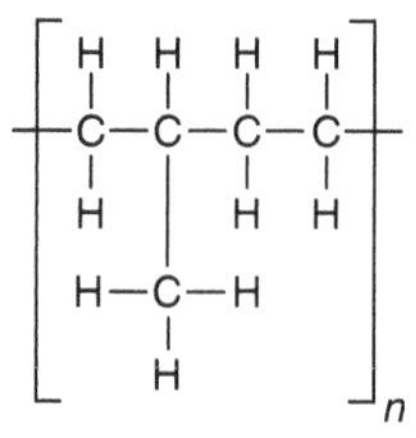

The homopolymers, being long chain high molecular weight molecules with a minimum of random orientations, have optimum chemical, thermal, and physical properties. For this reason, homopolymer material is preferred for difficult chemical, thermal, and physical conditions.

Copolymer PP is less brittle than the homopolymers and is able to withstand impact forces down to −20°F (−29°C), whereas, the homopolymer is extremely brittle below 40°F (4°C).

PP is not affected by most inorganic chemicals except the halogens and severe oxidizing conditions. It can be used with sulfur-bearing compounds, caustics, solvents, acids, and other organic chemicals. PP should not be used with oxidizing acids, detergents, low-boiling hydrocarbons, alcohols, aromatics, and some organic materials.

If exposed to sunlight, an ultraviolet absorber or screening agent should be in the formulation to protect it from degradation. Thermal oxidative degradation, particularly where copper is involved will pose a problem.

Refer to Table 2.32 for the compatibility of PP with select corrodents. Reference [1] provides a more detailed listing.

TABLE 2.32

Compatibility of PP with Selected Corrodents

| | Maximum Temperature | |
Chemical	°F	°C
Acetaldehyde	120	49
Acetamide	110	43
Acetic acid, 10%	220	104
Acetic acid, 50%	200	93
Acetic acid, 80%	200	93
Acetic acid, glacial	190	88
Acetic anhydride	100	38
Acetone	220	104
Acetyl chloride	X	X
Acrylic acid	X	X
Acrylonitrile	90	32
Adipic acid	100	38
Allyl alcohol	140	60
Allyl chloride	140	60
Alum	220	104
Aluminum acetate	100	38
Aluminum chloride, aqueous	200	93
Aluminum chloride, dry	220	104
Aluminum fluoride	200	93
Aluminum hydroxide	200	93
Aluminum nitrate	200	93
Aluminum oxychloride	220	104
Aluminum sulfate		
Ammonia gas	150	66
Ammonium bifluoride	200	93

(continued)

TABLE 2.32 *Continued*

Chemical	Maximum Temperature	
	°F	°C
Ammonium carbonate	220	104
Ammonium chloride, 10%	180	82
Ammonium chloride, 50%	180	82
Ammonium chloride, sat.	200	93
Ammonium fluoride, 10%	210	99
Ammonium fluoride, 25%	200	93
Ammonium hydroxide, 25%	200	93
Ammonium hydroxide, sat.	200	93
Ammonium nitrate	200	93
Ammonium persulfate	220	104
Ammonium phosphate	200	93
Ammonium sulfate, 10–40%	200	93
Ammonium sulfide	220	104
Ammonium sulfite	220	104
Amyl acetate	X	X
Amyl alcohol	200	93
Amyl chloride	X	X
Aniline	180	82
Antimony trichloride	180	82
Aqua regia 3:1	X	X
Barium carbonate	200	93
Barium chloride	220	104
Barium hydroxide	200	93
Barium sulfate	200	93
Barium sulfide	200	93
Benzaldehyde	80	27
Benzene	140	60
Benzenesulfonic acid, 10%	180	82
Benzoic acid	190	88
Benzyl alcohol	140	60
Benzyl chloride	80	27
Borax	210	99
Boric acid	220	104
Bromine gas, dry	X	X
Bromine gas, moist	X	X
Bromine, liquid	X	X
Butadiene	X	X
Butyl acetate	X	X
Butyl alcohol	200	93
n-Butylamine	90	32
Butyl phthalate	180	82
Butyric acid	180	82
Calcium bisulfide	210	99
Calcium bisulfite	210	99
Calcium carbonate	210	99
Calcium chlorate	220	104
Calcium chloride	220	104
Calcium hydroxide, 10%	200	93
Calcium hydroxide, sat.	220	104

(continued)

TABLE 2.32 *Continued*

Chemical	Maximum Temperature	
	°F	°C
Calcium hypochlorite	210	99
Calcium nitrate	210	99
Calcium oxide	220	104
Calcium sulfate	220	104
Caprylic acid	140	60
Carbon bisulfide	X	X
Carbon dioxide, dry	220	104
Carbon dioxide, wet	140	60
Carbon disulfide	X	X
Carbon monoxide	220	104
Carbon tetrachloride	X	X
Carbonic acid	220	104
Cellosolve	200	93
Chloroacetic acid, 50% water	80	27
Chloroacetic acid	180	82
Chlorine gas, dry	X	X
Chlorine gas, wet	X	X
Chlorine, liquid	X	X
Chlorobenzene	X	X
Chloroform	X	X
Chlorosulfonic acid	X	X
Chromic acid, 10%	140	60
Chromic acid, 50%	150	66
Chromyl chloride	140	60
Citric acid, 15%	220	104
Citric acid, conc.	220	104
Copper acetate	80	27
Copper carbonate	200	93
Copper chloride	200	93
Copper cyanide	200	93
Copper sulfate	200	93
Cresol	X	X
Cupric chloride, 5%	140	60
Cupric chloride, 50%	140	60
Cyclohexane	X	X
Cyclohexonol	150	66
Dibutyl phthalate	180	82
Dichloroacetic acid	100	38
Dichloroethane (ethylene dichloride)	80	27
Ethylene glycol	210	99
Ferric chloride	210	99
Ferric chloride, 50% in water	210	99
Ferrite nitrate, 10–50%	210	99
Ferrous chloride	210	99
Ferrous nitrate	210	99
Fluorine gas, dry	X	X
Fluorine gas, moist	X	X
Hydrobromic acid, dil.	230	110
Hydrobromic acid, 20%	200	93

(continued)

TABLE 2.32 *Continued*

Chemical	Maximum Temperature	
	°F	°C
Hydrobromic acid, 50%	190	88
Hydrochloric acid, 20%	220	104
Hydrochloric acid, 38%	200	93
Hydrocyanic acid, 10%	150	66
Hydrofluoric acid, 30%	180	82
Hydrofluoric acid, 70%	200	93
Hydrofluoric acid, 100%	200	93
Hypochlorous acid	140	60
Iodine solution, 10%	X	X
Ketones, general	110	43
Lactic acid, 25%	150	66
Lactic acid, conc.	150	66
Magnesium chloride	210	99
Malic acid	130	54
Manganese chloride	120	49
Methyl chloride	X	X
Methyl ethyl ketone	X	X
Methyl isobutyl ketone	80	27
Muriatic acid	200	93
Nitric acid, 5%	140	60
Nitric acid, 20%	140	60
Nitric acid, 70%	X	X
Nitric acid, anhydrous	X	X
Nitrous acid, conc.	X	X
Oleum	X	X
Perchloric acid, 10%	140	60
Perchloric acid, 70%	X	X
Phenol	180	82
Phosphoric acid, 50–80%	210	99
Picric acid	140	60
Potassium bromide, 30%	210	99
Salicylic acid	130	54
Silver bromide, 10%	170	77
Sodium carbonate	220	104
Sodium chloride	200	93
Sodium hydroxide, 10%	220	104
Sodium hydroxide, 50%	220	104
Sodium hydroxide, conc.	140	60
Sodium hypochlorite, 20%	120	49
Sodium hypochlorite, conc.	110	43
Sodium Sulfide, to 50%	190	88
Stannic chloride	150	66
Stannous chloride	200	93
Sulfuric acid, 10%	200	93
Sulfuric acid, 50%	200	93
Sulfuric acid, 70%	180	82
Sulfuric acid, 90%	180	82
Sulfuric acid, 98%	120	49

(continued)

TABLE 2.32 *Continued*

Chemical	Maximum Temperature	
	°F	°C
Sulfuric acid, 100%	X	X
Sulfuric acid, fuming	X	X
Sulfurous acid	180	82
Thionyl chloride	100	38
Toluene	X	X
Trichloroacetic acid	150	66
White liquor	220	104
Zinc chloride	200	93

The chemicals listed are in the pure state or in a saturated solution unless otherwise indicated. Compatibility is shown to the maximum allowable temperature for which data is available. Incompatibility is shown by an X. A blank space indicates that data is unavailable.

Source: From P.A. Schweitzer. 2004. *Corrosion Resistance Tables*, Vols. 1–4, 5th ed., New York: Marcel Dekker.

Polypropylene is widely used in engineering fabrics such as bale wraps, filter cloths, bags, ropes, and strapping. Piping and small tanks of polypropylene are widely used, and 90% of all battery casings are made of PP. Ignition resistant grades are available as a result of the addition of halogenated organic compounds. With this additive, polypropylene can be used in duct systems in the chemical industry. Because PP exhibits good flex life, it is useful in the construction of integral hinges. Polypropylene also has many textile applications such as carpet face and backing yarn, upholstery fabrics, and diaper cover stock. It is useful in outdoor clothing and sports clothes that are worn next to the body because its wicking qualities absorb body moisture and still leaves the wearer dry. PP also finds application in the automotive industry for interior trim and under-the-hood components. In appliance areas, it is used in washer agitators and dishwasher components. Consumer goods such as straws, toys and recreational items also make use of polypropylene.

2.24 Styrene–Acrylonitrile (SAN)

SAN is a copolymer of styrene and acrylonitrile having the following chemical structure:

$$
\left[\begin{array}{cc} \underset{|}{\overset{|}{C}}{-}\underset{|}{\overset{|}{C}} \\ \end{array}\right]\left[\begin{array}{cc} \underset{|}{\overset{|}{C}}{-}\underset{|}{\overset{|}{C}} \\ \end{array}\right]
$$

SAN is resistant to aliphatic hydrocarbons, not to aromatic and chlorinated hydrocarbons. It will be attacked by oxidizing agents and strong acids, and it will stress crack in the presence of certain organic compounds.

SAN will be degraded by UV light unless protective additives are incorporated into the formulation.

SAN finds application as food and beverage containers, dinnerware, housewares, appliances, interior refrigerator components, and toys. Industrial applications include fan blades and filter housings. Medical applications include tubing connectors and valves, labware, and urine bottles. The packaging industry makes use of SAN for cosmetic containers and displays.

2.25 Polyvinylidene Chloride (PVDC)

Polyvinylidene chloride is manufactured under the tradename Saran by Dow Chemical. It is a variant of PVC, having both chlorine atoms at the same end of the monomer instead of at the opposite ends as shown below:

$$\begin{array}{c} \text{Cl} \quad \text{H} \\ | \qquad | \\ -\text{C}-\text{C}- \\ | \qquad | \\ \text{Cl} \quad \text{H} \end{array}$$

Saron has improved strength, hardness, and chemical resistance over that of PVC. The operating temperature range is from 0 to 175°F (-18 to 80°C).

Saran is resistant to oxidants, mineral acids, and solvents. In applications such as plating solutions, chlorides and certain other chemicals, polyvinylidene chloride is superior to polypropylene and finds many applications in the handling of municipal water supplies and waste waters. Saran is also resistant to weathering and UV degradation. Refer to Table 2.33 for the compatibility of PVDC with selected corrodents. Reference [1] provides a more comprehensive listing.

Saran has found wide application in the plating industry and for handling deionized water, pharmaceuticals, food processing, and other applications where stream purity protection is critical. It also finds application as a lined piping system. Refer to Reference [2].

Saran is also used to manufacture auto seat covers, film, bristles, and prepared coatings.

TABLE 2.33

Compatibility of PVDC with Selected Corrodents

Chemical	Maximum Temperature	
	°F	°C
Acetaldehyde	150	66
Acetic acid, 10%	150	66
Acetic acid, 50%	130	54
Acetic acid, 80%	130	54
Acetic acid, glacial	140	60
Acetic anhydride	90	32
Acetone	90	32
Acetyl chloride	130	54
Acrylonitrile	90	32
Adipic acid	150	66
Allyl alcohol	80	27
Alum	180	82
Aluminum chloride, aqueous	150	66
Aluminum fluoride	150	66
Aluminum hydroxide	170	77
Aluminum nitrate	180	82
Aluminum oxychloride	140	60
Aluminum sulfate	180	82
Ammonia gas	X	X
Ammonium bifluoride	140	60
Ammonium carbonate	180	82
Ammonium chloride, sat.	160	71
Ammonium fluoride, 10%	90	32
Ammonium fluoride, 25%	90	32
Ammonium hydroxide, 25%	X	X
Ammonium hydroxide, sat.	X	X
Ammonium nitrate	120	49
Ammonium persulfate	90	32
Ammonium phosphate	150	66
Ammonium sulfate, 10–40%	120	49
Ammonium sulfide	80	27
Amyl acetate	120	49
Amyl alcohol	150	66
Amyl chloride	80	27
Aniline	X	X
Antimony trichloride	150	66
Aqua regia 3:1	120	49
Barium carbonate	180	82
Barium chloride	180	82
Barium hydroxide	180	82
Barium sulfate	180	82
Barium sulfide	150	66
Benzaldehyde	X	X
Benzene	X	X
Benzenesulfonic acid, 10%	120	49
Benzoic acid	120	49
Benzyl chloride	80	27

(continued)

TABLE 2.33 *Continued*

Chemical	Maximum Temperature °F	°C
Boris acid	170	77
Bromine, liquid	X	X
Butadiene	X	X
Butyl acetate	120	49
Butyl alcohol	150	66
Butyric acid	80	27
Calcium bisulfite	80	27
Calcium carbonate	180	82
Calcium chlorate	160	71
Calcium chloride	180	82
Calcium hydroxide, 10%	160	71
Calcium hydroxide, sat.	180	82
Calcium hypochlorite	120	49
Calcium nitrate	150	66
Calcium oxide	180	82
Calcium sulfate	180	82
Caprylic acid	90	32
Carbon bisulfide	90	32
Carbon dioxide, dry	180	82
Carbon dioxide, wet	80	27
Carbon disulfide	80	27
Carbon monoxide	180	82
Carbon tetrachloride	140	60
Carbonic acid	180	82
Cellosolve	80	27
Chloroacetic acid, 50% water	120	49
Chloroacetic acid	120	49
Chlorine gas, dry	80	27
Chlorine gas, wet	80	27
Chlorine, liquid	X	X
Chlorobenzene	80	27
Chloroform	X	X
Chlorosulfonic acid	X	X
Chromic acid, 10%	180	82
Chromic acid, 50%	180	82
Citric acid, 15%	180	82
Citric acid, conc.	180	82
Copper carbonate	180	82
Copper chloride	180	82
Copper cyanide	130	54
Copper sulfate	180	82
Cresol	150	66
Cupric chloride, 5%	160	71
Cupric chloride, 50%	170	77
Cyclohexane	120	49
Cyclohexanol	90	32
Dibutyl phthalate	180	82
Dichloroacetic acid	120	49

(continued)

TABLE 2.33 *Continued*

Chemical	Maximum Temperature	
	°F	°C
Dichloroethane (ethylene dichloride)	80	27
Ethylene glycol	180	82
Ferric chloride	140	60
Ferric chloride, 50% in water	140	60
Ferric nitrate, 10–50%	130	54
Ferrous chloride	130	54
Ferrous nitrate	80	27
Fluorine gas, dry	X	X
Fluorine gas, moist	X	X
Hydrobromic acid, dil.	120	49
Hydrobromic acid, 20%	120	49
Hydrobromic acid, 50%	130	54
Hydrochloric acid, 20%	180	82
Hydrochloric acid, 38%	180	82
Hydrocyanic acid, 10%	120	49
Hydrofluoric acid, 30%	160	71
Hydrofluoric acid, 100%	X	X
Hypochlorous acid	120	49
Ketones, general	90	32
Lactic acid, conc.	80	27
Magnesium chloride	180	82
Malic acid	80	27
Methyl chloride	80	27
Methyl ethyl ketone	X	X
Methyl isobutyl ketone	80	27
Muriatic acid	180	82
Nitric acid, 5%	90	32
Nitric acid, 20%	150	66
Nitric acid, 70%	X	X
Nitric acid, anhydrous	X	X
Oleum	X	X
Perchloric acid, 10%	130	54
Perchloric acid, 70%	120	49
Phenol	X	X
Phosphoric acid, 50–80%	130	54
Picric acid	120	49
Potassium bromide, 30%	110	43
Salicylic acid	130	54
Sodium carbonate	180	82
Sodium chloride	180	82
Sodium hydroxide, 0%	90	32
Sodium hydroxide, 50%	150	66
Sodium hydroxide, conc.	X	X
Sodium hypochlorite, 10%	130	54
Sodium hypochlorite, conc.	120	49
Sodium sulfide, to 50%	140	60
Stannic chloride	180	82
Stannous chloride	180	82

(continued)

TABLE 2.33 *Continued*

Chemical	Maximum Temperature	
	°F	°C
Sulfuric acid, 10%	120	49
Sulfuric acid, 50%	X	X
Sulfuric acid, 70%	X	X
Sulfuric acid, 90%	X	X
Sulfuric acid, 98%	X	X
Sulfuric acid, 100%	X	X
Sulfuric acid, fuming	X	X
Sulfurous acid	80	27
Thionyl chloride	X	X
Toluene	80	27
Trichloroacetic acid	80	27
Zinc chloride	170	77

The chemicals listed are in the pure state or in a saturated solution unless otherwise indicated. Compatibility is shown to the maximum allowable temperature for which data is available. Incompatibility is shown by an X. A blank space indicates that data is unavailable.

Source: From P.A. Schweitzer. 2004. *Corrosion Resistance Tables*, Vols. 1–4, 5th ed., New York: Marcel Dekker.

2.26 Polysulfone (PSF)

Polysulfone is an engineering polymer that can be used at elevated temperatures. It has the following chemical structure:

The linkages connecting the benzene rings are hydrolytically stable.

PSF has an operating temperature range of from −150 to 300°F (−101 to 149°C).

PSF is resistant to repeated sterilization by several techniques including steam autoclave, dry heat, ethylene oxide, certain chemicals, and radiation. It will stand exposure to soap, detergent solutions, and hydrocarbon oils even at elevated temperatures and under moderate stress levels. Polysulfone is unaffected by hydrolysis and has a very high resistance to mineral acids, alkali, and salt solutions. PSF is not resistant to polar organic solvents such as ketones, chlorinated hydrocarbons, and aromatic hydrocarbons.

Polysulfone has good weatherability, and it is not degraded by UV radiation. Refer to Table 2.34 for the compatibility of PSF with selected corrodents. Reference [1] provides a more comprehensive listing.

Polysulfone finds application as hot-water piping, lenses, iron handles, switches, and circuit breakers. Its rigidity and high temperature performance make it ideal for medical, microwave, and electronic application.

TABLE 2.34

Compatibility of PSF with Selected Corrodents

Chemical	Maximum Temperature	
	°F	°C
Acetaldehyde	X	
Acetic acid, 10%	200	93
Acetic acid, 50%	200	93
Acetic acid, 80%	200	93
Acetic acid, glacial	200	93
Acetone	X	
Acetyl chloride	X	
Aluminum chloride, aqueous	200	93
Aluminum chloride, dry	200	93
Aluminum fluoride	200	93
Aluminum oxychloride	150	66
Aluminum sulfate	200	93
Ammonia gas	X	
Ammonium carbonate	200	93
Ammonium chloride, 10%	200	93
Ammonium chloride, 50%	200	93
Ammonium chloride, sat.	200	93
Ammonium hydroxide, 25%	200	93
Ammonium hydroxide, sat.	200	93
Ammonium nitrate	200	93
Ammonium phosphate	200	93
Ammonium sulfate, to 40%	200	93
Amyl acetate	X	
Amyl alcohol	200	93
Aniline	X	
Aqua regia 3:1	X	
Barium carbonate	200	93
Barium chloride, 10%	200	93
Barium hydroxide	200	93
Barium sulfate	200	93
Benzaldehyde	X	
Benzene	X	
Benzoic acid	X	
Benzyl chloride	X	
Borax	200	93
Boric acid	200	93
Bromine gas, moist	200	93
Butyl acetate	X	

(continued)

TABLE 2.34 *Continued*

Chemical	Maximum Temperature	
	°F	°C
Butyl alcohol	200	93
n-Butylamine	X	
Calcium bisulfite	200	93
Calcium chloride	200	93
Calcium hypochlorite	200	93
Calcium nitrate	200	93
Calcium sulfate	200	93
Carbon bisulfide	X	
Carbon disulphide	X	
Carbon tetrachloride	X	
Carbonic acid	200	93
Cellosolve	X	
Chloroacetic acid	X	
Chlorine, liquid	X	
Chlorobenzene	X	
Chloroform	X	
Chlorosulfonic acid	X	
Chromic acid, 10%	140	60
Chromic acid, 50%	X	
Citric acid, 15%	100	38
Citric acid, 40%	80	27
Copper cyanide	200	93
Copper sulfate	200	93
Cresol	X	
Cupric chloride, 5%	200	93
Cupric chloride, 50%	200	93
Cyclohexane	200	93
Cyclohexanol	200	93
Dibutyl phthalate	180	82
Ethylene glycol	200	93
Ferric chloride	200	93
Ferric nitrate	200	93
Ferrous chloride	200	93
Hydrobromic acid, dil.	300	149
Hydrobromic acid, 20%	200	93
Hydrochloric acid, 20%	140	60
Hydrochloric acid, 38%	140	60
Hydrofluoric acid, 30%	80	27
Ketones, general	X	
Lactic acid, 25%	200	93
Lactic acid, conc.	200	93
Methyl chloride	X	
Methyl ethyl ketone	X	
Nitric acid, 5%	X	
Nitric acid, 20%	X	
Nitric acid, 70%	X	
Nitric acid, anhydrous	X	
Phosphoric acid, 50–80%	80	27

(continued)

TABLE 2.34 *Continued*

Chemical	Maximum Temperature	
	°F	°C
Potassium bromide, 30%	200	93
Sodium carbonate	200	93
Sodium chloride	200	93
Sodium hydroxide, 10%	200	93
Sodium hydroxide, 50%	200	93
Sodium hypochlorate, 20%	300	149
Sodium hypochlorite, conc.	300	149
Sodium sulfite, to 50%	200	93
Stannic chloride	200	93
Sulfuric acid, 10%	300	149
Sulfuric acid, 50%	300	149
Sulfuric acid, 70%	X	
Sulfuric acid, 90%	X	
Sulfuric acid, 98%	X	
Sulfuric acid, 100%	X	
Sulfuric acid, fuming	X	
Sulfurous acid	200	93
Toluene	X	

The chemicals listed are in the pure state or in a saturated solution unless otherwise indicated. Compatibility is shown to the maximum allowable temperature for which data is available. Incompatibility is shown by an X.

Source: From P.A. Schweitzer. 2004. *Corrosion Resistance Tables*, Vols. 1–4, 5th ed., New York: Marcel Dekker.

2.27 Polyvinyl Chloride (PVC)

Polyvinyl chloride is the most widely used of any of the thermoplasts. PVC is polymerized vinyl chloride that is produced from ethylene and anhydrous hydrochloric acid. The structure is:

$$\begin{array}{cccc} H & H & H & H \\ | & | & | & | \\ -C-C-C-C- \\ | & | & | & | \\ H & H & H & H \end{array}$$

Two types of PVC are produced, normal impact (type 1) and high impact (type 2). Type 1 is a rigid, unplasticized PVC having normal impact with optimum chemical resistance. Type 2 has optimum impact resistance and reduced chemical resistance.

Type 1 PVC (unplasticized) resists attack by most acids and strong alkalies, gasoline, kerosene, aliphatic alcohols, and hydrocarbons. It is particularly useful in the handling of hydrochloric acid. The chemical resistance of type 2 PVC to oxidizing and highly alkaline material is reduced.

PVC may be attacked by aromatics, chlorinated organic compounds, and lacquer solvents. PVC is resistant to all normal atmospheric pollutants including weather and UV degradation.

Refer to Table 2.35 for the compatibility of type 2 PVC with selected corrodents and Table 2.36 for type 1 PVC. Reference [1] provides a more comprehensive listing.

TABLE 2.35

Compatibility of Type 2 PVC with Selected Corrodents

	Maximum Temperature	
Chemical	°F	°C
Acetaldehyde	X	X
Acetamide	X	X
Acetic acid, 10%	100	38
Acetic acid, 50%	90	32
Acetic acid, 80%	X	X
Acetic acid, glacial	X	X
Acetic anhydride	X	X
Acetone	X	X
Acetyl chloride	X	X
Acrylic acid	X	X
Acrylonitrile	X	X
Adipic acid	140	60
Allyl alcohol	90	32
Allyl chloride	X	X
Alum	140	60
Aluminum acetate	100	38
Aluminum chloride, aqueous	140	60
Aluminum fluoride	140	60
Aluminum hydroxide	140	60
Aluminum nitrate	140	60
Aluminum oxychloride	140	60
Aluminum sulfate	140	60
Ammonia gas	140	60
Ammonium bifluoride	90	32
Ammonium carbonate	140	60
Ammonium chloride, 10%	140	60
Ammonium chloride, 50%	140	60
Ammonium chloride, sat.	140	60
Ammonium fluoride, 10%	90	32
Ammonium fluoride, 25%	90	32
Ammonium hydroxide, 25%	140	60
Ammonium hydroxide, sat.	140	60
Ammonium nitrate	140	60
Ammonium persulfate	140	60

(continued)

TABLE 2.35 *Continued*

Chemical	Maximum Temperature	
	°F	°C
Ammonium phosphate	140	60
Ammonium sulfate, 10–40%	140	60
Ammonium sulfide	140	60
Amyl acetate	X	X
Amyl alcohol	X	X
Amyl chloride	X	X
Aniline	X	X
Antimony trichloride	140	60
Aqua regia 3:1	X	X
Barium carbonate	140	60
Barium chloride	140	60
Barium hydroxide	140	60
Barium sulfate	140	60
Barium sulfide	140	60
Benzaldehyde	X	X
Benzene	X	X
Benzenesulfonic acid, 10%	140	60
Benzoic acid	140	60
Benzyl alcohol	X	X
Borax	140	60
Boric acid	140	60
Bromine gas, dry	X	X
Bromine gas, moist	X	X
Bromine, liquid	X	X
Butadiene	60	16
Butyl acetate	X	X
Butyl alcohol	X	X
n-Butylamine	X	X
Butyric acid	X	X
Calcium bisulfide	140	60
Calcium bisulfite	140	60
Calcium carbonate	140	60
Calcium chlorate	140	60
Calcium chloride	140	60
Calcium hydroxide, 10%	140	60
Calcium hydroxide, sat.	140	60
Calcium hypochlorite	140	60
Calcium nitrate	140	60
Calcium oxide	140	60
Calcium sulfate	140	60
Calcium bisulfide	X	X
Carbon dioxide, dry	140	60
Carbon dioxide, wet	140	60
Carbon disulfide	X	X
Carbon monoxide	140	60
Carbon tetrachloride	X	X
Carbonic acid	140	60
Cellosolve	X	X

(continued)

TABLE 2.35 *Continued*

Chemical	Maximum Temperature	
	°F	°C
Chloroacetic acid	105	40
Chlorine gas, dry	140	60
Chlorine gas, wet	X	X
Chlorine, liquid	X	X
Chlorobenzene	X	X
Chloroform	X	X
Chlorosulfonic acid	60	16
Chromic acid, 10%	140	60
Chromic acid, 50%	X	X
Citric acid, 15%	140	60
Citric acid, conc.	140	60
Copper carbonate	140	60
Copper chloride	140	60
Copper cyanide	140	60
Copper sulfate	140	60
Cresol	X	X
Cyclohexanol	X	X
Dichloroacetic acid	120	49
Dichloroethane (ethylene dichloride)	X	X
Ethylene glycol	140	60
Ferric chloride	140	60
Ferric nitrate, 10–50%	140	60
Ferrous chloride	140	60
Ferrous nitrate	140	60
Fluorine gas, dry	X	X
Fluorine gas, moist	X	X
Hydrobromic acid, dil.	140	60
Hydrobromic acid, 20%	140	60
Hydrobromic acid, 50%	140	60
Hydrochloric acid, 20%	140	60
Hydrochloric acid, 38%	140	60
Hydrocyanic acid, 10%	140	60
Hydrofluoric acid, 30%	120	49
Hydrofluoric acid, 70%	68	20
Hypochlorous acid	140	60
Ketones, general	X	X
Lactic acid, 25%	140	60
Lactic acid, conc.	80	27
Magnesium chloride	140	60
Malic acid	140	60
Methyl chloride	X	X
Methyl ethyl ketone	X	X
Methyl isobutyl ketone	X	X
Muriatic acid	140	60
Nitric acid, 5%	100	38
Nitric acid, 20%	140	60
Nitric acid, 70%	70	140
Nitric acid, anhydrous	X	X

(continued)

TABLE 2.35 *Continued*

Chemical	Maximum Temperature	
	°F	**°C**
Nitric acid, conc.	60	16
Oleum	X	X
Perchloric acid, 10%	60	16
Perchloric acid, 70%	60	16
Phenol	X	X
Phosphoric acid, 50–80%	140	60
Picric acid	X	X
Potassium bromide, 30%	140	60
Salicylic acid	X	X
Silver bromide, 10%	105	40
Sodium carbonate	140	60
Sodium chloride	140	60
Sodium hydroxide, 10%	140	60
Sodium hydroxide, 50%	140	60
Sodium hydroxide, conc.	140	60
Sodium hypochlorite, 20%	140	60
Sodium hypochlorite, conc.	140	60
Sodium sulfide, to 50%	140	60
Stannic chloride	140	60
Stannous chloride	140	60
Sulfuric acid, 10%	140	60
Sulfuric acid, 50%	140	60
Sulfuric acid, 70%	140	60
Sulfuric acid, 90%	X	X
Sulfuric acid, 98%	X	X
Sulfuric acid, 100%	X	X
Sulfuric acid fuming	X	X
Sulfurous acid	140	60
Thionyl chloride	X	X
Toluene	X	X
Trichloroacetic acid	X	X
White liquor	140	60
Zinc chloride	140	60

The chemicals listed are in the pure state or in a saturated solution unless otherwise indicated. Compatibility is shown to the maximum allowable temperature for which data is available. Incompatibility is shown by an X. A blank space indicates that data is unavailable.

Source: From P.A. Schweitzer. 2004. *Corrosion Resistance Tables*, Vols. 1–4, 5th ed., New York: Marcel Dekker.

The primary application for PVC include water, gas, vent, drain, and corrosive chemical piping, electrical conduit, and wire insulation. It is also used as a liner. Reference [2] provides details of PVC piping.

The automotive industry makes use of PVC for exterior applications as body-side moldings. Its ability to be pigmented and its excellent weathering ability to retain color makes it ideal for this application.

TABLE 2.36

Compatibility of Type 1 PVC with Selected Corrodents

	Maximum Temperature	
Chemical	°F	°C
Acetaldehyde	X	
Acetamide	X	
Acetic acid, 10%	140	60
Acetic acid, 50%	140	60
Acetic acid, 80%	140	60
Acetic acid glacial	130	54
Acetic anhydride	X	
Acetone	X	
Acetyl chloride	X	
Acrylic acid	X	
Acrylonitrile	X	
Adipic acid	140	60
Allyl alcohol	90	32
Allyl chloride	X	
Alum	140	60
Aluminum acetate	100	38
Aluminum chloride, aqueous	140	60
Aluminum chloride, dry	140	60
Aluminum fluoride	140	60
Aluminum hydroxide	140	60
Aluminum nitrate	140	60
Aluminum oxychloride	140	60
Aluminum sulfate	140	60
Ammonia gas	140	60
Ammonium bifluoride	90	32
Ammonium carbonate	140	60
Ammonium chloride, 10%	140	60
Ammonium chloride, 50%	140	60
Ammonium chloride, sat.	140	60
Ammonium fluoride, 1%	140	60
Ammonium fluoride, 25%	140	60
Ammonium hydroxide, 25%	140	60
Ammonium hydroxide, sat.	140	60
Ammonium nitrate	140	60
Ammonium persulfate	140	60
Ammonium phosphate	140	60
Ammonium sulfate, 10–40%	140	60
Ammonium sulfide	140	60
Ammonium sulfite	120	49
Amyl acetate	X	
Amyl alcohol	140	60
Amyl chloride	X	
Aniline	X	
Antimony trichloride	140	60
Aqua regia 3:1	X	
Barium carbonate	140	60

(continued)

TABLE 2.36 *Continued*

Chemical	Maximum Temperature	
	°F	°C
Barium chloride	140	60
Barium hydroxide	140	60
Barium sulfate	140	60
Barium sulfide	140	60
Benzaldehyde	X	
Benzene	X	
Benzenesulfonic acid	140	60
Benzoic acid	140	60
Benzyl alcohol	X	
Benzyl chloride	X	
Borax	140	60
Boric acid	140	60
Bromine gas, dry	X	
Bromine gas, moist	X	
Bromine, liquid	X	
Butadiene	140	60
Butyl acetate	X	
Butyl alcohol	X	
n-Butylamine	X	
Butyric acid	X	
Calcium bisulfide	140	60
Calcium bisulfite	140	60
Calcium carbonate	140	60
Calcium chlorate	140	60
Calcium chloride	140	60
Calcium hydroxide, sat.	140	60
Calcium hypochlorite	140	60
Calcium nitrate	140	60
Calcium oxide	140	60
Calcium sulfate	140	60
Caprylic acid	120	49
Carbon bisulfide	X	
Carbon dioxide, dry	140	60
Carbon dioxide, wet	140	60
Carbon disulfide	X	
Carbon monoxide	140	60
Carbon tetrachloride	X	
Carbonic acid	140	60
Cellosolve	X	
Chloroacetic acid, 50% water	X	
Chloroacetic acid	140	60
Chlorine gas, dry	140	60
Chlorine gas, wet	X	
Chlorine, liquid	X	
Chlorobenzene	X	
Chloroform	X	
Chlorosulfonic acid	X	
Chromic acid, 10%	140	60

(continued)

TABLE 2.36 *Continued*

Chemical	Maximum Temperature	
	°F	°C
Chromic acid, 50%	X	
Chromyl chloride	120	49
Citric acid, 15%	140	60
Citric acid, conc.	140	60
Copper acetate	80	27
Copper carbonate	140	60
Copper chloride	140	60
Copper cyanide	140	60
Copper sulfate	140	60
Cresol	120	49
Cupric chloride	140	60
Cupric chloride, 50%	150	66
Cyclohexane	80	27
Cyclohexanol	X	
Dibutyl phthalate	80	27
Dichloroacetic acid	100	38
Dichloroethane	X	
Ethylene glycol	140	60
Ferric chloride	140	60
Ferric chloride, 50% water	140	60
Ferric nitrate, 10–50%	140	60
Fluorine gas, dry	X	
Fluorine gas, moist	X	
Hydrobromic acid dil.	140	60
Hydrobromic acid, 20%	140	60
Hydrobromic acid, 50%	140	60
Hydrochloric acid, 20%	140	60
Hydrochloric acid, 38%	140	60
Hydrocyanic acid	140	60
Hydrofluoric acid, 30%	130	54
Hydrofluoric acid, 70%	X	
Hypochlorous acid	140	60
Iodine solution, 10%	100	38
Ketones, general	X	
Lactic acid, 25%	140	60
Lactic acid, conc.	80	27
Magnesium chloride	140	60
Malic acid	140	60
Manganese chloride	90	32
Methyl chloride	X	
Methyl ethyl ketone	X	
Methyl isobutyl ketone	X	
Muriatic acid	160	71
Nitric acid, 5%	140	60
Nitric acid, 20%	140	60
Nitric acid, 70%	140	60
Nitric acid, anhydrous	X	
Nitrous acid, conc.	80	27

(continued)

TABLE 2.36 *Continued*

Chemical	Maximum Temperature	
	°F	°C
Oleum	X	
Perchloric acid, 10%	140	60
Perchloric acid, 70%	X	
Phenol	X	
Phosphoric acid, 50–80%	140	60
Picric acid	X	
Potassium bromide, 30%	140	60
Salicylic acid	140	60
Silver bromide, 10%	140	60
Sodium carbonate	140	60
Sodium chloride	140	60
Sodium hydroxide, 10%	140	60
Sodium hydroxide, 50%	140	60
Sodium hydroxide, conc.	140	60
Sodium hydroxide, 20%	140	60
Sodium hypochlorite, conc.	140	60
Sodium sulfide, to 50%	140	60
Stannic chloride	140	60
Stannous chloride	140	60
Sulfuric acid, 10%	140	60
Sulfuric acid, 50%	140	60
Sulfuric acid, 70%	140	60
Sulfuric acid, 90%	140	60
Sulfuric acid, 98%	X	
Sulfuric acid, 100%	X	
Sulfuric acid, fuming	X	
Sulfurous acid	140	60
Thionyl chloride	X	
Toluene	X	
Trichloroacetic acid	90	32
White liquor	140	60
Zinc chloride	140	60

The chemicals listed are in the pure state or in a saturated solution unless otherwise indicated. Compatibility is shown to the maximum allowable temperature for which data is available. Incompatibility is shown by an X.

Source: From P.A. Schweitzer. 2004. *Corrosion Resistance Tables*, Vols. 1–4, 5th ed., New York: Marcel Dekker.

2.28   Chlorinated Polyvinyl Chloride (CPVC)

When acetylene and hydrochloric acid are reacted to produce polyvinyl chloride, the chlorination is approximately 56.8%. Further chlorination of

PVC to approximately 67% produces CPVC whose chemical structure is:

$$-\overset{\displaystyle \underset{|}{H}}{\underset{|}{C}}-\overset{\displaystyle \underset{|}{H}}{\underset{|}{C}}-\overset{\displaystyle \underset{|}{Cl}}{\underset{|}{C}}-\overset{\displaystyle \underset{|}{H}}{\underset{|}{C}}-$$

While PVC is limited to a maximum operating temperature of 140°F (60°C), CPVC has a maximum operating temperature of 190°F (82°C). There are special formulations of CPVC that permit operating temperatures of 200°F (93°C).

There are many similarities in the chemical resistance of CPVC and PVC. However, care must be exercised because there are differences. Overall, the corrosion resistance of CPVC is somewhat inferior to that of PVC. In general, PVC is inert to most mineral acids, bases, salts, and paraffinic hydrocarbons.

CVPC is not recommended for use with most polar organic materials including various solvents, chlorinated or aromatic hydrocarbons, esters, and ketones.

It is ideally suited for handling hot water and/or steam condensate. Refer to Table 2.37 for the compatibility of CPVC with selected corrodents. Reference [1] provides a more comprehensive listing.

TABLE 2.37

Compatibility of CPVC with Selected Corrodents

Chemical	Maximum Temperature	
	°F	°C
Acetaldehyde	X	X
Acetic acid, 10%	90	32
Acetic acid, 50%	X	X
Acetic acid, 80%	X	X
Acetic acid, glacial	X	X
Acetic anhydride	X	X
Acetone	X	X
Acetyl chloride	X	X
Acrylic acid	X	X
Acrylonitrile	X	X
Adipic acid	200	93
Allyl alcohol, 96%	200	90
Allyl chloride	X	X
Alum	200	93
Aluminum acetate	100	38
Aluminum chloride, aqueous	200	93
Aluminum chloride, dry	180	82
Aluminum fluoride	200	93
Aluminum hydroxide	200	93
Aluminum nitrate	200	93
Aluminum oxychloride	200	93

(continued)

TABLE 2.37 *Continued*

Chemical	Maximum Temperature	
	°F	°C
Aluminum sulfate	200	93
Ammonia gas, dry	200	93
Ammonium bifluoride	140	60
Ammonium carbonate	200	93
Ammonium chloride, 10%	180	82
Ammonium chloride, 50%	180	82
Ammonium chloride, sat.	200	93
Ammonium fluoride, 10%	200	93
Ammonium fluoride, 25%	200	93
Ammonium hydroxide, 25%	X	X
Ammonium hydroxide, sat.	X	X
Ammonium nitrate	200	93
Ammonium persulfate	200	93
Ammonium phosphate	200	93
Ammonium sulfate, 10–40%	200	93
Ammonium sulfide	200	93
Ammonium sulfite	160	71
Amyl acetate	X	X
Amyl alcohol	130	54
Amyl chloride	X	X
Aniline	X	X
Antimony trichloride	200	93
Aqua regia 3:1	80	27
Barium carbonate	200	93
Barium chloride	180	82
Barium hydroxide	180	82
Barium sulfate	180	82
Barium sulfide	180	82
Benzaldehyde	X	X
Benzene	X	X
Benzenesulfonic acid, 10%	180	82
Benzoic acid	200	93
Benzyl alcohol	X	X
Benzyl chloride	X	X
Borax	200	93
Boric acid	210	99
Bromine gas, dry	X	X
Bromine gas, moist	X	X
Bromine, liquid	X	X
Butadiene	150	66
Butyl acetate	X	X
Butyl alcohol	140	60
n-Butylamine	X	X
Butyric acid	140	60
Calcium bisulfide	180	82
Calcium bisulfite	210	99
Calcium carbonate	210	99
Calcium chlorate	180	82

(continued)

TABLE 2.37   *Continued*

Chemical	Maximum Temperature	
	°F	°C
Calcium chloride	180	82
Calcium hydroxide, 10%	170	77
Calcium hydroxide, sat.	210	99
Calcium hypochlorite	200	93
Calcium nitrate	180	82
Calcium oxide	180	82
Calcium sulfate	180	82
Caprylic acid	180	82
Carbon bisulfide	X	X
Carbon dioxide, dry	210	99
Carbon dioxide, wet	160	71
Carbon disulfide	X	X
Carbon monoxide	210	99
Carbon tetrachloride	X	X
Carbonic acid	180	82
Cellosolve	180	82
Chloroacetic acid, 50% water	100	38
Chloroacetic acid	X	X
Chlorine gas, dry	140	60
Chlorine gas, wet	X	X
Chlorine, liquid	X	X
Chlorobenzene	X	X
Chloroform	X	X
Chlorosulfonic acid	X	X
Chromic acid, 10%	210	99
Chromic acid, 50%	210	99
Chromyl chloride	180	82
Citric acid, 15%	180	82
Citric acid, conc.	180	82
Copper acetate	80	27
Copper carbonate	180	82
Copper chloride	210	99
Copper cyanide	180	82
Copper sulfate	210	99
Cresol	X	X
Cupric chloride, 5%	180	82
Cupric chloride, 50%	180	82
Cyclohexane	X	X
Cyclohexanol	X	X
Dichloroacetic acid, 20%	100	38
Dichloroethane (ethylene dichloride)	X	X
Ethylene glycol	210	99
Ferric chloride	210	99
Ferric chloride, 50% in water	180	82
Ferric nitrate, 10–50%	180	82
Ferrous chloride	210	99
Ferrous nitrate	180	82
Fluorine gas, dry	X	X

(continued)

TABLE 2.37 *Continued*

Chemical	Maximum Temperature	
	°F	°C
Fluorine gas, moist	80	27
Hydrobromic acid, dil.	130	54
Hydrobromic acid, 20%	180	82
Hydrobromic acid, 50%	190	88
Hydrochloric acid, 20%	180	82
Hydrochloric acid, 38%	170	77
Hydrocyanic acid, 10%	80	27
Hydrofluoric acid, 30%	X	X
Hydrofluoric acid, 70%	90	32
Hydrofluoric acid, 100%	X	X
Hypochlorous acid	180	82
Ketones, general	X	X
Lactic acid, 25%	180	82
Lactic acid, conc.	100	38
Magnesium chloride	230	110
Malic acid	180	82
Manganese chloride	180	82
Methyl chloride	X	X
Methyl ethyl ketone	X	X
Methyl isobutyl ketone	X	X
Muriatic acid	170	77
Nitric acid, 5%	180	82
Nitric acid, 20%	160	71
Nitric acid, 70%	180	82
Nitric acid, anhydrous	X	X
Nitrous acid, conc.	80	27
Oleum	X	X
Perchloric acid, 10%	180	82
Perchloric acid, 70%	180	82
Phenol	140	60
Phosphoric acid, 50–80%	180	82
Picric acid	X	X
Potassium bromide, 30%	180	82
Salicylic acid	X	X
Silver bromide, 10%	170	77
Sodium carbonate	210	99
Sodium chloride	210	99
Sodium hydroxide, 10%	190	88
Sodium hydroxide, 50%	180	82
Sodium hydroxide, conc.	190	88
Sodium hypochlorite, 20%	190	88
Sodium hypochlorite, conc.	180	82
Sodium sulfide, to 50%	180	82
Stannic chloride	180	82
Stannous chloride	180	82
Sulfuric acid, 10%	180	82
Sulfuric acid, 50%	180	82
Sulfuric acid, 70%	200	93

(continued)

TABLE 2.37 *Continued*

Chemical	Maximum Temperature	
	°F	°C
Sulfuric acid, 90%	X	X
Sulfuric acid, 98%	X	X
Sulfuric acid, 100%	X	X
Sulfuric acid, fuming	X	X
Sulfurous acid	180	82
Thionyl chloride	X	X
Toluene	X	X
Trichloroacetic acid, 20%	140	60
White liquor	180	82
Zinc chloride	180	82

The chemicals listed are in the pure state or in a saturated solution unless otherwise indicated. Compatibility is shown to the maximum allowable temperature for which data is available. Incompatibility is shown by an X.

Source: From P.A. Schweitzer. 2004. *Corrosion Resistance Tables*, Vols. 1–4, 5th ed., New York: Marcel Dekker. With permission.

The primary applications of PVC include hot water and/or steam condensate piping, corrosive chemical piping at elevated temperatures, valves, fume ducts, internal column packing, and miscellaneous fabricated items. Reference [2] provides detailed information on CPVC piping.

2.29   Chlorinated Polyether (CPE)

Chlorinated polyether is sold under the tradename of Penton. It has a maximum operating temperature of 225°F (107°C). CPE is resistant to most acids and bases, oxidizing agents, and common solvents. It has the advantage of being able to resist acids at elevated temperatures. It is not resistant to nitric acid above 10% concentration. Refer to Table 2.38 for the compatibility of CPE with selected corrodents. Reference [1] provides a more comprehensive listing.

CPE finds application as bearing retainers, tanks, tank linings, and in process equipment.

2.30   Polyacrylonitrile (PAN)

PAN is a member of the olefin family. In general, it is somewhat similar to the other olefins such as polyethylene and polypropylene in terms of appearance, general chemical characteristics, and electrical properties.

PAN has the following structural formula:

$$-\overset{\displaystyle H}{\underset{\displaystyle H}{C}}-\overset{\displaystyle H}{\underset{\displaystyle C=N}{C}}-$$

PAN has corrosion resistance properties similar to those of polypropylene and polyethylene.

TABLE 2.38

Compatibility of CPE with Selected Corrodents

Chemical	Maximum Temperature	
	°F	°C
Acetaldehyde	130	54
Acetic acid, 10%	250	121
Acetic acid, 50%	250	121
Acetic acid, 80%	250	121
Acetic acid, glacial	250	121
Acetic anhydride	150	66
Acetone	90	32
Acetyl chloride	X	
Adipic acid	250	121
Allyl alcohol	250	121
Allyl chloride	90	32
Alum	250	121
Aluminum chloride, aqueous	250	121
Aluminum fluoride	250	121
Aluminum hydroxide	250	121
Aluminum oxychloride	220	104
Aluminum sulfate	250	121
Ammonia gas, dry	220	104
Ammonium bifluoride	220	104
Ammonium carbonate	250	121
Ammonium chloride, sat.	250	121
Ammonium fluoride, 10%	250	121
Ammonium fluoride, 25%	250	121
Ammonium hydroxide, 25%	250	121
Ammonium hydroxide, sat.	250	121
Ammonium nitrate	250	121
Ammonium persulfate	180	82
Ammonium phosphate	250	121
Ammonium sulfate, 10–40%	250	121
Ammonium sulfide	250	121
Amyl acetate	180	82
Amyl alcohol	220	104
Amyl chloride	220	104
Aniline	150	66
Antimony trichloride	220	104

(continued)

TABLE 2.38 *Continued*

Chemical	Maximum Temperature	
	°F	°C
Aqua regia 3:1	80	27
Barium carbonate	250	121
Barium chloride	250	121
Barium hydroxide	250	121
Barium sulfate	250	121
Barium sulfide	220	104
Benzaldehyde	80	27
Benzenesulfonic acid, 10%	220	104
Benzoic acid	250	121
Benzyl alcohol		
Benzyl chloride	80	27
Borax	140	60
Boric acid	250	121
Bromine	140	60
Bromine, liquid	X	
Butadiene	250	121
Butyl acetate	130	54
Butyl alcohol	220	104
Butyric acid	230	110
Calcium bisulfate	250	121
Calcium carbonate	250	121
Calcium chlorate	150	66
Calcium chloride	250	121
Calcium hydroxide	250	121
Calcium hypochlorite	180	82
Calcium nitrate	250	121
Calcium oxide	250	121
Calcium sulfate	250	121
Carbon bisulfide	X	
Carbon dioxide, dry	250	121
Carbon dioxide, wet	250	121
Carbon disulfide	X	
Carbon monoxide	250	121
Carbon tetrachloride	200	93
Carbonic acid	250	121
Cellosolve	220	104
Chloroacetic acid	220	104
Chlorine gas, dry	100	38
Chlorine gas, wet	80	27
Chlorine, liquid	X	
Chlorobenzene	150	66
Chloroform	80	27
Chlorosulfonic acid	X	
Chromic acid, 10%	250	121
Chromic acid, 50%	250	121
Citric acid, 15%	250	121
Citric acid, 25%	250	121
Copper chloride	250	121
Copper cyanide	250	121
Copper sulfate	250	121
Cresol	140	60

(continued)

TABLE 2.38 *Continued*

Chemical	Maximum Temperature	
	°F	°C
Cyclohexane	220	104
Cyclohexanol	220	60
Dichloroacetic acid		
Dichloroethane	X	
Ethylene glycol	220	104
Ferric chloride	250	121
Ferric nitrate	250	121
Ferrous chloride	250	121
Hydrobromic acid, 20%	250	121
Hydrochloric acid, 20%	250	121
Hydrochloric acid, 38%	300	149
Hydrochloric acid, 70%		
Hydrochloric acid, 100%		
Hypochlorous acid	150	66
Ketones, general	X	
Lactic acid, 25%	250	121
Magnesium chloride	250	121
Malic acid	250	121
Methyl chloride	230	110
Methyl ethyl ketone	90	32
Methyl isobutyl ketone	80	27
Muriatic acid		
Nitric acid, 5%	180	82
Nitric acid, 20%	80	27
Nitric acid, 70%	80	27
Nitric acid, anhydrous	X	
Oleum		
Perchloric acid, 10%	140	60
Perchloric acid, 70%	X	
Phenol		
Phosgene gas, dry		
Phosphoric acid, 50–80%	250	121
Picric acid	150	66
Potassium bromide, 30%	250	121
Sodium carbonate	250	121
Sodium chloride	250	121
Sodium hydroxide, 10%	180	82
Sodium hydroxide, conc.	180	82
Sodium hypochlorite, 20%	180	82
Sodium hypochlorite, conc.	200	93
Sodium sulfide, to 50%	230	110
Stannic chloride	250	121
Stannous chloride	250	121
Sulfuric acid, 10%	230	110
Sulfuric acid, 50%	230	110
Sulfuric acid, 70%	230	110
Sulfuric acid, 90%	130	54
Sulfuric acid, 98%	X	
Sulfuric acid, 100%	X	
Sulfuric acid, fuming		
Sulfurous acid	200	93

(continued)

TABLE 2.38 *Continued*

| | Maximum Temperature | |
Chemical	°F	°C
Thionyl chloride	X	
Toluene	200	93
White liquor	200	93
Zinc chloride	250	121

The chemical listed are in the pure state or in a saturated solution unless otherwise indicated. Compatibility is shown to the maximum allowable temperature for which data is available. Incompatibility is shown by an X. A blank space indicates that data is unavailable.

Source: From P.A. Schweitzer. 2004. *Corrosion Resistance Tables*, Vols. 1–4, 5th ed., New York: Marcel Dekker.

Polyacrylonitrile is used to produce molded appliance parts, automotive parts, garden hose, vending machine tubing, chemical apparatus, typewriter cases, bags, luggage shells, and trim.

2.31 Polyurethane (PUR)

Polyurethanes are produced from either polyethers or polyesters. Those produced from polyether are more resistant to hydrolysis and have higher resilience, good energy-absorbing characteristics, good hysteresis characteristics, and good all-around chemical resistance. The polyester-based urethanes are generally stiffer and will have higher compression and tensile moduli, higher tear strength and cut resistance, higher operating temperature, lower compression set, optimum abrasion resistance, and good fuel and oil resistance. Refer to Figure 2.2 for the structural formula.

Polyurethanes exhibit excellent resistance to oxygen aging, but they have limited life in high humidity and high temperature applications. Water affects PUR in two ways, temporary plasticization and permanent degradation. Moisture plasticization results in a slight reduction in hardness and tensile strength. When the absorbed water is removed, the original properties are restored. Hydrolytic degradation causes permanent reduction in physical and electrical properties.

Because polyurethane is a polar material, it is resistant to nonpolar organic fluids such as oils, fuels, and greases, but it will be readily attacked and even dissolved by polar organic fluids such as dimethylformamide and dimethyl sulfoxide.

Table 2.39 provides the compatibility of PUR with selected corrodents. Reference [1] provides a more comprehensive listing.

Polyurethanes are used where the properties of good abrasion resistance and low coefficient of friction are required. They are also used

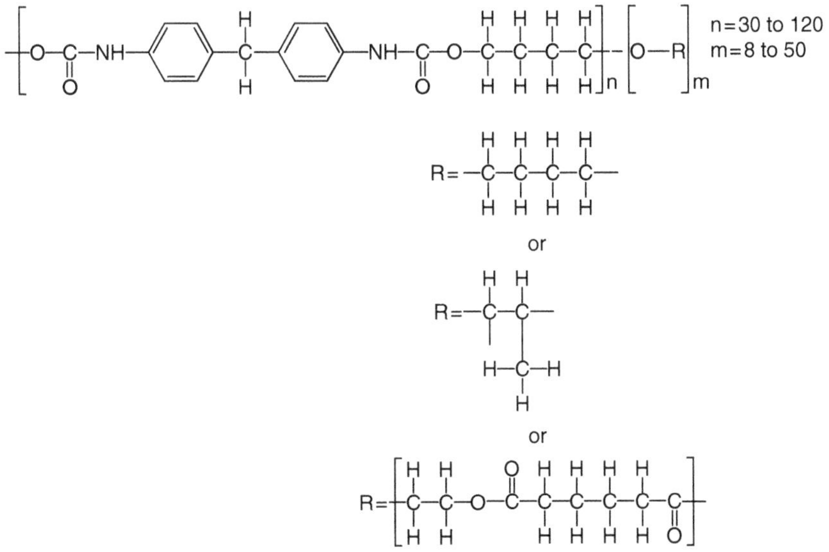

FIGURE 2.2
Chemical structure of PUR.

TABLE 2.39

Compatibility of PUR with Selected Corrodents

Chemical	
Acetaldehyde	X
Acetamine	X
Acetic acid, 10%	X
Acetic acid, 50%	X
Acetic acid, 80%	X
Acetic acid, glacial	X
Acetic anhydride	X
Acetone	X
Acetyl chloride	X
Ammonium carbonate	R
Ammonium chloride, 10%	R
Ammonium chloride, 50%	R
Ammonium chloride, sat.	R
Ammonium hydroxide, 25%	R
Ammonium hydroxide, sat.	R
Ammonium persulfate	X
Amyl acetate	X
Amyl alcohol	X
Aniline	X
Aqua regia 3:1	X
Barium chloride	R
Barium hydroxide	R

(continued)

TABLE 2.39 *Continued*

Chemical	
Barium sulfide	R
Benzaldehyde	X
Benzene	X
Benzenesulfonic acid	X
Benzoic acid	X
Benzyl alcohol	X
Benzyl chloride	X
Borax	R
Boric acid	R
Bromine, liquid	X
Butadiene	X
Butyl acetate	X
Carbon bisulfide	R
Calcium chloride	R
Calcium hydroxide, 10%	R
Calcium hydroxide, sat.	R
Calcium hypochlorite	X
Calcium nitrate	R
Carbon dioxide, dry	R
Carbon dioxide, wet	R
Carbon monoxide	R
Carbon tetrachloride	X
Carbonic acid	R
Chloroacetic acid	X
Chlorine gas, dry	X
Chlorine gas, wet	X
Chlorobenzene	X
Chloroform	X
Chlorosulfonic acid	X
Chromic acid, 10%	X
Chromic acid, 50%	X
Copper chloride	R
Copper cyanide	R
Copper sulfate	R
Cyclohexane	R
Cresol	X
Ethylene glycol	R
Ferric chloride	R
Ferric chloride, 50%	R
Ferric nitrate, 10–50%	R
Hydrochloric acid, 20%	R
Hydrochloric acid, 38%	X
Magnesium chloride	R
Methyl chloride	X
Methyl ethyl ketone	X
Methyl isobutyl ketone	X
Nitric acid, 5%	X
Nitric acid, 20%	X
Nitric acid, 70%	X

(continued)

TABLE 2.39 *Continued*

Chemical	
Nitric acid, anhydrous	X
Oleum	X
Perchloric acid, 10%	X
Perchloric acid, 70%	X
Phenol	X
Potassium bromide, 30%	R
Sodium chloride	R
Sodium hydroxide, 50%	R
Sodium hypochlorite, conc.	X
Sulfuric acid, 10%	X
Sulfuric acid, 50%	X
Sulfuric acid, 70%	X
Sulfuric acid, 90%	X
Sulfuric acid, 98%	X
Sulfuric acid, 100%	X
Toluene	X

The chemicals listed are in the pure state or in a saturated solution unless otherwise indicated. Compatibility at 90°F/32°C is shown by an R. incompatibility is shown by an X.

for high-impact car panels and other parts. Polyurethanes are also used for toys, gears, bushings, pulleys, golf balls, ski-goggle frames, shoe components, and ski boots. The medical field makes use of PUR for diagnostic devices, tubing, and catheters.

2.32 Polybutylene Terephthalate (PBT)

PBT is also known as a thermoplastic polyester. These thermoplastic polyesters are highly crystalline with a melting point of approximately 430°F (221°C). It has a structural formula as follows:

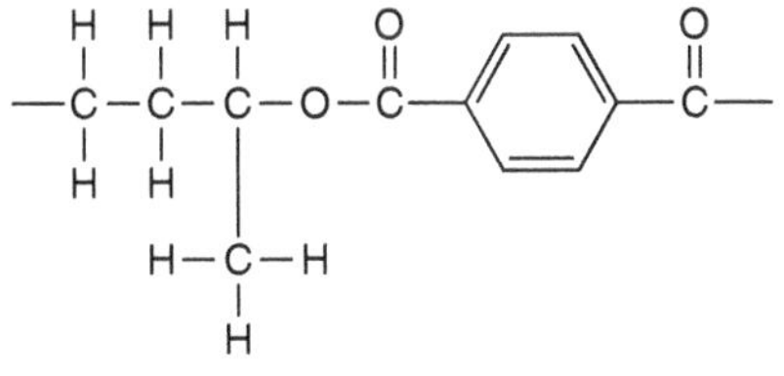

PBT exhibits good chemical resistance, in general, to dilute mineral acids, aliphatic hydrocarbons, aromatic hydrocarbons, ketones, and esters with limited resistance to hot water and washing soda. It is not resistant to chlorinated hydrocarbons and alkalies.

PBT has good weatherability and is resistant to UV degradation. Refer to Table 2.40 for the compatibility of PBT with selected corrodents.

TABLE 2.40

Compatibility of PBT with Selected Corrodents

Chemical	
Acetic acid, 5%	R
Acetic acid, 10%	R
Acetone	R
Ammonia, 10%	X
Benzene	R
Butyl acetate	R
Calcium chloride	R
Calcium disulfide	R
Carbon tetrachloride	R
Chlorobenzene	X
Chloroform	X
Citric acid, 10%	R
Diesel oil	R
Dioxane	R
Edible oil	R
Ethanol	R
Ether	R
Ethyl acetate	R
Ethylene chloride	X
Ethylene glycol	R
Formic acid	R
Fruit juice	R
Fuel oil	R
Gasoline	R
Glycerine	R
Heptane/hexane	R
Hydrochloric acid, 38%	X
Hydrochloric acid, 2%	R
Hydrogen peroxide, 30%	R
Hydrogen peroxide, 0–5%	R
Ink	R
Linseed oil	R
Methanol	R
Methyl ethyl ketone	R
Methylene chloride	X
Motor oil	R
Nitric acid, 2%	R
Paraffin oil	R
Phosphoric acid, 10%	R
Potassium hydroxide, 50%	X
Potassium dichromate, 10%	R
Potassium permanganate, 10%	R
Silicone oil	R
Soap solution	R
Sodium bisulfate	R
Sodium carbonate	R
Sodium chloride, 10%	R
Sodium hydroxide, 50%	X

(continued)

TABLE 2.40 *Continued*

Chemical	
Sodium nitrate	R
Sodium thiosulfate	R
Sulfuric acid, 98%	X
Sulfuric acid, 2%	R
Toluene	R
Vaseline	R
Water, cold	R
Water, hot	R
Wax, molten	R
Xylene	R

R, material resistant at 73°F/20°C; X, material not resistant.

The unreinforced resins are used in housings that require excellent impact in moving parts such as gears, bearings, pulleys, and writing instruments. Flame retardant formulations find application as television, radio, electronics, business machines, and pump components.

Reinforced resins find application in the automotive, electrical, electronic, and general industrial areas.

2.33 Acetals

Acetals are a group of high-performance engineering polymers that resemble the polyamide somewhat in appearance but not in properties. The general repeating chemical structure is:

$$\left[\begin{array}{c} H \\ | \\ -C-O- \\ | \\ H \end{array} \right]_n$$

There are two basic types of acetals. DuPont produces the homopolymers whereas Celanese produces the copolymers.

Neither the homopolymer nor the copolymer is resistant to strong mineral acids, but the copolymers are resistant to strong bases. The acetals will be degraded by UV light and gamma rays. They also exhibit poor resistance to oxidizing agents but, in general, are resistant to the common solvents. Refer to Table 2.41 for the compatibility of acetals with selected corrodents.

Homopolymer materials are usually implied when no identification of acetal type is made.

TABLE 2.41

Compatibility of Acetals with Selected Corrodents

Chemical	Maximum Temperature	
	°F	°C
Acetaldehyde	70	23
Acetic acid, 5%	70	23
Acetic acid, 10%	X	
Acetone	70	23
Acid mine water	100	38
Ammonia, wet	X	
Ammonium hydroxide, conc.	X	
Aniline	70	23
Aqua regia	X	
Benzene	70	23
Benzyl chloride	70	23
n-Butylamine	X	
Calcium bisulfite	X	
Calcium chloride	X	
Calcium chlorate, 5%	70	23
Calcium hydroxide, 10%	70	23
Calcium hydroxide, 20%	X	
Calcium hypochlorite, 2%	X	
Calcium nitrate	X	
Calcium sulfate	X	
Calcium disulfide	70	23
Carbon tetrachloride	X	
Carbonic acid	70	23
Cellosolve	70	23
Chloric acid	X	
Chlorine, dry	X	
Chlorine, moist	70	23
Chloroacetic acid	X	
Chlorobenzene, dry	70	23
Chloroform	X	
Chlorosulfonic acid	X	
Chromic acid	X	
Copper nitrate	X	
Cupric chloride	X	
Cyclohexanne	70	23
Cyclohexanol	70	23
Cyclohexanne	70	23
Detergents	70	23
Dichloroethane	70	23
Diethyl ether	70	23
Dimethyl formamide	140	60
Dimethyl sulfoxide	70	23
Dioxane	140	60
Ethanolamine	70	23
Ethers	70	23
Ethyl acetate, 10%	200	93
Ethyl chloride, wet	70	23

(continued)

TABLE 2.41 *Continued*

Chemical	Maximum Temperature	
	°F	°C
Ethyl ether	70	23
Ethylene chloride	70	23
Ethylenediamine	70	23
Fatty acid	95	35
Ferric chloride	X	
Ferric nitrate	X	
Ferric sulfate	X	
Ferrous chloride	X	
Ferrous sulfate, 10%	X	
Formic acid, 3%	X	
Heptane	70	23
Hexane	70	23
Hydrochloric acid, 1–20%	X	
Hydrofluoric acid, 4%	X	
Hydrogen peroxide	X	
Hydrogen sulfide, moist	X	
Kerosene	140	60
Lactic acid	X	
Linseed oil	70	23
Magnesium chloride, 10%	70	23
Magnesium hydroxide, 10%	70	23
Magnesium sulfate, 10%	70	23
Magnesium sulfate	70	23
Mercury	70	23
Methyl acetate	X	
Methyl alcohol	70	23
Methylene chloride	70	23
Milk	70	23
Tetrahydrofuran	70	23
Thionyl chloride	X	
Turpentine	140	60
Mineral oil	70	23
Motor oil	150	71
Monoethanolamine	X	
Nitric acid, 1%	X	
Nitric acid	X	
Oxalic acid	X	
Paraffin	70	23
Perchloroethylene	70	23
Perchloric acid, 10%	X	
Phosphoric acid	X	
Pyridine	70	23
Silver nitrate	70	23
Sodium bicarbonate, 50%	70	23
Sodium carbonate, 20%	180	82
Sodium carbonate	70	23
Sodium chlorate, 10%	70	23
Sodium chloride, 10%	180	82

(continued)

TABLE 2.41 *Continued*

Chemical	Maximum Temperature	
	°F	°C
Sodium chloride	70	23
Sodium cyanide, 10%	70	23
Sodium hydroxide, 10%	70	23
Sodium hypochlorite	X	
Sodium nitrate, 50%	70	23
Sodium thiosulfate	70	23
Sulfuric acid, 10%	X	
Sulfuric acid	X	
Sulfurous acid, 10%	X	
Trichloroethylene	X	
Wax	200	93

The chemicals listed are in the pure state or in a saturated solution unless otherwise indicated. Compatibility is shown to the maximum allowable temperature for which data is available. Incompatibility is shown by an X.

Acetals are used in the automotive industry for fuel-system components and moving parts such as gears, structural components in the suspension system, and bushings. Applications also include appliance housings and plumbing fixtures.

━━━━━━━━

References

1. P.A. Schweitzer. 2004. *Corrosion Resistance Tables*, Vols. 1–4, 5th ed., New York: Marcel Dekker.
2. P.A. Schweitzer. 1994. *Corrosion Resistant Piping Systems*, New York: Marcel Dekker.
3. P.A. Schweitzer. 1996. *Corrosion Engineering Handbook*, New York: Marcel Dekker.
4. P.A. Schweitzer. 2001. *Corrosion Resistant Linings and Coatings*, New York: Marcel Dekker.

3

Thermoset Polymers

Thermoset polymers assume a permanent shape or set once cured. Once set, they cannot be reshaped. They are formed by a large amount of cross-linking of linear prepolymers (a small amount of cross-linking will produce elastomers) or by direct formation of networks by the reaction of two monomers. The latter is the more prominent of the two methods. It is a stepwise or condensation method that has been defined as "the reaction of two monomers to produce a third plus a by-product, usually water or alcohol." Because in some cases a by-product is not produced, this definition is no longer exactly correct. The reaction is now referred to as a "stepwise" polymerization. When the reaction results in a by-product, it is called a "condensation reaction." Table 3.1 lists the principal thermoset polymers.

Although fewer in number than the thermoplastic polymers, thermosets comprise approximately 14% of the total polymer market. Compared to the thermoplasts, they are more brittle, stronger, harder, and generally more temperature resistant. Table 3.2 gives the operating temperature range of the thermoset polymers. In addition, they offer the advantages of better dimensional stability, creep resistance, chemical resistance, and good electrical properties. Their disadvantages lie in the fact that most are more difficult to process and more expensive.

Phenolics represent about 43% of the thermoset market, making them the most widely used. They are relatively inexpensive and are readily molded with good stiffness. Most contain wood or glass-flour fillers and, on occasion, glass fibers.

For molded products, usually formed by compression or transfer molding, mineral or cellulose fibers are often used as low-cost general purpose fillers, and glass fiber fillers are often used for optimum strength or dimensional stability.

3.1 Corrosion of Thermosets

Unreinforced, unfilled thermoset polymers can corrode by several mechanisms. The type of corrosion can be divided into two main categories: physical and chemical.

TABLE 3.1

Principal Thermoset Polymers

Alkyds
Amino resins
Diallyl phthalates (allyls)
Epoxies
Furans
Melamines
Phenol formaldehyde
Phenolics
Polybutadienes
Polyurethanes
Silicones
Unsaturated polyesters
Ureas
Vinyl esters

Physical corrosion is the interaction of a thermoset polymer with its environment so that its properties are altered but no chemical reactions take place. The diffusion of a liquid into the polymer is a typical example. In many cases, physical corrosion is reversible: after the liquid is removed, the original properties are restored.

When a polymer absorbs a liquid or gas, that results in plasticization or swelling of the thermoset network, physical corrosion has taken place. For a cross-linked thermoset, swelling caused by solvent absorption will be at a maximum when the solvent and polymer solubility parameters are exactly matched.

TABLE 3.2

Operating Temperature Range of Themoset Polymers

Polymer	Allowable Temperature (°F/°C)	
	Minimum	Maximum
Epoxy	−423/−252	300/150
Polyesters:		
Bisphenol A fumarate		250–300/120–150
Halogenated		250–300/120–150
Hydrogenated bisphenol-A– bisphenol-A		250–300/120–150
Isophthalic		150/70
Terephthalic		250–300/120–150
Vinyl ester		200–280/98–140
Furan		300–400/150–200
Phenolic		230/110
Silicone	−423/−252	500/260

Temperatures might have to be modified depending on the corrodent.

Chemical corrosion takes place when the bonds in the thermoset are broken by means of a chemical reaction with the polymers' environment. There may be more than one form of chemical corrosion taking place at the same time. Chemical corrosion is usually not reversible.

As result of chemical corrosion, the polymer itself may be affected in one or more ways. For example, the polymer may be embrittled, softened, charred, crazed, delaminated, discolored, blistered, or swollen. All thermosets will be attacked in essentially the same manner. However, certain chemically resistant types suffer negligible attack or exhibit significantly lower rates of attack under a wide range of severely corrosive conditions. This is the result of the unique molecular structure of the resins, including built-in properties of ester groups.

Cure of the resin plays an important part in the chemical resistance of the thermoset. Improper curing will result in a loss of corrosion resistance properties. Construction of the laminate and the type of reinforcing used also affect the corrosion resistance of the laminate. The degree and nature of the bond between the resin and the reinforcement also plays an important role.

The various modes of attack affect the strength of the laminate in different ways, depending upon the environment and other service conditions and the mechanisms or combination of mechanisms that are at work.

Some environments may weaken primary and/or secondary polymer linkages with resulting depolymerization. Other environments may cause swelling or microcracking, whereas others may hydrolyze ester groupings or linkages. In certain environments, repolymerization can occur with a resultant change in structure. Other results may be chain scission and decreases in molecular weight or simple solvent action. Attack or absorption at the interface between the reinforcing material and the resin will result in weakening.

In general, chemical attack on thermoset polymers is a "go/no go" situation. With an improper environment, attack on the reinforced polymer will occur in a relatively short time. Experience has indicated that if an installation has been soundly engineered and has operated successfully for 12 months, in all probability, it will continue to operate successfully for a substantial period of time.

Thermoset polymers are not capable of handling concentrated sulfuric acid (93%) and concentrated nitric acid. Pyrolysis or charring of the resin quickly occurs so that within a few hours, the laminate is destroyed. Tests show that polyesters and vinyl esters can handle up to 70% sulfuric acid for long periods of time.

The attack of aqueous solutions on reinforced polymers occurs through hydrolysis, with the water degrading bonds in the backbone of the resin molecules. The ester linkage is the most susceptible. The attack of solvents is of a different nature. The solvent penetrates the resin matrix of the polymer through spaces between the polymer chains causing the laminate surface to swell, soften, and crack. In the first stages of solvent attack, the following occur:

1. Softening; decrease in hardness by a substantial amount.
2. Swelling; the laminate swells considerably.
3. Weight; pronounced weight gain. Anything over 2% is cause for concern.

There are many degrees of attack:

1. If there is less than a 2% weight gain and considerable retention in hardness over a 12 month period, plus little or no swelling, the laminate will do very well.
2. If hardness is retained at a high level, weight gain has stabilized, and swelling has stabilized at the 12 month level, then the resin may do fairly well in limited service.
3. Bisphenol resin laminate in contact with toluene experiences a 19% weight gain in one month, a swelling of 19%, and hardness drops from 43 to 0, typical of a material undergoing total failure. As the solvent attack continues:
   a. Softening: hardness drops to zero.
   b. Swelling: the absorption of the molecule continues, producing mechanical stresses that cause fractures in the laminate.

This may occur in the liquid or vapor zone.

Organic compounds with carbon–carbon unsaturated double bonds, such as carbon disulfide, are powerful swelling solvents and show greater swelling action than their saturated counterparts. Smaller solvent molecules can penetrate a polymer matrix more effectively. The degree of similarity between solvent and resin is important. Slightly polar resins, such as polyesters and the vinyl esters, are attacked by mildly polar solvents.

Generally, saturated, long-chain organic molecules, such as the straight chain hydrocarbons, are handled well by polyesters. This is why polyester gasoline tanks are so successful.

The polymer's ability to resist attack is improved by

1. The cross-link density of the resin.
2. The ability of the resin to pack into a tight structure.
3. The heat distortion temperature of the resin, which is strongly related to its solvent resistance. The higher the heat distortion temperature, the better the solvent resistance. The latter is not true of all resin systems.

In general, the more brittle resins possess poor solvent resistance whereas the more resilient resins can withstand a greater degree of solvent absorption.

Orthophthalic, isophthalic, bisphenol, and chlorinated or brominated polyesters exhibit poor resistance to such solvents as acetone, carbon disulfide, toluene, trichloroethylene, trichloroethane, and methyl ethyl ketone. The vinyl esters show improved solvent resistance. Heat-cured epoxies exhibit better solvent resistance. However, the furan resins offer the best all-around solvent resistance. They excel in this area. Furan resins are capable of handling solvents in combination with acids and bases.

Stress corrosion is another factor to be considered. The failure rate of glass-reinforced composites can be significant. This is particularly true of composites exposed to the combination of acid and stress. Weakening of the glass fibers upon exposure to acid is believed to be caused by ion exchange between the acid and glass.

Under stress, an initial fiber fracture occurs that is specifically a tensile type of failure. If the resin matrix surrounding the failed fiber fractures, the acid is allowed to attack the next available fiber, which subsequently fractures. The process continues until total failure occurs.

3.2 Joining of Thermosets

Thermoset polymers cannot be solvent cemented or fusion welded. However, they can be adhesive bonded using epoxies, thermosetting acrylics, and urethanes. Prior to bonding, the surfaces must be abraded and solvent cleaned.

3.3 Ultraviolet Light Stability

Thermoset resins will be degraded by ultraviolet light. In view of this, it is necessary that UV protection be provided. Polymers installed outdoors will suffer severe degradation in a year or two from ultraviolet exposure. The glass filament will fray. After the initial fraying, the phenomenon will stop far short of having any particular effect on the strength of the material. This can be prevented by adding an ultraviolet inhibitor to the resin. Another approach is to apply a 3-mil-thick coat of black epoxy paint that will also effectively screen out the ultraviolet rays.

With the exception of UV light stability, thermoset polymers resist weathering, being resistant to normal atmospheric pollutants. Refer to Table 3.3 for the atmospheric resistance of fiberglass reinforced thermoset polymers.

3.4 Reinforcing Materials

Many different reinforcing fibers are used in laminates and reinforced polymers. Which fiber is selected depends upon the cost, the properties

TABLE 3.3

Atmospheric Resistance of Fiberglass-Reinforced Thermoset Polymers

Polymer	UV[a] Degradation	Moisture[b] Absorption	Weathering	Ozone	SO$_2$	NO$_2$	H$_2$S
Epoxy	R	0.03	R	R	R	R	R
Polyesters							
Bisphenol A–fumarate	RS	0.20	R	R	R	R	R
Halogenated	RS	0.20	R	R	R	R	R
Hydrogenated bisphenol-A–bisphenol-A	RS	0.20	R	R	R[c]	R	
Isophthalic	RS	0.20	R	R	R	R	R
Terephthalic	RS	0.20	R	R	R	R	R
Vinyl ester	R	0.20	R	R	R	R	R
Furan	R	2.65	R	R	R	R	R
Phenolic							
Silicone	R	0.02–0.06	R	R	R	R	

[a] R, resistant; RS, resistant only if stabilized with UV protector; X, not resistant.
[b] Water absorption rate 24 h at 73°F/23°C (%).
[c] SO$_3$ will cause severe attack.

required, and the nature or the resin system. Fiberglass is the most often uses reinforcement material. Table 3.4 lists the common reinforcements and the properties they impart to the reinforced polymer.

3.4.1 Glass Fibers

Glass fibers are formed continuously from a melt in a special fiber-forming furnace. There are six formulations that are produced. The most common is E glass. This glass resists moisture and results in products with excellent electrical properties. C glass is designed to be used where optimum chemical resistance is required. D glass has very good electrical properties, particularly the dielectric constant, and is used in electronic applications. S glass is used for its high strength and stiffness while R glass is a lower cost fiber than S glass.

3.4.1.1 E Glass

As discussed previously, E glass is an electrical grade borosilicate glass. It possesses excellent water resistance, strength, and low elongation; and it is reasonable in cost. Practically all glass mat, continuous filaments, and woven roving come from this source.

3.4.1.2 C Glass

This is a calcium aluminosilicate glass widely used for surfacing mats, glass flakes, or flake glass linings, and for acid resistant cloths. C glass has poor

TABLE 3.4

Properties Imparted to Polymer by Reinforcing Materials

Reinforcing Fiber	Mechanical Strength	Electrical Properties	Impact Resistance	Corrosion Resistance	Machining and Punching	Heat Resistance	Moisture Resistance	Abrasion Resistance	Low Cost	Stiffness
Glass strands	X		X	X		X	X		X	X
Glass fabric	X	X	X	X		X	X			X
Glass mat			X	X		X	X		X	X
Asbestos		X	X			X				
Paper		X	X		X				X	
Cotton/linen	X	X	X		X				X	
Nylon		X	X	X				X	X	
Short inorganic fibers	X		X							
Organic fibers	X	X	X	X	X			X		
Ribbons		X			X		X			
Polyethylene	X		X				X			X
Metals	X		X			X	X			X
Aramid	X	X	X			X		X		X
Boron	X		X			X				X
Carbon/ graphite	X		X			X				X
Ceramic	X		X							X

water resistance and carries a premium cost. C veil has been used as a surfacing mat for many years when the standard specification for corrosion resistant equipment was 10-mil C veil. The use of 20-mil C veil and brittle resins such as the bisphenols should be avoided as they are subject to impact and handling damage. Until the development of synthetic veils, 10-mill C glass was nearly universally used. It is available in thicknesses of 10, 15, 20, and 30 mils.

3.4.1.3 S Glass

Because of its exceptional strength, S glass is widely used in the aerospace industry. It has excellent resistance to acids and water, but it is several times as expensive as E glass. Because of its cost, it is not used in the corrosion industry. It is comparable in strength to aramid fiber.

3.4.1.4 Glass Filaments

Glass filaments are available in a variety of diameters as shown in Table 3.5. Filaments in the range of diameters *G–T* are normally used to reinforce polymers. These filaments are formed into strands with 204, 400, 800, 1000, 2000, 3000, and 4000 filaments to a strand. The filaments are bonded into a

TABLE 3.5

Glass Fiber Diameters

Filament	Diameter Range $(\times 10^5$ in.)	Nominal Diameter $(\times 10^5$ in.)
B	10.0–14.9	12.3
C	15.0–19.9	17.5
D	20.0–24.9	22.5
DE	23.0–27.9	25.0
E	25.0–29.9	27.5
F	30.0–34.9	32.5
G	35.0–39.9	36.0
G	35.0–39.0	37.5
H	40.0–44.9	42.5
J	45.0–49.9	47.5
K	50.0–54.9	52.5
L	55.0–59.9	57.5
M	60.0–64.9	62.5
N	65.0–69.9	67.5
P	70.0–74.9	72.5
Q	75.0–79.9	77.5
R	80.0–84.9	82.5
S	85.0–89.9	87.5
T	90.0–94.9	92.5
U	95.0–99.9	97.5
V	100.0–104.9	102.5
W	105.0–109.9	107.5

strand using a sizing agent applied to the filament. The sizing agent also gives them environmental and abrasive protection. Coupling agents, such as silanes, chrome complexes, and polymers, are added to the finished products to improve adhesion of the resin matrix to the glass fibers. These strands are then used to manufacture the various types of glass reinforcements. Glass fabrics, glass mats, and chopped strands are the most common reinforcements used to reinforce polymers.

3.4.1.5 Chopped Strands

Strands of glass, (usually E glass) are mechanically cut into lengths from 0.25 to 2 in. and are used to reinforce molding compounds. The longer lengths are used with thermoset resins whereas the short fibers are used with thermoplastic polymers.

3.4.1.6 Glass Mats

Glass strands are cut and dipped into a moving conveyer belt where a polymer binder is applied to hold the mat together. Mat weights vary from 0.75 to 3 oz./ft.^{2} in widths up to 10 ft.

Continuous strand mats are produced by continuously depositing the uncut strands in a swirling pattern onto a moving belt where a binder is applied. Continuous strand mats exhibit better physical properties than cut strand mats, but the material is less homogenous. Continuous mats vary in weight from 0.75 to 4.5 oz./ft.^{2} and are available in widths up to 6 ft.

Woven roving is a mat fabric made by weaving multiple strands collected into a roving and into a coarse fabric. The physical properties of woven roving are intermediate between mats and fabrics. Various thicknesses are available in widths up to 10 ft. These constructions are used in low-pressure laminations and poltrusions.

3.4.1.7 Glass Fabrics

Reinforced polymers make use of many different glass fabrics. E glass is used in most and filament laminates of D, G, H, and K are common. Strands of glass filaments are plied into yarns and woven into fabrics on looms. The machine direction of the loom is called the warp whereas the cross section is the welt (also called "woof" or "fill"). The number of yarns can be varied in both warp and welt to control the weight, thickness, appearance, and strength of fabric.

3.4.2 Polyester

Polyester is primarily used for surfacing mat for the resin-rich inner surface of filament wound or custom contact-molded structures. It may also be used in conjunction with C glass surfacing mats.

The tendency of C glass to bridge and form voids is reduced or eliminated by overwinding a C glass surfacing mat with a polyester mat under tension. Nexus (registered trademark of Burlington Industries) surfacing veils are also used on exterior surfaces of poltruded products. The Nexus surfacing veil possesses a relatively high degree of elongation that makes it very compatible with the higher elongation resins and reduces the risk of checking, crazing, and cracking in temperature-cycling operations.

The Nexus surfacing veil exhibits excellent resistance to alcohols, bleaching agents, water, hydrocarbons, and aqueous solutions of most weak acids at boiling. Being a polyester derivative, it is not resistant to strong acids such as 93% sulfuric.

3.4.3 Carbon Fiber

Carbon fiber is available in mat form, typically 0.2, 0.5, 1, and 2 oz./yd.^{2}. It is also available in a 0.5 oz./yd.^{2} mat as a blend with 33% glass fiber. Other blends are also available such as 25% carbon filler—75% glass fiber, and 50% aluminized glass—50% carbon fiber.

Carbon fibers are also available in continuous roving and as chopped fibers in sizes 1/8–2 in.

The carbon fiber mat, either alone or supplemented with a ground carbon or graphite filler, provides in-depth grounding systems and static control in hazardous areas where static sparks may result in fires or explosions.

3.4.4 Aramid Fibers

Aramid is a generic name for aromatic polyamide (nylon). It is sold under the tradename Kevlar, manufactured by DuPont. The fiber is used in the same manner as glass fibers to reinforce polymers. Because of its great tensile strength and consistency coupled with low density, it essentially revolutionized pressure vessel technology. Aramid composites, though still widely used for pressure vessels, have been largely replaced by very high strength graphite fibers.

Aramid fibers are woven into fabrics for reinforcing polymers. The fabrics range in size from 2-mil (1 oz./yd.^{2}) to thick 30-mil (6 oz./yd.^{2}) cloths. Kevlar is available in surfacing mats and cloth form. Surfacing mats can be had in weights of 0.4 and 1 oz./yd.^{2}.

Aramid fiber finds varied applications in aircraft and aerospace components because of its low density. This light weight is also advantageous in the fabrication of laminated canoes and kayaks. Other aramid fiber sport applications include downhill skis, tennis racquets, and golf shafts.

Aramid reinforcement is also available as paper and sold under tradenames of Nomex by DuPont and TP Technora from Teijin America. Aramid paper is used in circuit boards to improve crack resistance and to impart a smooth surface. The paper is limited by such drawbacks as a high coefficient of thermal expansion, poor resin adhesion, and difficult drilling and machining characteristics.

3.4.5 Polyethylene Fibers

Allied Signal, Inc. produces a high-strength, high-modulus polyethylene fiber under the tradename Spectra. The polyethylene is an ultrahigh molecular weight polymer, up to 5 million, compared to conventional polyethylene that has a molecular weight of about 200,000. Two Spectra fibers are produced, the 900 and 1000 grades. The 1000 grade has higher strength and modulus. Because the density of Spectra is the lowest of all of the fibers, it finds many applications in aerospace laminates such as wing tips and heliocopter seats. Spectra laminates must not be exposed to temperatures above 250°F (121°C). Table 3.6 compares the corrosion resistance of Spectra and Aramid fibers.

3.4.6 Paper

Kraft paper is widely used as a reinforcement. When saturated with phenolics, it is made into a common printed wiring board. When combined with melamine, it becomes a decorative high-pressure laminate used in furniture, countertops, and wall panels. Paper reinforcement is inexpensive and easy to machine, drill, and punch. It imparts good electrical properties, but it is sensitive to moisture and cannot withstand high temperatures.

TABLE 3.6

Corrosion Resistance of Spectra and Aramid Fibers

| | % Retention of Fiber Tensile Strength | | | |
| | Spectra | | Aramid | |
Chemical	6 Months	2 Years	6 Months	2 Years
Seawater	100	100	100	98
Hydraulic fluid	100	100	100	87
Kerosene	100	100	100	97
Gasoline	100	100	93	X
Toluene	100	96	72	X
Acetic acid, glacial	100	100	82	X
Hydrochloric acid 1 M	100	100	40	X
Sodium hydroxide 5 M	100	100	42	X
Ammonium hydroxide, 29%	100	100	70	X
Perchlorethylene	100	100	75	X
Detergent solution, 10%	100	100	91	X
Clorox	91	73	0	0

X, samples not tested due to physical deterioration.

3.4.7 Cotton and Linen

Fabrics of cotton or linen, when impregnated with phenolic resin, are used in several grades of laminates. They have good impact strength and abrasion resistance, and they machine well. Laminates with these fabrics have better water resistance than paper-based laminates. These laminates are used for gears and pulleys.

3.5 Polyesters

Polyesters were first prepared in 1929 by William H. Carothers. The manufacturing techniques he developed are still in use today. It was not until the late 1930s that Ellis discovered that polyesters' cross-linking increased around 30 times in the presence of unsaturated monomers. Large-scale commercialization of unsaturated polyesters resulted from the 1942 discovery by the U.S. Rubber Company that adding glass fibers to polyesters greatly improved their physical properties. Widespread use of polyesters had been hampered because of their inherent brittleness. The first successful commercial use of glass-reinforced polyesters was the production of radomes for military aircraft during World War II. The first fiberglass-reinforced boat hulls were made in 1946, and fiberglass is still a major application for unsaturated polyesters.

Of all the reinforced thermoset polymers used for corrosion resistance, unsaturated polyesters are the most widely used. Unsaturated polyesters are produced via a condensation reaction between a dibasic organic acid or anhydride and a difunctional alcohol as shown in Figure 3.1. At least one of these components contributes sites of unsaturation to the oligomer chain. The oligomer or prepolymer is then dissolved in an unsaturated monomer such as styrene. To initiate cross-linking, a free radical source, such as an organic peroxide, is added to the liquid resin. The cross-linking reaction is a free-radical copolymerization between the resin oligomer and the unsaturated monomer. The types and ratio of components used to manufacture the oligomers, the manufacturing procedure, and the molecular weight of the oligomer will determine the properties of the cured polyester.

Maleic anhydride or fumaric acid provides the source of unsaturation for almost all of the polyester resins currently produced with maleic anhydride being the most common.

$$HOn-R-OH + nHO-\overset{\overset{O}{\|}}{C}-R-\overset{\overset{O}{\|}}{C}-OH \longrightarrow \left[R-O-\overset{\overset{O}{\|}}{C}-R-\overset{\underset{\|}{C}}{\underset{O}{\|}}-O\right]_n + nH_2O$$

FIGURE 3.1
Condensation reaction.

Propylene glycol is the most common diol used because it yields polymers with the best overall properties. Figure 3.2 lists the more common components used in the manufacture of unsaturated polyesters.

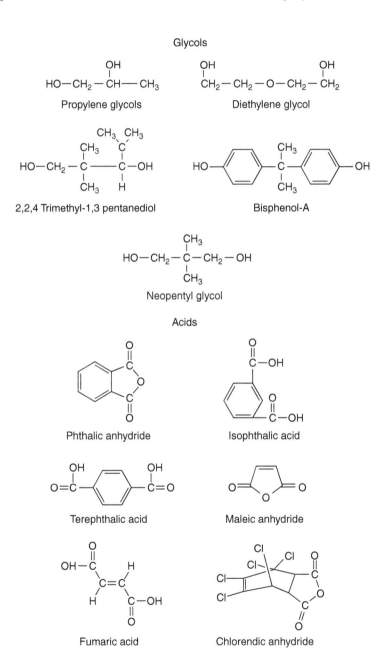

FIGURE 3.2
Components of unsaturated polyesters.

The primary unsaturated polyesters are:

General purpose (orthophthalic)
Isophthalic
Bisphenol A fumarate
Hydrogenated bisphenol A
Halogenated polyester
Terephthalate (PET)

The strength of a laminate is primarily determined by the type and amount of reinforcing used. The specific polyester resin will determine operating temperatures and corrosion resistance.

3.5.1   General Purpose Polyesters

These resins have the following formula:

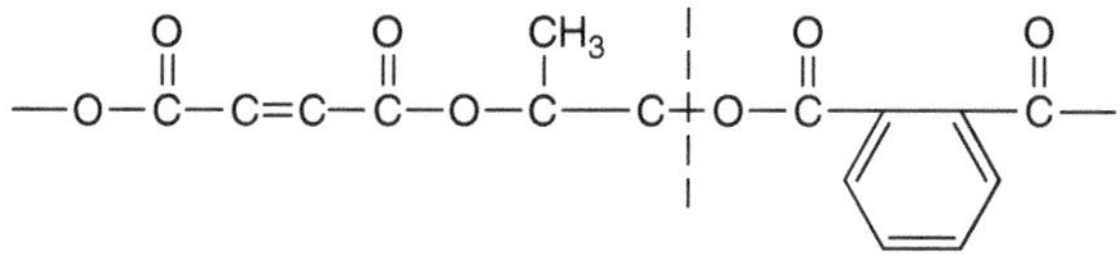

General purpose polyesters are based on phthalic anhydride as the saturated monomer and are the lowest-cost class of resin. These general purpose resins are not normally recommended for use in corrosive service. They are adequate for use with nonoxidizing mineral acids and corrodents that are relatively mild. Tests have indicated that general purpose resins will provide satisfactory service with the following materials up to 125°F (52°C):

Acids		
Acetic acid, 10%	Fatty acids	Oleic acid
Citric acid	Lactic acid, 1%	Benzoic acid
		Boric acid
Salts		
Aluminum sulfate	Copper sulfate	Magnesium sulfate
Ammonium chloride	Ferric chloride	Nickel chloride
Ammonium sulfate, 10%	Ferric nitrate	Nickel nitrate
Calcium chloride (sat.)	Ferric sulfate	Nickel sulfate
Calcium sulfate	Ferrous chloride	Potassium chloride
	Magnesium chloride	Potassium sulfate
		Sodium chloride, 10%
Solvents		
Amyl alcohol	Glycerine	Kerosene
		Naphtha

General purpose resins are not satisfactory in contact with the following:

Oxidizing acids.
Alkaline solutions such as calcium hydroxide, sodium hydroxide, sodium carbonate. Bleach solutions such as 5% sodium hypochlorite. Solvents such as carbon disulfide, carbon tetrachloride, and gasoline.

This resin predominates in boat building. It exhibits excellent resistance to all types of water, including seawater. For the most part, these resins find application in building panels, boats, radomes, television satellite dishes, tote boxes, fishing poles, building materials, car and truck bodies, and any application where exposure is confined to ambient conditions.

3.5.2 Isophthalic Polyesters

Isophthalic polyesters have the following formula:

These polyester resins use isophthalic acid in place of phthalic anhydride as the saturated monomer. This increases the cost to produce, but it improves physical properties and corrosion resistance.

The standard corrosion grade isophthalic polyesters are made with 1:1 molar ratio of isophthalic acid to maleic anhydride or fumaric acid with propylene glycol. In general, the properties of isophthalic polyesters are superior to the lower-cost general purpose polyesters, not only in chemical resistance but also in physical properties. They can be formulated to be fire retardant or to protect against the effects of ultraviolet light.

Isophthalic polyesters have a relatively wide range of corrosion resistance. They are satisfactory for use up to 125°F (52°C) in acids such as 10% acetic, benzoic, boric, citric, oleic, 25% phosphoric, tartaric, 10–25% sulfuric, and fatty acids.

Most organic salts are also compatible with isophthalic polyesters. Solvents such as amyl alcohols, ethylene glycol, formaldehyde, gasoline, kerosine, and naphtha are also compatible.

The isophthalic polyester resins are not resistant to acetone, amyl acetate, benzene, carbon disulfide, solutions of alkaline, salts of potassium and sodium, hot distilled water, or higher concentrations of oxidizing acids.

Refer to Table 3.7 for the compatibility of isophthalic polyester with selected corrodents. Reference [1] provides a more comprehensive listing.

TABLE 3.7

Compatibility of Isophthalic Polyester with Selected Corrodents

Chemical	Maximum Temperature	
	°F	°C
Acetaldehyde	X	X
Acetic acid, 10%	180	82
Acetic acid, 50%	110	43
Acetic acid, 80%	X	X
Acetic acid, glacial	X	X
Acetic anhydride	X	X
Acetone	X	X
Acetyl chloride	X	X
Acrylic acid	X	X
Acrylonitrile	X	X
Adipic acid	220	104
Allyl alcohol	X	X
Allyl chloride	X	X
Alum	250	121
Aluminum chloride, aqueous	180	82
Aluminum chloride, dry	170	77
Aluminum fluoride, 10%	140	60
Aluminum hydroxide	160	71
Aluminum nitrate	160	71
Aluminum sulfate	180	82
Ammonia gas	90	32
Ammonium carbonate	X	X
Ammonium chloride, 10%	160	71
Ammonium chloride, 50%	160	71
Ammonium chloride, sat.	180	32
Ammonium fluoride, 10%	90	32
Ammonium fluoride, 25%	90	32
Ammonium hydroxide, 25%	X	X
Ammonium hydroxide, sat.	X	X
Ammonium nitrate	160	71
Ammonium persulfate	160	71
Ammonium phosphate	160	71
Ammonium sulfate, 10%	180	82
Ammonium sulfide	X	X
Ammonium sulfite	X	X
Amyl acetate	X	X
Amyl alcohol	160	71
Amyl chloride	X	X
Aniline	X	X
Antimony trichloride	160	71
Aqua regia 3:1	X	X
Barium carbonate	190	88
Barium chloride	140	60
Barium hydroxide	X	X
Barium sulfate	160	71
Barium sulfide	90	32
Benzaldehyde	X	X
Benzene	X	X

(continued)

TABLE 3.7 *Continued*

Chemical	Maximum Temperature	
	°F	°C
Benzenesulfonic acid, 10%	180	82
Benzoic acid	180	82
Benzyl alcohol	X	X
Benzyl chloride	X	X
Borax	140	60
Boric acid	180	82
Bromine gas, dry	X	X
Bromine gas, moist	X	X
Bromine, liquid	X	X
Butyl acetate	X	X
Butyl alcohol	80	27
n-Butylamine	X	X
Butyric acid, 25%	129	49
Calcium bisulfide	160	71
Calcium bisulfite	150	66
Calcium carbonate	160	71
Calcium chlorate	160	71
Calcium chloride	180	82
Calcium hydroxide, 10%	160	71
Calcium hydroxide, sat.	160	71
Calcium hypochlorite, 10%	120	49
Calcium nitrate	140	60
Calcium oxide	160	71
Calcium sulfate	160	71
Caprylic acid	160	71
Carbon bisulfide	X	X
Carbon dioxide, dry	160	71
Carbon dioxide, wet	160	71
Carbon disulfide	X	X
Carbon monoxide	160	71
Carbon tetrachloride	X	X
Carbonic acid	160	71
Cellosolve	X	X
Chloroacetic acid, 50% water	X	X
Chloroacetic acid to 25%	150	66
Chlorine gas, dry	160	71
Chlorine gas, wet	160	71
Chlorine, liquid	X	X
Chlorobenzene	X	X
Chloroform	X	X
Chlorosulfonic acid	X	X
Chromic acid, 10%	X	X
Chromic acid, 50%	X	X
Chromyl chloride	140	60
Citric acid, 15%	160	71
Citric acid, conc.	200	93
Copper acetate	160	71
Copper chloride	180	82
Copper cyanide	160	71

(continued)

TABLE 3.7 *Continued*

Chemical	Maximum Temperature	
	°F	°C
Copper sulfate	200	93
Cresol	X	X
Cupric chloride, 5%	170	77
Cupric chloride, 50%	170	77
Cyclohexane	80	27
Dichloroacetic acid	X	X
Dichloroethane (ethylene dichloride)	X	X
Ethylene glycol	120	49
Ferric chloride	180	82
Ferric chloride, 50% in water	160	71
Ferric nitrate, 10–50%	180	82
Ferrous chloride	180	82
Ferrous nitrate	160	71
Fluorine gas, dry	X	X
Fluorine gas, moist	X	X
Hydrobromic acid, dil.	120	49
Hydrobromic acid, 20%	140	60
Hydrobromic acid, 50%	140	60
Hydrochloric acid, 20%	160	71
Hydrochloric acid, 38%	160	71
Hydrocyanic acid, 10%	90	32
Hydrofluoric acid, 30%	X	X
Hydrofluoric acid, 70%	X	X
Hydrofluoric acid, 100%	X	X
Hypochlorous acid	90	32
Ketones, general	X	X
Lactic acid, 25%	160	71
Lactic acid, conc.	160	71
Magnesium chloride	180	82
Malic acid	90	32
Methyl ethyl ketone	X	X
Methyl isobutyl ketone	X	X
Muriatic acid	160	71
Nitric acid, 5%	120	49
Nitric acid, 20%	X	X
Nitric acid, 70%	X	X
Nitric acid, anhydrous	X	X
Nitrous acid, conc.	120	49
Oleum	X	X
Perchloric acid, 10%	X	X
Perchloric acid, 70%	X	X
Phenol	X	X
Phosphoric acid, 50–80%	180	82
Picric acid	X	X
Potassium bromide, 30%	160	71
Salicylic acid	100	38
Sodium carbonate, 20%	90	32
Sodium chloride	200	93
Sodium hydroxide, 10%	X	X

(continued)

TABLE 3.7 *Continued*

Chemical	Maximum Temperature	
	°F	°C
Sodium hydroxide, 50%	X	X
Sodium hydroxide, conc.	X	X
Sodium hypochlorite, 20%	X	X
Sodium hypochlorite, conc.	X	X
Sodium sulfide to 50%	X	X
Stannic chloride	180	82
Stannous chloride	180	82
Sulfuric acid, 10%	160	71
Sulfuric acid, 50%	150	66
Sulfuric acid, 70%	X	X
Sulfuric acid, 90%	X	X
Sulfuric acid, 98%	X	X
Sulfuric acid, 100%	X	X
Sulfuric acid, fuming	X	X
Sulfurous acid	X	X
Thionyl chloride	X	X
Toluene	110	43
Trichloroacetic acid, 50%	170	77
White liquor	X	X
Zinc chloride	180	82

The chemicals listed are in the pure state or in saturated solution unless otherwise indicated. Compatibility is shown to the maximum allowable temperature for which data are available. Incompatibility is shown by an X. A blank space indicates that data are unavailable.

3.5.2.1 *Typical Applications*

Isophthalic polyesters provide corrosion resistance in a wide variety of end uses. Recent excavations of 25-year-old FRP underground gasoline storage tanks offer evidence to the resin's resistance to internal chemical attack or external attack by water in the surrounding soil. Isopolyester pipes and containment vaults also provide protection against failure from corrosion by acid media. Reference [2] provides information on FRP piping systems.

Isopolyesters used in above-ground tanks and tank linings for chemical storage provide similar protection. In food-contact applications, these resins withstand acids and corrosive salts encountered in foods and food handling.

The range of successful commercial applications include support structures, ladders, grating, platforms, and guard rails. Currently, there are more than 100 standard structural shapes available from which platforms, supports, structures, and decking can be fabricated. These materials offer several advantages, including corrosion resistance, nonconductivity, high strength, light weight, and dimensional stability.

The water resistance properties of the isophthlates also provide advantages in appearance and performance in cast countertops as well as pool, tub, and spa applications. Other applications include their use in cooling towers and use as rail cars to transport automobiles.

3.5.3 Bisphenol A Fumarate Polyesters

This is a premium corrosion-resistant grade resin. It costs approximately twice as much as a general purpose resin and approximately one-third more than an isophthalic resin. The structural formula is shown in Figure 3.3.

Standard bisphenol-A fumarate resins are derived from the propylene glycol or oxide diether of bisphenol-A and fumaric acid. The aromatic structure provided by the bisphenol-A provides several benefits. Thermal stability is improved, and the heat distortion point of the resin is mainly raised from the more rigid nature of the aromatic structure. The number of interior chain ester groups is reduced so the resistance to hydrolysis and saponification is increased. Bisphenol A fumarate polyesters have the best hydrolysis resistance of any commercial unsaturated polyester.

The bisphenol polyesters are superior in their corrosion resistant properties to the isophthalic polyesters. They exhibit good performance with moderate alkaline solutions of bleaching agents. The bisphenol polyesters will break down in highly concentrated acids or alkalies. These resins can be used in the handling of the following materials:

	Acids to 200°F (93°C)	
Acetic	Fatty acids	Stearic
Benzoic	Hydrochloric, 10%	Sulfonic, 30%
Boric	Lactic	Tannic
Butyric	Maleic	Tartaric
Chloroacetic, 15%	Oleic	Trichloroacetic, 50%
Chromic, 5%	Oxalic	Rayon spin bath
Citric	Phosphoric, 80%	

	Salt Solution to 200°F (93°C)	
All aluminum salts	Most plating solutions	Iron salts
Most ammonium salts	Copper salts	Zinc salts
Calcium salts		

	Alkalies	
Ammonium hydroxide, 5% to 160°F (71°C)	Chlorine dioxide, 15% to 200°F (93°C)	Sodium chlorite to 200°F (93°C)
Calcium hydroxide, 25% to 160°F (71°C)	Potassium hydroxide, 25% to 160°F (71°C)	Sodium hydrosulfate to 200°F (93°C)
Calcium hypochlorite, 20% to 200°F (93°C)		

	Solvents	
Isophthalics are resistant to most all solvents plus	Alcohols at ambient temperatures	Linseed oil
Sour crude oil	Glycerine	

	Gases to 200°F (93°C)	
Carbon dioxide	Chlorine, wet	Sulfur trioxide
Carbon monoxide	Sulfur dioxide, dry	Rayon waste gases 150°F (66°C)
Chlorine, dry	Sulfur dioxide, wet	

FIGURE 3.3
Chemical structure of bisphenol A fumarate resins.

Solvents such as benzene, carbon disulfide, ether, methyl ethyl ketone, toluene, xylene, trichloroethylene, and trichloroethane will attack the resin. Sulfuric acid above 70%, sodium hydroxide, and 30% chromic acid will also attack the resin. Refer to Table 3.8 for the compatibility of bisphenol-A fumarate polyester resin with selected corrodents. Table 3.9 lists the compatibility of hydrogenated bisphenol-A fumarate polyesters with selected corrodents. Reference [1] provides a more comprehensive listing.

TABLE 3.8

Compatibility of Bisphenol A–Fumarate Polyester with Selected Corrodents

| | Maximum Temperature | |
Chemical	°F	°C
Acetaldehyde	X	X
Acetic acid, 10%	220	104
Acetic acid, 50%	160	171
Acetic acid, 80%	160	171
Acetic acid, glacial	X	X
Acetic anhydride	110	43
Acetone	X	X
Acetyl chloride	X	X
Acrylic acid	100	38
Acrylonitrile	X	X
Adipic acid	220	104
Allyl alcohol	X	X
Allyl chloride	X	X
Alum	220	104
Aluminum chloride, aqueous	200	93
Aluminum fluoride, 10%	90	32
Aluminum hydroxide	160	71
Aluminum nitrate	200	93
Aluminum sulfate	200	93
Ammonia gas	200	93
Ammonium carbonate	90	32
Ammonium chloride, 10%	200	93
Ammonium chloride, 50%	220	104

(continued)

TABLE 3.8 *Continued*

Chemical	Maximum Temperature	
	°F	°C
Ammonium chloride, sat.	220	104
Ammonium fluoride, 10%	180	82
Ammonium fluoride, 25%	120	49
Ammonium hydroxide, 25%	100	38
Ammonium hydroxide, 20%	140	60
Ammonium nitrate	220	104
Ammonium persulfate	180	82
Ammonium phosphate	80	27
Ammonium sulfate, 10–40%	220	104
Ammonium sulfide	110	43
Ammonium sulfite	80	27
Amyl acetate	80	27
Amyl alcohol	200	93
Amyl chloride	X	X
Aniline	X	X
Antimony trichloride	220	104
Aqua regia 3:1	X	X
Barium carbonate	200	93
Barium chloride	220	104
Barium hydroxide	150	66
Barium sulfate	220	104
Barium sulfide	140	60
Benzaldehyde	X	X
Benzene	X	X
Benzenesulfonic acid, 10%	200	93
Benzoic acid	180	82
Benzyl alcohol	X	X
Benzyl chloride	X	X
Borax	220	104
Boric acid	220	104
Bromine gas, dry	90	32
Bromine gas, moist	100	38
Bromine, liquid	X	X
Butyl acetate	80	27
Butyl alcohol	80	27
n-Butylamine	X	X
Butyric acid	220	104
Calcium bisulfite	180	82
Calcium carbonate	210	99
Calcium chlorate	200	93
Calcium chloride	220	104
Calcium hydroxide, 10%	180	82
Calcium hydroxide, sat.	160	71
Calcium hypochlorite, 10%	80	27
Calcium nitrate	220	104
Calcium sulfate	220	104
Caprylic acid	160	71
Carbon bisulfide	X	X

(continued)

TABLE 3.8 *Continued*

Chemical	Maximum Temperature	
	°F	°C
Carbon dioxide, dry	350	177
Carbon dioxide, wet	210	99
Carbon disulfide	X	X
Carbon monoxide	350	177
Carbon tetrachloride	110	43
Carbonic acid	90	32
Cellosolve	140	60
Chloroacetic acid, 50% water	140	60
Chloroacetic acid to 25%	80	27
Chlorine gas, dry	200	93
Chlorine gas, wet	200	93
Chlorine, liquid	X	X
Chlorobenzene	X	X
Chloroform	X	X
Chlorosulfonic acid	X	X
Chromic acid, 10%	X	X
Chromic acid, 50%	X	X
Chromyl chloride	150	66
Citric acid, 15%	220	104
Citric acid, conc.	220	104
Copper acetate	180	82
Copper chloride	220	104
Copper cyanide	220	104
Copper sulfate	220	104
Cresol	X	X
Cyclohexane	X	X
Dichloroacetic acid	100	38
Dichloroethane	X	X
(ethylene dichloride)	220	104
Ethylene glycol	220	104
Ferric chloride	220	104
Ferric chloride, 50% in water	220	104
Ferric nitrate, 10–50%	220	104
Ferrous chloride	220	104
Ferrous nitrate	220	104
Fluorine gas, moist		
Hydrobromic acid, dil.	220	104
Hydrobromic acid, 20%	220	104
Hydrobromic acid, 50%	160	71
Hydrochloric acid, 20%	190	88
Hydrochloric acid, 38%	X	X
Hydrocyanic acid, 10%	200	93
Hydrofluoric acid, 30%	90	32
Hypochlorous acid, 20%	90	32
Iodine solution, 10%	200	93
Lactic acid, 25%	210	99
Lactic acid, conc.	220	104
Magnesium chloride	220	104

(continued)

TABLE 3.8 *Continued*

Chemical	Maximum Temperature	
	°F	°C
Malic acid	160	71
Methyl ethyl ketone	X	X
Methyl isobutyl ketone	X	X
Muriatic acid	130	54
Nitric acid, 5%	160	71
Nitric acid, 20%	100	38
Nitric acid, 70%	X	X
Nitric acid, anhydrous	X	X
Oleum	X	X
Phenol	X	X
Phosphoric acid, 50–80%	220	104
Picric acid	110	43
Potassium bromide, 30%	200	93
Salicylic acid	150	66
Sodium carbonate	160	71
Sodium chloride	220	104
Sodium hydroxide, 10%	130	54
Sodium hydroxide, 50%	220	104
Sodium hydroxide, conc.	200	93
Sodium hypochlorite, 20%	X	X
Sodium sulfide to 50%	210	99
Stannic chloride	200	93
Stannous chloride	220	104
Sulfuric acid, 10%	220	104
Sulfuric acid, 50%	220	104
Sulfuric acid, 70%	160	71
Sulfuric acid, 90%	X	X
Sulfuric acid, 98%	X	X
Sulfuric acid, 100%	X	X
Sulfuric acid, fuming	X	X
Sulfurous acid	110	43
Thionyl chloride	X	X
Toluene	X	X
Trichloroacetic acid, 50%	180	82
White liquor	180	82
Zinc chloride	250	121

The chemicals listed are in the pure state or in a saturated solution unless otherwise indicated. Compatibility is shown to the maximum allowable temperature for which data are available. Incompatibility is shown by an X. A blank space indicates that data are unavailable.

Source: From P.A. Schweitzer. 2004. *Corrosion Resistance Tables*, Vols. 1–4, 5th ed., New York: Marcel Dekker.

TABLE 3.9

Compatibility of Hydrogenated Bisphenol A–Bisphenol A
Polyester with Selected Corrodents

Chemical	Maximum Temperature	
	°F	°C
Acetic acid, 10%	200	93
Acetic acid, 50%	160	71
Acetic anhydride	X	X
Acetone	X	X
Acetyl chloride	X	X
Acrylonitrile	X	X
Aluminum acetate		
Aluminum chloride, aqueous	200	93
Aluminum fluoride	X	X
Aluminum sulfate	200	93
Ammonium chloride, sat.	200	93
Ammonium nitrate	200	93
Ammonium persulfate	200	93
Ammonium sulfide	100	38
Amyl acetate	X	X
Amyl alcohol	200	93
Amyl chloride	90	32
Aniline	X	X
Antimony trichloride	80	27
Aqua regia 3:1	X	X
Barium carbonate	180	82
Barium chloride	200	93
Benzaldehyde	X	X
Benzene	X	X
Benzoic acid	210	99
Benzyl alcohol	X	X
Benzyl chloride	X	X
Boric acid	210	99
Bromine, liquid	X	X
Butyl acetate	X	X
n-Butylamine	X	X
Butyric acid	X	X
Calcium bisulfide	120	49
Calcium chlorate	210	99
Calcium chloride	210	99
Calcium hypochlorite, 10%	180	82
Carbon bisulfide	X	X
Carbon disulfide	X	X
Carbon tetrachloride	X	X
Chloroacetic acid, 50% water	90	32
Chlorine gas, dry	210	99
Chlorine gas, wet	210	99
Chloroform	X	X
Chromic acid, 50%	X	X
Citric acid, 15%	200	93
Citric acid, conc.	210	99

(continued)

TABLE 3.9 *Continued*

Chemical	Maximum Temperature	
	°F	°C
Copper acetate	210	99
Copper chloride	210	99
Copper cyanide	210	99
Copper sulfate	210	99
Cresol	X	X
Cyclohexane	210	99
Dichloroethane (ethylene dichloride)	X	X
Ferric chloride	210	99
Ferric chloride, 50% in water	200	93
Ferric nitrate, 10–50%	200	93
Ferrous chloride	210	99
Ferrous nitrate	210	99
Hydrobromic acid, 20%	90	32
Hydrobromic acid, 50%	90	32
Hydrochloric acid, 20%	180	82
Hydrochloric acid, 38%	190	88
Hydrocyanic acid, 10%	X	X
Hydrofluoric acid, 30%	X	X
Hydrofluoric acid, 70%	X	X
Hydrofluoric acid, 100%	X	X
Hypochlorous acid, 50%	210	99
Lactic acid, 25%	210	99
Lactic acid, conc.	210	99
Magnesium chloride	210	99
Methyl ethyl ketone	X	X
Methyl isobutyl ketone	X	X
Muriatic acid	190	88
Nitric acid, 5%	90	32
Oleum	X	X
Perchloric acid, 10%	X	X
Perchloric acid, 70%	X	X
Phenol	X	X
Phosphoric acid, 50–80%	210	99
Sodium carbonate, 10%	100	38
Sodium chloride	210	99
Sodium hydroxide, 10%	100	38
Sodium hydroxide, 50%	X	X
Sodium hydroxide, conc.	X	X
Sodium hypochlorite, 10%	160	71
Sulfuric acid, 10%	210	99
Sulfuric acid, 50%	210	99
Sulfuric acid 70 %	90	32
Sulfuric acid, 90%	X	X
Sulfuric acid, 98%	X	X
Sulfuric acid, 100%	X	X
Sulfuric acid, fuming	X	X
Sulfurous acid 25 %	210	99
Toluene	90	32

(continued)

TABLE 3.9 *Continued*

Chemical	Maximum Temperature	
	°F	°C
Trichloroacetic acid	90	32
Zinc chloride	200	93

The chemicals listed are in the pure state or in a saturated solution unless otherwise indicated. Compatibility is shown to the maximum allowable temperature for which data are available. Incompatibility is shown by an X. A blank space indicates that data are unavailable.

Source: From P.A. Schweitzer. 2004. *Corrosion Resistance Tables*, Vols. 1–4, 5th ed., New York: Marcel Dekker.

3.5.3.1 Typical Applications

Bisphenol A resins find applications similar to the isophthalic resins but in areas requiring greater resistance to corrosion.

3.5.4 Halogenated Polyesters

Halogenated resins consist of chlorinated or brominated polymers. The chlorinated polyester resins cured at room temperature and reinforced with fiberglass possess unique physical, mechanical, and corrosion resistance properties. These are also known as chlorendic polyesters.

These resins have a very high heat distortion point, and the laminates, show very high retention of physical strength at elevated temperature upsets as in flue gas desulfurization scrubbers, some of which may reach a temperature of 400°F (204°C). These laminates are routinely used as chimney liners at temperatures of 240–280°F (116–138°C).

These resins have the highest heat resistance of any chemically resistant polyester. They are also inherently fire retardant. A noncombustible rating of 20 can be achieved, making this the safest possible polyester for stacks, hoods, fans, ducts, or wherever a fire hazard might exist. This fire retardancy is achieved by the addition of antimony trioxide.

Excellent resistance is exhibited in contact with oxidizing acids and solutions such as 35% nitric acid at elevated temperatures, 70% nitric acid at room temperature, 40% chromic acid, chlorine water, wet chlorine, and 15% hypochlorites. They also resist neutral and acid salts, nonoxidizing acids, organic acids, mercaptans, ketones, aldehydes, alcohols, glycols, organic esters, and fats and oils. Table 3.10 is a general application guide.

These polymers are not resistant to highly alkaline solutions of sodium hydroxide, concentrated sulfuric acid, alkaline solutions with pH greater than 10, aliphatic, primary and aromatic amines, amides, and other alkaline organics, phenols, and acid halides. Table 3.11 provides the compatibility of halogenated polyesters with selected corrodents. Reference [1] provides a more comprehensive listing.

TABLE 3.10

General Application Guide for Chlorinated Polyester Laminates

Environment	Comments
Acid halides	Not recommended
Acids, mineral nonoxidizing	Resistant to 250°F/121°C
Acids, organic	Resistant to 250°F/121°C; glacial acetic acid to 120°F/49°C
Alcohols	Resistant to 180°F/82°C
Aldehydes	Resistant to 180°F/82°C
Alkaline solutions pH >10	Not recommended for continuous exposure
Amines, aliphatic, primary aromatic	Can cause severe attack
Amides, other alkaline organics	Can cause severe attack
Esters, organic	Resistant to 180°F/82°C
Fats and oils	Resistant to 200°F/95°C
Glycols	Resistant to 180°F/82°C
Ketones	Resistant to 180°F/82°C
Mercaptans	Resistant to 180°F/82°C
Phenol	Not recommended
Salts, acid	Resistant to 250°F/121°C
Salts, neutral	Resistant to 250°F/121°C
Water, demineralized, distilled, deionized, steam and condensate	Resistant to 212°F/100°C; Lowest absorption of any polyester

TABLE 3.11

Compatibility of Halogenated Polyester with Selected Corrodents

Chemical	Maximum Temperature	
	°F	°C
Acetaldehyde	X	X
Acetic acid, 10%	140	60
Acetic acid, 50%	90	32
Acetic acid, glacial	110	43
Acetic anhydride	100	38
Acetone	X	X
Acetyl chloride	X	X
Acrylic acid	X	X
Acrylonitrile	X	X
Adipic acid	220	104
Allyl alcohol	X	X
Allyl chloride	X	X
Alum, 10%	200	93
Aluminum chloride, aqueous	120	49
Aluminum fluoride, 10%	90	32
Aluminum hydroxide	170	77
Aluminum nitrate	160	71
Aluminum oxychloride		
Aluminum sulfate	250	121

(continued)

TABLE 3.11 *Continued*

Chemical	Maximum Temperature	
	°F	°C
Ammonia gas	150	66
Ammonium carbonate	140	60
Ammonium chloride, 10%	200	93
Ammonium chloride, 50%	200	93
Ammonium chloride, sat.	200	93
Ammonium fluoride, 10%	140	60
Ammonium fluoride, 25%	140	60
Ammonium hydroxide, 25%	90	32
Ammonium hydroxide, sat.	90	32
Ammonium nitrate	200	93
Ammonium persulfate	140	60
Ammonium phosphate	150	66
Ammonium sulfate, 10–40%	200	93
Ammonium sulfide	120	49
Ammonium sulfite	100	38
Amyl acetate	190	85
Amyl alcohol	200	93
Amyl chloride	X	X
Aniline	120	49
Antimony trichloride, 50%	200	93
Aqua regia 3:1	X	X
Barium Carbonate	250	121
Barium chloride	250	121
Barium hydroxide	X	X
Barium sulfate	180	82
Barium sulfide	X	X
Benzaldehyde	X	X
Benzene	90	32
Benzenesulfonic acid, 10%	120	49
Benzoic acid	250	121
Benzyl alcohol	X	X
Benzyl chloride	X	X
Borax	190	88
Boric acid	180	82
Bromine gas, dry	100	38
Bromine gas, moist	100	38
Bromine, liquid	X	X
Butyl acetate	80	27
Butyl alcohol	100	38
n-Butylamine	X	X
Butyric acid, 20%	200	93
Calcium bisulfide	X	X
Calcium bisulfite	150	66
Calcium carbonate	210	99
Calcium chlorate	250	121
Calcium chloride	250	121
Calcium hydroxide, sat.	X	X
Calcium Hypochlorite, 20%	80	27

(continued)

TABLE 3.11 *Continued*

Chemical	Maximum Temperature	
	°F	°C
Calcium nitrate	220	104
Calcium oxide	150	66
Calcium sulfate	250	121
Caprylic acid	140	60
Carbon bisulfide	X	X
Carbon dioxide, dry	250	121
Carbon dioxide, wet	250	121
Carbon disulfide	X	X
Carbon monoxide	170	77
Carbon tetrachloride	120	49
Carbonic acid	160	71
Cellosolve	80	27
Chloroacetic acid, 50% water	100	38
Chloroacetic acid, 25%	90	32
Chlorine gas, dry	200	93
Chlorine gas, wet	220	104
Chlorine, liquid	X	X
Chlorobenzene	X	X
Chloroform	X	X
Chlorosulfonic acid	X	X
Chromic acid, 10%	180	82
Chromic acid, 50%	140	60
Chromyl chloride	210	99
Citric acid, 15%	250	121
Citric acid, conc.	250	121
Copper acetate	210	99
Copper chloride	250	121
Copper cyanide	250	121
Copper sulfate	250	121
Cresol	X	X
Cyclohexane	140	60
Dibutyl phthalate	100	38
Dichloroacetic acid	100	38
Dichloroethane (ethylene dichloride)	X	X
Ethylene glycol	250	121
Ferric chloride	250	121
Ferric chloride, 50% in water	250	121
Ferric nitrate, 10–50%	250	121
Ferrous chloride	250	121
Ferrous nitrate	160	71
Hydrobromic acid, dil.	200	93
Hydrobromic acid, 20%	160	71
Hydrobromic acid, 50%	200	93
Hydrochloric acid, 20%	230	110
Hydrochloric acid, 38%	180	82
Hydrocyanic acid, 10%	150	66
Hydrofluoric acid, 30%	120	49
Hydrochlorous acid, 10%	100	38

(continued)

TABLE 3.11 *Continued*

Chemical	Maximum Temperature	
	°F	°C
Lactic acid, 25%	200	93
Lactic acid, conc.	200	93
Magnesium Chloride	250	121
Malic acid, 10%	90	32
Methyl chloride	80	27
Methyl ethyl ketone	X	X
Methyl isobutyl ketone	80	27
Muriatic acid	190	88
Nitric acid, 5%	210	99
Nitric acid, 20%	80	27
Nitric acid, 70%	80	27
Nitrous acid, conc.	90	32
Oleum	X	X
Perchloric acid, 10%	90	32
Perchloric acid, 70%	90	32
Phenol, 5%	90	32
Phosphoric acid, 50–80%	250	121
Picric acid	100	38
Potassium bromide, 30%	230	110
Salicylic acid	130	54
Sodium carbonate, 10%	190	88
Sodium chloride	250	121
Sodium hydroxide, 10%	110	43
Sodium hydroxide, 50%	X	X
Sodium hydroxide, conc.	X	X
Sodium hypochlorite, 20%	X	X
Sodium hypochlorite, conc.	X	X
Sodium sulfide to 50%	X	X
Stannic chloride	80	27
Stannous Chloride	250	121
Sulfuric acid, 10%	260	127
Sulfuric acid, 50%	200	93
Sulfuric acid, 70%	190	88
Sulfuric acid, 90%	X	X
Sulfuric acid, 98%	X	X
Sulfuric acid, 100%	X	X
Sulfuric acid, fuming	X	X
Sulfurous acid, 10%	80	27
Thionyl chloride	X	X
Toluene	110	43
Trichloroacetic acid, 50%	200	93
White liquor	X	X
Zinc chloridex	200	93

The chemicals listed are in the pure state or in a saturated solution unless otherwise indicated. Compatibility is shown to the maximum allowable temperature for which data are available. Incompatibility is shown by an X. A blank space indicates that the data are unavailable.

Source: From P.A. Schweitzer. 2004. *Corrosion Resistance Tables*, Vols. 1–4, 5th ed., New York: Marcel Dekker.

3.5.4.1 Typical Applications

Halogenated polyesters are widely used in the pulp and paper industry in bleach atmospheres where they outperform stainless steel and high-nickel alloys. Applications are also found for ductwork, fans, and other areas where potential fire hazards may be present. They are also used for high-temperature applications such as chimney liners, chemical storage tanks, and chemical piping among other applications.

3.5.5 Terephthalate Polyesters (PET)

Terephthalate polyesters are based on terephthalic acid, the para isomer of phthalic acid. The properties of cured terephthalic-based polyesters are similar to those of isophthalic polyesters, with the terephthalics having higher heat distortion temperatures and being somewhat softer at equal saturation levels.

The PET's corrosion resistance is fairly similar to the isophthalics. Testing has indicated that the benzene resistance of comparably formulated resins is lower for PET versus isophthalic polyesters. This trend is also followed where retention of flexural modulus is elevated for various terephthalic resins versus the standard grade isophthalic resin. The PET's loss of properties in gasoline is greater than the isophthalics at the same level of unsaturation, but as the unsaturation increases, the gasoline resistance reverses with the PET performing better. The trend was seen only at unsaturated acid levels of greater than 50 mol%. This was achieved with a reversal performance in 10% sodium hydroxide where the PET with lower unsaturation was better than the isophthalic level. This follows the general trend for thermosets that as crosslink density increases, solvent resistance increases.

Refer to Table 3.12 for the compatibility of terephthalate polyester with selected corrodents. Reference [1] provides a more comprehensive listing.

TABLE 3.12

Compatibility of Polyester Terephthalate (PET) with Selected Corrodents

Chemical	Maximum Temperature	
	°F	°C
Acetic acid, 10%	300	149
Acetic acid, 50%	300	149
Acetic anhydride	X	X
Acetone	X	X
Acetyl chloride	X	X
Acrylonitrile	80	26
Aluminum chloride aqueous	170	77
Aluminum sulfate	300	149
Ammonium chloride, sat.	170	77

(continued)

TABLE 3.12 *Continued*

Chemical	Maximum Temperature	
	°F	°C
Ammonium nitrate	140	70
Ammonium persulfate	180	82
Amyl acetate	80	26
Amyl alcohol	250	121
Aniline	X	X
Antimony trichloride	250	121
Aqua regia 3:1	80	26
Barium carbonate	250	121
Barium chloride	250	121
Benzaldehyde	X	X
Benzene	X	X
Benzoic acid	250	121
Benzyl alcohol	80	27
Benzyl chloride	250	121
Boric acid	200	93
Bromine liquid	80	27
Butyl acetate	250	121
Butyric acid	250	121
Calcium chloride	250	121
Calcium hypochlorite	250	121
Carbon tetrachloride	250	121
Chloroacetic acid, 50%	X	X
Chlorine gas, dry	80	27
Chlorine gas, wet	80	27
Chloroform	250	121
Chromic acid, 50%	250	121
Citric acid, 15%	250	121
Citric acid, conc.	150	66
Copper chloride	170	77
Copper sulfate	170	77
Cresol	X	X
Cyclohexane	80	27
Dichloroethane	X	X
Ferric chloride	250	121
Ferric nitrate, 10–50%	170	77
Ferrous chloride	250	121
Hydrobromic acid, 20%	250	121
Hydrobromic acid, 50%	250	121
Hydrochloric acid, 20%	250	121
Hydrochloric acid, 38%	90	32
Hydrocyanic acid, 10%	80	27
Hydrochloric acid, 30%	X	X
Hydrofluoric acid, 70%	X	X
Hydrofluoric acid, 100%	X	X
Lactic acid, 25%	250	121
Lactic acid, conc.	250	121
Magnesium chloride	250	121
Methyl ethyl ketone	250	121

(continued)

TABLE 3.12　*Continued*

Chemical	Maximum Temperature	
	°F	°C
Methyl isobutyl ketone	X	X
Muriatic acid	90	32
Nitric acid, 5%	150	66
Perchloric acid, 10%	X	X
Perchloric acid, 70%	X	X
Phenol	X	X
Phosphoric acid, 50–80%	250	121
Sodium carbonate, 10%	250	121
Sodium chloride	250	121
Sodium hydroxide, 10%	150	66
Sodium hydroxide, 50%	X	X
Sodium hydroxide, conc.	X	X
Sodium hypochlorite, 20%	80	27
Sulfuric acid, 10%	160	71
Sulfuric acid, 50%	140	60
Sulfuric acid, 70%	X	X
Sulfuric acid, 90%	X	X
Sulfuric acid, 98%	X	X
Sulfuric acid, 100%	X	X
Sulfuric acid, fuming	X	X
Toluene	250	121
Trichloroacetic acid	250	121
Zinc chloride	250	121

The chemicals listed are in the pure state or in a saturated solution unless otherwise indicated. Compatibility is shown to the maximum allowable temperature for which data are available. Incompatibility is shown by an X. A blank space indicates that data are unavailable.

Source: From P.A. Schweitzer. 2004. *Corrosion Resistance Tables*, Vols. 1–4, 5th ed., New York: Marcel Dekker.

3.5.5.1　Typical Applications

The terephthalates are used in applications similar to those of the isophthalates but where corrosive conditions and temperatures are more moderate.

3.6　Epoxy Polyesters

Epoxy-based thermosets are the most widely used and versatile thermosets. They dominated the reinforcing piping field until the introduction of the vinyl esters, and they are still widely used. They find usage in many applications, including adhesives, coatings, encapsulants, tooling compounds, composites, and molding compounds. Their versatility is due to the wide latitude in properties that can be achieved by formulation.

A large variety of epoxy resins, modifiers, and curing agents are available that permit the epoxy formulator to tailor the epoxy system to meet the needs of each application.

3.6.1 Resin Types

The epoxide or oxirane functionality is a three membered carbon-oxygen-carbon ring. The simplest 1,2-epoxide is ethylene oxide:

$$\overset{\displaystyle O}{\overset{\diagup\diagdown}{CH_2-CH_2-}}$$

A common term used in naming epoxy resins is the term "glycidyl." The terminology for the glycidyl group:

$$\overset{\displaystyle O}{\overset{\diagup\diagdown}{CH_2-CHCH_2-}}$$

comes from the trivially named glycidyl

$$\overset{\displaystyle O}{\overset{\diagup\diagdown}{CH_2-CHCH_2OH-}}$$

and glycidic acid:

$$\overset{\displaystyle O}{\overset{\diagup\diagdown}{CH_2-CHCOOH}}$$

 Figure 3.4 illustrates the structures of the most common epoxy resins in use today. Two or more epoxide groups per molecule are required to form a cross-linked network.

The most widely used thermoset epoxy is the diglycidyl ether of bisphenol A (DGEBA). It is available in both liquid and solid forms. The chemical structure is shown in Figure 3.4.

By controlling operating conditions and varying the ratio of epichlorohydrin to bisphenol A, products of different molecular weight can be produced. For liquid resins, the n in the structured formula is generally less than 1; for solid resins, n is 2 or greater. Solids with very high meeting points have n values as high as 20.

The novolacs are another class of epoxy resins. They are produced by reacting a novolac resin, usually formed by the reaction of o-cresol or phenol and formaldehyde with epichlorohydrin. Figure 3.5 shows the general structure. These materials are used as transfer molding powders, electrical laminates, and parts where superior thermal properties and high resistance to solvents and chemicals are required.

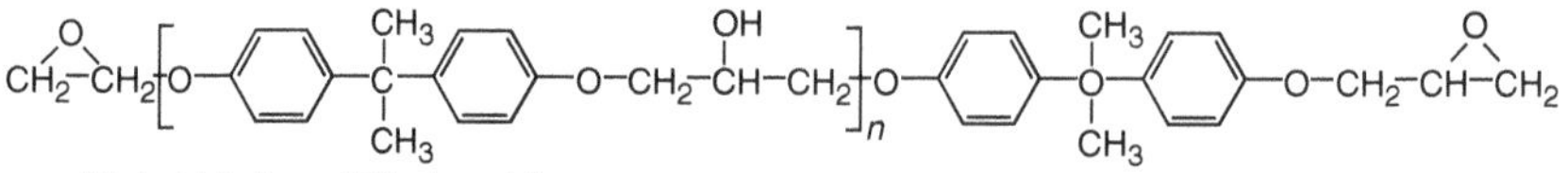

Diglycidyl ether of Bisphenol-A

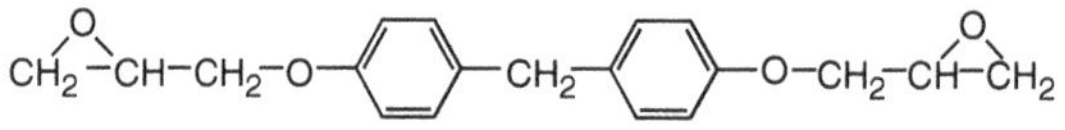

Diglycidyl ether of Bisphenol-F

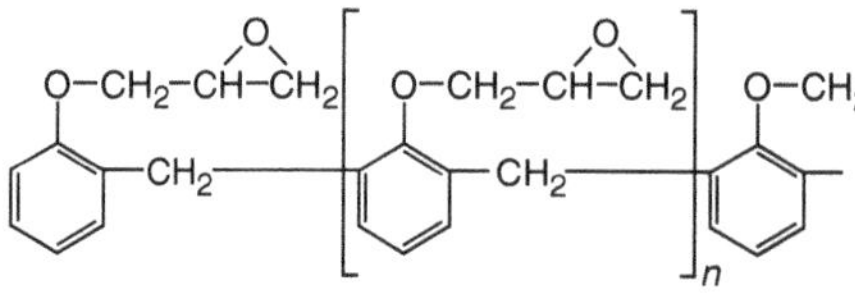

Glycidyl ether of phenolic novolac

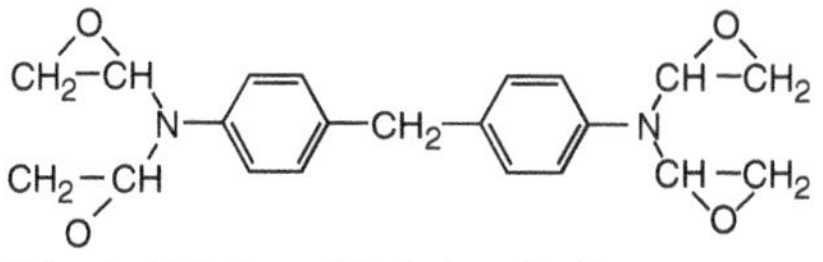

Tetraglycidyl ether of Methylenedianiline

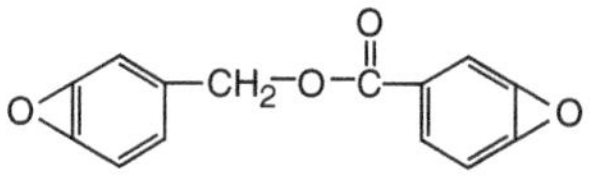

3,4 epoxycyclohexylmethyl-3,4epoxycyclohexane carboxylate

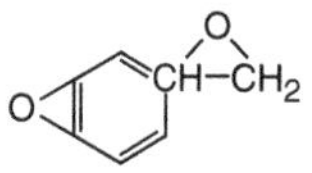

Vinyl cyclohexane diepoxide

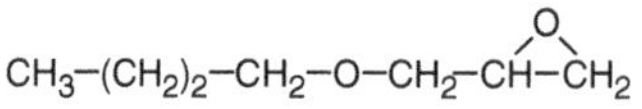

Butyl glycidyl ether

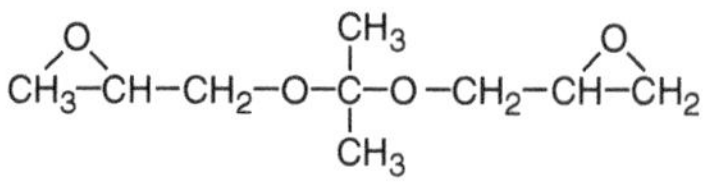

Diglycidyl ether of neopentyl glycol

FIGURE 3.4
Common commercial epoxy resins.

3.6.2 Curing

Epoxy resins must be cured with cross-linking agents (hardeners) or catalysts to develop desired properties. Cross-linking takes place at the epoxy and hydroxyl groups that are the reaction sites. Useful agents are amines, anhydrides, aldehyde condensation products, and Lewis acid catalysts. To achieve a balance of application properties and initial handling

FIGURE 3.5
General structure of novolac epoxy resins.

characteristics, careful selection of the proper curing agent is required. The primary types of curing agents are aromatic amines, aliphatic amines, catalytic curing agents, and acid anhydrides.

3.6.2.1 Aromatic Amines

Aromatic amines usually require an elevated temperature cure. Epoxies cured with aromatic amines usually have a longer working life than epoxies cured with aliphatic amines. These curing agents are relatively difficult to use because they are solids and must be melted into the epoxy. However, the allowable operating temperature for epoxies cured with aromatic amines are higher than those epoxies cured with aliphatic amines.

3.6.2.2 Aliphatic Amines

These are widely used because the curing of the epoxies takes place at room temperature. High exothermic temperatures develop during the curing reaction that limit the mass of material that can be cured. The electrical and physical properties of these aliphatic-cured resins had the greatest tendency toward degradation of electrical and physical properties at elevated temperatures. Typical aliphatic amines used include diethylene triamine (DETA) and triethylene tetramine (TETA).

3.6.2.3 Catalytic Curing Agents

Catalytic curing agents require a temperature of 200°F (93°C) or higher to react. These epoxy formulations exhibit a longer working life than the aliphatic amine cured epoxies. The exothermic reaction may be critically affected by the mass of the resin mixture. Typical materials used include piperidine, boron trifluoride ethylamine complex, and benzyl dimethyl-amine (BDMA).

3.6.2.4 Acid Anhydrides

These curing agents are becoming more widely used because they are easy to work with, have minimum toxicity problems compared with amines, and offer optimum high-temperature properties to the cured resin. Typical acid anhydrides used include nadic methyl anhydride (NMA), dodecenyl succenic anhydride (DDSA), hexahydrophthalic anhydride (HHPA), and alkendic anyhydride.

3.6.3   Corrosion Resistance

The epoxy resin family exhibits good resistance to alkalies, non-oxidizing acids, and many solvents. Typically, epoxies are compatible with the following materials at 200°F (93°C) unless otherwise noted:

Acids		
Acetic acid, 10% (150°F (66°C))	Fatty acids	Rayon spin bath
Benzoic acid	Hydrochloric acid, 10%	Oxalic acid
Butyric acid	Sulfamic acid, 20% to 180°F (82°C)	
Bases		
Sodium hydroxide, 50% to 180°F (82°C)	Calcium hydroxide	Magnesium hydroxide
Sodium sulfide, 10%	Trisodium phosphate	
Salts		
Aluminum	Magnesium	Sodium
Calcium	Metallic salts	Most ammonium salts
Iron	Potassium	
Solvents		
Alcohols	Isopropyl to 150°F (66°C)	Naphtha
Methyl	Benzene to 150°F (66°C)	Toluene
Ethyl	Ethylacetate to 150°F (66°C)	Xylene
Miscellaneous		
Distilled water	Sour crude oil	Diesel fuel
Seawater	Jet fuel	Black liquor
White liquor	Gasoline	
Epoxies are not satisfactory for use with		
Bromine water	Fluorine	Sulfuric acid above, 70%
Chromic acid	Methylene chloride	Wet chlorine gas
Bleaches	Hydrogen peroxide	Wet sulfur dioxide

Refer to Table 3.13 for the compatibility of epoxies with selected corrodents. Reference [1] provides a more comprehensive listing.

3.6.4   Typical Applications

Epoxy resins find many applications in the chemical process industry as piping. Refer to Reference [2] for more details. They are also widely used in the electronics field because of the wide variety of formulations possible. Formulations range from flexible to rigid in the cured state and from thin liquids to thick pastes and molding powders in the uncured state. Embedding applications (potting, casting, encapsulating, and impregnating) in molded parts and laminated construction are the predominant uses.

TABLE 3.13

Compatibility of Epoxy with Selected Corrodents

Chemical	Maximum Temperature	
	°F	°C
Acetaldehyde	150	66
Acetamide	90	32
Acetic acid, 10%	190	88
Acetic acid, 50%	110	43
Acetic acid, 80%	110	43
Acetic anhydride	X	X
Acetone	110	43
Acetyl chloride	X	X
Acrylic acid	X	X
Acrylonitrile	90	32
Adipic acid	250	121
Allyl alcohol	X	X
Allyl chloride	140	60
Alum	300	149
Aluminum chloride, aqueous, 1%	300	149
Aluminum chloride, dry	90	32
Aluminum fluoride	180	82
Aluminum hydroxide	180	82
Aluminum nitrate	250	121
Aluminum sulfate	300	149
Ammonia gas, dry	210	99
Ammonium bifluoride	90	32
Ammonium carbonate	140	60
Ammonium chloride, sat.	180	82
Ammonium fluoride, 25%	150	60
Ammonium hydroxide, 25%	140	60
Ammonium hydroxide, sat.	150	66
Ammonium nitrate, 25%	250	121
Ammonium persulfate	250	121
Ammonium phosphate	140	60
Ammonium sulfate, 10–40%	300	149
Ammonium sulfite	100	38
Amyl acetate	80	27
Amyl alcohol	140	60
Amyl chloride	80	27
Aniline	150	66
Antimony trichloride	180	82
Aqua regia 3:1	X	X
Barium carbonate	240	116
Barium chloride	250	121
Barium hydroxide, 10%	200	93
Barium sulfate	250	121
Barium sulfide	300	149
Benzaldehyde	X	X
Benzene	160	71
Benzenesulfonic acid, 10%	160	71
Benzoic acid	200	93

(continued)

TABLE 3.13 *Continued*

Chemical	Maximum Temperature	
	°F	°C
Benzyl alcohol	X	X
Benzyl chloride	60	16
Borax	250	121
Boric acid, 4%	200	93
Bromine gas, dry	X	X
Bromine gas, moist	X	X
Bromine, liquid	X	X
Butadiene	100	38
Butyl acetate	170	77
Butyl alcohol	140	60
n-Butylamine	X	X
Butyric acid	210	99
Calcium bisulfide		
Calcium bisulfite	200	93
Calcium carbonate	300	149
Calcium chlorate	200	93
Calcium chloride, 37.5%	190	88
Calcium hydroxide, sat.	180	82
Calcium hypochlorite, 70%	150	66
Calcium nitrate	250	121
Calcium sulfate	250	121
Calcium acid	X	X
Carbon bisulfide	100	38
Carbon dioxide, dry	200	93
Carbon disulfide	100	38
Carbon monoxide	80	27
Carbon tetrachloride	170	77
Carbonic acid	200	93
Cellosolve	140	60
Chloroacetic acid, 92% water	150	66
Chloroacetic acid	X	X
Chlorine gas, dry	150	66
Chlorine gas, wet	X	X
Chlorobenzene	150	66
Chloroform	110	43
Chlorosulfonic acid	X	X
Chromic acid, 10%	110	43
Chromic acid, 50%	X	X
Citric acid, 15%	190	88
Citric acid, 32%	190	88
Copper acetate	200	93
Copper carbonate	150	66
Copper chloride	250	121
Copper cyanide	150	66
Copper sulfate, 17%	210	99
Cresol	100	38
Cupric chloride, 5%	80	27
Cupric chloride, 50%	80	27

(*continued*)

TABLE 3.13 *Continued*

Chemical	Maximum Temperature	
	°F	°C
Cyclohexane	90	32
Cyclohexanol	80	27
Dichloroacetic acid	X	X
Dichloroethane (ethylene dichloride)	X	X
Ethylene glycol	300	149
Ferric chloride	300	149
Ferric chloride, 50% in water	250	121
Ferric nitrate, 10–50%	250	121
Ferrous chloride	250	121
Ferrous nitrate		
Fluorine gas, dry	90	32
Hydrobromic acid, dil.	180	82
Hydrobromic acid, 20%	180	82
Hydrobromic acid, 50%	110	43
Hydrochloric acid, 20%	200	93
Hydrochloric acid, 38%	140	60
Hydro cyanic acid, 10%	160	71
Hydrofluoric acid, 30%	X	X
Hydrofluoric acid, 70%	X	X
Hydrofluoric acid, 100%	X	X
Hypochlorous acid	200	93
Ketones, general	X	X
Lactic acid, 25%	220	104
Lactic acid, conc.	200	93
Magnesium chloride	190	88
Methyl chloride	X	X
Methyl ethyl ketone	90	32
Methyl isobutyl ketone	140	60
Muriatic acid	140	60
Nitric acid, 5%	160	71
Nitric acid, 20%	100	38
Nitric acid, 70%	X	X
Nitric acid, anhydrous	X	X
Nitrous acid, conc.	X	X
Oleum	X	X
Perchloric acid, 10%	90	32
Perchloric acid, 70%	80	27
Phenol	X	X
Phosphoric acid, 50–80%	110	43
Picric acid	80	27
Potassium bromide, 30%	200	93
Salicylic acid	140	60
Sodium carbonate	300	149
Sodium chloride	210	99
Sodium hydroxide, 10%	190	88
Sodium hydroxide, 50%	200	93
Sodium hypochlorite, 20%	X	X
Sodium hypochlorite, conc.	X	X

(continued)

TABLE 3.13 *Continued*

Chemical	Maximum Temperature	
	°F	°C
Sodium sulfide to 10%	250	121
Stannic chloride	200	93
Stannous chloride	160	71
Sulfuric acid, 10%	140	60
Sulfuric acid, 50%	110	43
Sulfuric acid, 70%	110	43
Sulfuric acid, 90%	X	X
Sulfuric acid, 98%	X	X
Sulfuric acid, 100%	X	X
Sulfuric acid, fuming	X	X
Sulfurous acid, 20%	240	116
Thionyl chloride	X	X
Toluene	150	66
Trichloroacetic acid	X	X
White liquor	90	32
Zinc chloride	250	121

The chemicals listed are in the pure state or in a saturated solution unless otherwise indicated. Compatibility is shown to the maximum allowable temperature for which data are available. Incompatibility is shown by an X. A blank space indicates that the data are unavailable.

Source: From P.A. Schweitzer. 2004. *Corrosion Resistance Tables*, Vols. 1–4, 5th ed., New York: Marcel Dekker.

3.7 Vinyl Esters

The vinyl ester class of resins was developed during the late 1950s and early 1960s. Vinyl esters were first used as dental fillings, replacing acrylic materials that were being used at that time. Over the next several years, changes in the molecular structure of the vinyl esters produced resins that found extensive use in corrosion-resistant equipment.

Present-day vinyl esters possess several advantages over unsaturated polyesters. They provide improved toughness in the cured polymer while maintaining good thermal stability and physical properties at elevated temperatures. This improved toughness permits the resins to be used in castings as well as in reinforced products. The structural formulas for typical vinyl ester resins are shown in Figure 3.6. These resins have improved bonding to inorganic fillers and reinforcements as a result of the internal hydroxyl group. Composites produced from vinyl esters have improved bonding to the reinforcements and improved damage resistance.

Vinyl ester resins are also available in halogenated modifications for ductwork and stack construction where fire retardance and ignition

FIGURE 3.6
Structural formula for typical vinyl ester resins.

resistance are major concerns. Vinyl esters have a number of basic advantages including:

1. They cure rapidly and give high early strength and superior creep resistance as a result of their molecular structure.
2. They provide excellent fiber wet-out and good adhesion to the glass fiber. In many cases, they are similar to the amine-cured epoxies but less than the heat-cured epoxies.
3. Vinyl ester laminates have slightly higher strengths than polyesters, but they are not as high as the heat-cured epoxies.
4. The vinyl esters have better impact resistance and greater tolerance to cyclic temperatures, pressure fluctuations, and mechanical shock than the chlorendic and bisphenol polyester resins. This results in a tough laminate that is resistant to cracking and crazing.
5. Because of the basic structure of the vinyl ester molecule, it is more resistant to hydrolysis and oxidation or halogenation than the polyesters.

The vinyl ester resins have an upper temperature limit of approximately 225°F (107°C). Exceptions to this are the vinyl ester resins having a novolac backbone. These resins can be used at 325–350°F (163–177°C).

In general, vinyl esters can be used to handle most hot, highly chlorinated, and acid mixtures at elevated temperatures. They also provide excellent resistance to strong mineral acids and bleaching solutions. Vinyl esters excel in alkaline and bleach environments and are used extensively in the very corrosive conditions found in the pulp and paper industry.

The family of vinyl esters includes a wide variety of formulations. As a result, there can be differences in the compatibility of formulations among manufacturers. When one checks compatibility in a table, one must keep in mind that all formulations may not act as shown. An indication that vinyl ester is compatible generally means that at least one formulation is compatible. This is the case in Table 3.14 that shows the compatibility of

TABLE 3.14

Compatibility of Vinyl Ester with Selected Corrodents

Chemical	Maximum Temperature	
	°F	°C
Acetaldehyde	X	X
Acetamide		
Acetic acid, 10%	200	93
Acetic acid, 50%	180	82
Acetic acid, 80%	150	66
Acetic acid, glacial	150	66
Acetic anhydride	100	38
Acetone	X	X
Acetyl chloride	X	X
Acrylic acid	100	38
Acrylonitrile	X	X
Adipic acid	180	82
Allyl alcohol	90	32
Allyl chloride	90	32
Alum	240	116
Aluminum acetate	210	99
Aluminum chloride, aqueous	260	127
Aluminum chloride, dry	140	60
Aluminum fluoride	100	38
Aluminum hydroxide	200	93
Aluminum nitrate	200	93
Aluminum oxychloride		
Aluminum sulfate	250	121
Ammonia gas	100	38
Ammonia bifluoride	150	66
Ammonium carbonate	150	66
Ammonium chloride, 10%	200	93
Ammonium chloride, 50%	200	93
Ammonium chloride, sat.	200	93
Ammonium fluoride, 10%	140	60
Ammonium fluoride, 25%	140	60
Ammonium hydroxide, 25%	100	38
Ammonium hydroxide, sat.	130	54
Ammonium nitrate	250	121
Ammonium persulfate	180	82
Ammonium phosphate	200	93
Ammonium sulfate, 10–40%	220	104
Ammonium sulfide	120	49
Ammonium sulfite	220	104

(continued)

TABLE 3.14 *Continued*

Chemical	Maximum Temperature	
	°F	°C
Amyl acetate	110	38
Amyl alcohol	210	99
Amyl chloride	120	49
Aniline	X	X
Antimony trichloride	160	71
Aqua regia 3:1	X	X
Barium carbonate	260	127
Barium chloride	200	93
Barium hydroxide	150	66
Barium sulfate	200	93
Barium sulfide	180	82
Benzaldehyde	X	X
Benzene	X	X
Benzenesulfonic acid, 10%	200	93
Benzoic acid	180	82
Benzyl alcohol	100	38
Benzyl chloride	90	32
Borax	210	99
Boric acid	200	93
Bromine gas, dry	100	38
Bromine gas, moist	100	38
Bromine, liquid	X	X
Butadiene		
Butyl acetate	80	27
Butyl alcohol	120	49
n-Butylamine	X	X
Butyric acid	130	54
Calcium bisulfide		
Calcium bisulfite	180	82
Calcium carbonate	180	82
Calcium chlorate	260	127
Calcium chloride	180	82
Calcium hydroxide, 10%	180	82
Calcium hydroxide, sat.	180	82
Calcium hypochlorite	180	82
Calcium nitrate	210	99
Calcium oxide	160	71
Calcium sulfate	250	116
Caprylic acid	220	104
Carbon bisulfide	X	X
Carbon dioxide, dry	200	93
Carbon dioxide, wet	220	104
Carbon disulfide	X	X
Carbon monoxide	350	177
Carbon tetrachloride	180	82
Carbonic acid	120	49
Cellosolve	140	60
Chloroacetic acid, 50% water	150	66

(continued)

TABLE 3.14 *Continued*

Chemical	Maximum Temperature	
	°F	°C
Chloroacetic acid	200	93
Chlorine gas, dry	250	121
Chlorine gas, wet	250	121
Chlorine, liquid	X	X
Chlorobenzene	110	43
Chloroform	X	X
Chlorosulfonic acid	X	X
Chromic acid, 10%	150	66
Chromic acid, 50%	X	X
Chromyl chloride	210	99
Citric acid, 15%	210	99
Citric acid, conc.	210	99
Copper acetate	210	99
Copper carbonate		
Copper chloride	220	104
Copper cyanide	210	99
Copper sulfate	240	116
Cresol	X	X
Cupric chloride, 5%	260	127
Cupric chloride, 50%	220	104
Cyclohexane	150	66
Cyclohexanol	150	66
Dibutyl phthalate	200	93
Dichloroacetic acid	100	38
Dichloroethane (ethylene dichloride)	110	43
Ethylene glycol	210	99
Ferric chloride	210	99
Ferric chloride, 50% in water	210	99
Ferric nitrate, 10–50%	200	93
Ferrous chloride	200	93
Ferrous nitrate	200	93
Fluorine gas, dry	X	X
Fluorine gas, moist	X	X
Hydrobromic acid, dil.	180	82
Hydrobromic acid, 20%	180	82
Hydrobromic acid, 50%	200	93
Hydrochloric acid, 20%	220	104
Hydrochloric acid, 38%	180	82
Hydrocyanic acid, 10%	160	71
Hydrofluoric acid, 30%	X	X
Hydrofluoric acid, 70%	X	X
Hydrofluoric acid, 100%	X	X
Hypochlorous acid	150	66
Iodine solution, 10%	150	66
Ketones, general	X	X
Lactic acid, 25%	210	99
Lactic acid, conc.	200	93
Magnesium chloride	260	127

(continued)

TABLE 3.14 *Continued*

Chemical	Maximum Temperature	
	°F	°C
Malic acid, 10%	140	60
Manganese chloride	210	99
Methyl chloride		
Methyl ethyl ketone	X	X
Methyl isobutyl ketone	X	X
Muriatic acid	180	82
Nitric acid, 5%	180	82
Nitric acid, 20%	150	66
Nitric acid, 70%	X	X
Nitric acid, anhydrous	X	X
Nitrous acid, 10%	150	66
Oleum	X	X
Perchloric acid, 10%	150	66
Perchloric acid, 70%	X	X
Phenol	X	X
Phosphoric acid, 50–80%	210	99
Picric acid	200	93
Potassium bromide, 30%	160	71
Salicylic acid	150	66
Silver bromide, 10%		
Sodium carbonate	180	82
Sodium chloride	180	82
Sodium hydroxide, 10%	170	77
Sodium hydroxide, 50%	220	104
Sodium hydroxide, conc.		
Sodium hypochlorite, 20%	180	82
Sodium hypochlorite, conc.	100	38
Sodium sulfide to 50%	220	104
Stannic chloride	210	99
Stannous chloride	200	93
Sulfuric acid, 10%	200	93
Sulfuric acid, 50%	210	99
Sulfuric acid, 70%	180	82
Sulfuric acid, 90%	X	X
Sulfuric acid, 98%	X	X
Sulfuric acid, 100%	X	X
Sulfuric acid, fuming	X	X
Sulfurous acid, 10%	120	49
Thionyl chloride	X	X
Toluene	120	49
Trichloroacetic acid, 50%	210	99
White liquor	180	82
Zinc chloride	180	82

The chemicals listed are in the pure state or in a saturated solution unless otherwise indicated. Compatibility is shown to the maximum allowable temperature for which data are available. Incompatibility is shown by an X. A blank space indicates that data are unavailable.

Source: From P.A. Schweitzer. 2004. *Corrosion Resistance Tables*, Vols. 1–4, 5th ed., New York: Marcel Dekker.

vinyl ester laminates with selected corrodents. The resin manufacturer must be consulted to verify the resistance. Reference [1] provides a more comprehensive listing.

3.7.1 Typical Applications

Vinyl ester laminates are used in industrial equipment and scrubbers such as absorption towers, process vessels, storage tanks, piping, hood scrubbers, ducts, and exhaust stacks, which all handling corrosive materials. Reference [2] provides detailed information concerning vinyl esters as piping materials.

3.8 Furans

Furan polymers are derivatives of furfuryl alcohol and furfural. With an acid catalyst, polymerization occurs by the condensation route, which generates heat and a by-product of water. The exotherm must be controlled to prevent the water vapor from blistering and cracking the laminate. Furan resin catalysts should have exotherms above 65°F (18°C) but not over 85°F (30°C).

All furan laminates must be postured to drive out the reaction condensate to achieve optimum properties. Curing for a fresh laminate should start with an initial temperature of 50°F (66°C) for 4 h, which is slowly raised to 180°F (82°C) for 8 h of curing. Too fast a cure can result in a blistered or cracked laminate. A final Barcol hardness of 40–45 is necessary to develop optimum laminate properties. The structural formula for the resin is:

Furan thermosets are noted for their excellent resistance to solvents. They are considered to have the best overall chemical resistance of all the thermosets. They also exhibit excellent resistance to strong concentrated mineral acids, caustics, and combinations of solvents with acids and bases.

Because there are different formulations of furans, the supplier should be consulted as to the compatibility of a particular resin with the corrodents to be encountered.

Typical materials with which furan resins are compatible are:

	Solvents	
Acetone	Styrene	Methanol
Benzene	Toluene	Methyl ethyl ketone
Carbon disulfide	Ethanol	Trichloroethylene
Chlorobenzene	Ethyl acetate	Xylene
Perchloroethylene		
	Acids	
Acetic acid	Nitric acid, 5%	Sulfuric acid, 60% to 150°F (66°C)
Hydrochloric acid	Phosphoric acid	
	Bases	
Diethylamine	Sodium sulfide	Sodium hydroxide, 50%
Sodium carbonate		
	Water	
Demineralized	Distilled	
	Others	
Pulp mill liquor		

Furan resins are not satisfactory for use with oxidizing media such as chromic or nitric acids, peroxides, hypochlorites, chlorine, phenol, and concentrated sulfuric acid.

Refer to Table 3.15 for the compatibility of furan resins with selected corrodents. Reference [1] provides a more comprehensive listing.

TABLE 3.15

Compatibility of Furan Resins with Selected Corrodents

Chemical	Maximum Temperature	
	°F	°C
Acetaldehyde	X	X
Acetic acid, 10%	212	100
Acetic acid, 50%	160	71
Acetic acid, 80%	80	27
Acetic acid, glacial	80	27
Acetic anhydride	80	27
Acetone	80	27
Acetyl chloride	200	93
Acrylic acid	80	27
Acrylonitrile	80	27
Adipic acid, 25%	280	138
Allyl alcohol	300	149
Allyl chloride	300	149
Alum, 5%	140	60
Aluminum chloride, aqueous	300	149
Aluminum chloride, dry	300	149
Aluminum fluoride	280	138
Aluminum hydroxide	260	127
Aluminum sulfate	160	71
Ammonium carbonate	240	116

(continued)

TABLE 3.15   *Continued*

Chemical	Maximum Temperature	
	°F	°C
Ammonium hydroxide, 25%	250	121
Ammonium hydroxide, sat.	200	93
Ammonium nitrate	250	121
Ammonium persulfate	260	127
Ammonium phosphate	260	127
Ammonium sulfate, 10–40%	260	127
Ammonium sulfide	260	127
Ammonium sulfite	240	116
Amyl acetate	260	127
Amyl alcohol	278	137
Amyl chloride	X	X
Aniline	80	27
Antimony trichloride	250	121
Aqua regia 3:1	X	X
Barium carbonate	240	116
Barium chloride	260	127
Barium hydroxide	260	127
Barium sulfide	260	127
Benzaldehyde	80	27
Benzene	160	71
Benzenesulfonic acid, 10%	160	71
Benzoic acid	260	127
Benzyl alcohol	80	27
Benzyl chloride	140	60
Borax	140	60
Boric acid	300	149
Bromine gas, dry	X	X
Bromine gas, moist	X	X
Bromine, liquid, 3% max	300	149
Butadiene		
Butyl acetate	260	127
Butyl alcohol	212	100
n-Butylamine	X	X
Butyric acid	260	127
Calcium bisulfite	260	127
Calcium chloride	160	71
Calcium hydroxide, sat.	260	127
Calcium hypochlorite	X	X
Calcium nitrate	260	127
Calcium oxide		
Calcium sulfate	260	127
Caprylic acid	250	121
Carbon bisulfide	160	71
Carbon dioxide, dry	90	32
Carbon dioxide, wet	80	27
Carbon disulfide	260	127
Carbon tetrachloride	212	100
Cellosolve	240	116

(continued)

TABLE 3.15 *Continued*

Chemical	Maximum Temperature	
	°F	°C
Chloroacetic acid, 50% water	100	38
Chloroacetic acid	240	116
Chlorine gas, dry	260	127
Chlorine gas, wet	260	127
Chlorine, liquid	X	X
Chlorobenzene	260	127
Chloroform	X	X
Chlorosulfonic acid	260	127
Chromic acid, 10%	X	X
Chromic acid, 50%	X	X
Chromyl chloride	250	121
Citric acid, 15%	250	121
Citric acid, conc.	250	121
Copper acetate	260	127
Copper carbonate		
Copper chloride	260	127
Copper cyanide	240	116
Copper sulfate	300	149
Cresol	260	127
Cupric chloride, 5%	300	149
Cupric chloride, 50%	300	149
Cyclohexane	140	60
Cyclohexanol		
Dichloroacetic acid	X	X
Dichloroethane (ethylene dichloride)	250	121
Ethylene glycol	160	71
Ferric chloride	260	127
Ferric chloride, 50% in water	160	71
Ferric nitrate, 10–50%	160	71
Ferrous chloride	160	71
Ferrous nitrate		
Fluorine gas, dry	X	X
Fluorine gas, moist	X	X
Hydrobromic acid, dil.	212	100
Hydrobromic acid, 20%	212	100
Hydrobromic acid, 50%	212	100
Hydrochloric acid, 20%	212	100
Hydrochloric acid, 38%	80	27
Hydrocyanic acid, 10%	160	71
Hydrofluoric acid, 30%	230	110
Hydrofluoric acid, 70%	140	60
Hydrofluoric acid, 100%	140	60
Hypochlorous acid	X	X
Iodine solution, 10%	X	X
Ketones, general	100	38
Lactic acid, 25%	212	100
Lactic acid, conc.	160	71
Magnesium chloride	260	127

(continued)

TABLE 3.15 *Continued*

Chemical	Maximum Temperature	
	°F	°C
Malic acid, 10%	260	127
Manganese chloride	200	93
Methyl chloride	120	49
Methyl ethyl ketone	80	27
Methyl isobutyl ketone	160	71
Muriatic acid	80	27
Nitric acid, 5%	X	X
Nitric acid, 20%	X	X
Nitric acid, 70%	X	X
Nitric acid, anhydrous	X	X
Nitrous acid, conc.	X	X
Oleum	190	88
Perchloric acid, 10%	X	X
Perchloric acid, 70%	260	127
Phenol	X	X
Phosphoric acid, 50%	212	100
Picric acid		
Potassium bromide, 30%	260	127
Salicylic acid	260	127
Silver bromide, 10%		
Sodium carbonate	212	100
Sodium chloride	260	127
Sodium hydroxide, 10%	X	X
Sodium hydroxide, 50%	X	X
Sodium hydroxide, conc.	X	X
Sodium hypochlorite, 15%	X	X
Sodium hypochlorite, conc.	X	X
Sodium sulfide to 10%	260	127
Stannic chloride	260	127
Stannous chloride	250	121
Sulfuric acid, 10%	160	71
Sulfuric acid, 50%	80	27
Sulfuric acid, 70%	80	27
Sulfuric acid, 90%	X	X
Sulfuric acid, 98%	X	X
Sulfuric acid, 100%	X	X
Sulfuric acid, fuming	X	X
Sulfurous acid	160	71
Thionyl chloride	X	X
Toluene	212	100
Trichloroacetic acid, 30%	80	27
White liquor	140	60
Zinc chloride	160	71

The chemicals listed are in the pure state or in a saturated solution unless otherwise indicated. Compatibility is shown to the maximum allowable temperature for which data are available. Incompatibility is shown by an X. A blank space indicates that data are unavailable.

Source: From P.A. Schweitzer. 2004. *Corrosion Resistance Tables*, Vols. 1–4, 5th ed., New York: Marcel Dekker.

3.8.1 Typical Applications

Furan resins cost approximately 30–50% more than polyester resins and do not have as good of an impact resistance as the polyesters. They find application as piping, tanks, and special equipment such as scrubbing columns. Refer to Reference [2] for more information regarding furan piping systems.

3.9 Phenolics

These are the oldest commercial classes of polymers in use today. Although first discovered in 1876, it was not until, after the "heat and pressure" patent was applied for in 1907 by Leo H. Bakeland, that the development and application of phenolic molding compounds became economical.

Phenolic resin precursors are formed by a condensation reaction and water is formed as a by-product. The structural formula is:

Phenolic compounds are available in a large number of variations, depending on the nature of the reactants, the ratios used, and the catalysts, plasticizers, lubricants, fillers, and pigments employed. An extremely large number of phenolic materials are available as a result of the many resin and filler combinations.

Phenolic resins exhibit resistance to most organic solvents, especially aromatics and chlorinated solvents. Organic polar solvents capable of hydrogen bonding, alcohols and ketones, can attack phenolics. Although phenolics have an aromatic character, the phenolic hydroxyls provide sites for hydrogen bonding and attack by caustics. Phenolics are not suitable for use in strong alkaline environments. Strong mineral acids also attack the phenolics, and acids such as nitric, chromic, and hydrochloric cause severe degradation. Sulfuric and phosphoric acids may be suitable under some conditions. There is some loss of properties when phenolics are in contact with organic acids such as acetic, formic, and oxalic.

Refer to Table 3.16 for the compatibility of phenolics with, selected corrodents. Reference [1] provides a more comprehensive listing.

3.9.1 Typical Applications

In addition to molding compounds, phenolics are used to bond friction materials for automotive brake linings, clutch parts, and transmission bands. They serve as binders for core material in furniture, the water-resistant adhesive for exterior grade plywood, binders for wood particle boards, and

TABLE 3.16

Compatibility of Phenolics with Selected Corrodents

	Maximum Temperature	
Chemical	°F	°C
Acetaldehyde		
Acetamide		
Acetic acid, 10%	212	100
Acetic acid, 50%		
Acetic acid, 80%		
Acetic acid, glacial	70	21
Acetic anhydride	70	21
Acetone	X	X
Acetyl chloride		
Acrylic acid		
Acrylonitrile		
Adipic acid		
Allyl alcohol		
Allyl chloride		
Alum		
Aluminum acetate		
Aluminum chloride, aqueous	90	32
Aluminum chloride, dry		
Aluminum fluoride		
Aluminum hydroxide		
Aluminum nitrate		
Aluminum oxychloride		
Aluminum sulfate	300	149
Ammonia gas	90	32
Ammonia bifluoride		
Ammonium carbonate	90	32
Ammonium chloride, 10%	80	27
Ammonium chloride, 50%	80	27
Ammonium chloride, sat.	80	27
Ammonium fluoride, 10%		
Ammonium fluoride, 25%		
Ammonium hydroxide, 25%	X	X
Ammonium hydroxide, sat.	X	X
Ammonium nitrate	160	71
Ammonium persulfate		
Ammonium phosphate		
Ammonium sulfate, 10–40%	300	149
Ammonium sulfide		
Ammonium sulfite		
Amyl acetate		
Amyl alcohol		
Amyl chloride		
Aniline	X	X
Antimony trichloride		
Aqua regia 3:1		
Barium carbonate		
Barium chloride		

(continued)

TABLE 3.16 *Continued*

Chemical	Maximum Temperature	
	°F	°C
Barium hydroxide		
Barium sulfate		
Barium sulfide		
Benzaldehyde	70	21
Benzene	160	71
Benzenesulfonic acid, 10%	70	21
Benzoic acid		
Benzyl alcohol		
Benzyl chloride	70	21
Borax		
Boric acid		
Bromine gas, dry		
Bromine gas, moist		
Bromine, liquid		
Butadiene		
Butyl acetate	X	X
Butyl alcohol		
n-Butylamine		
Butyl phthalate	160	71
Butyric acid, 25%		
Calcium bisulfide		
Calcium bisulfite		
Calcium carbonate		
Calcium chlorate		
Calcium chloride	300	149
Calcium hydroxide, 10%		
Calcium hydroxide, sat.		
Calcium hypochlorite, 10%	X	X
Calcium nitrate		
Calcium oxide		
Calcium sulfate		
Caprylic acid		
Carbon bisulfide		
Carbon dioxide, dry	300	149
Carbon dioxide, wet	300	149
Carbon disulfide		
Carbon monoxide		
Carbon tetrachloride	200	93
Carbonic acid	200	93
Cellosolve		
Chloroacetic acid, 50% water		
Chloroacetic acid		
Chlorine gas, dry		
Chlorine gas, wet	X	X
Chlorine, liquid	X	X
Chlorobenzene	260	127
Chloroform	160	71
Chlorosulfonic acid		
Chromic acid, 10%	X	X

(continued)

TABLE 3.16   *Continued*

Chemical	Maximum Temperature	
	°F	°C
Chromic acid, 50%	X	X
Chromyl chloride		
Citric acid, 15%	160	71
Citric acid, conc.	160	71
Copper acetate		
Copper carbonate		
Copper chloride		
Copper cyanide		
Copper sulfate	300	149
Cresol		
Cupric chloride, 5%		
Cupric chloride, 50%		
Cyclohexane		
Cyclohexanol		
Dibutyl phthalate		
Dichloroacetic acid		
Dichloroethane (ethylene dichloride)		
Ethylene glycol	70	21
Ferric chloride	300	149
Ferric chloride, 50% in water	300	149
Ferric nitrate, 10–50%		
Ferrous chloride, 40%		
Ferrous nitrate		
Fluorine gas, dry		
Fluorine gas, moist		
Hydrobromic acid, dil.	200	93
Hydrobromic acid, 20%	200	93
Hydrobromic acid, 50%	200	93
Hydrochloric acid, 20%	300	149
Hydrochloric acid, 38%	300	149
Hydrocyanic acid, 10%		
Hydrofluoric acid, 30%	X	X
Hydrofluoric acid, 60%	X	X
Hydrofluoric acid, 100%	X	X
Hypochlorous acid		
Iodine solution, 10%		
Ketones, general		
Lactic acid, 25%	160	71
Lactic acid, conc.		
Magnesium chloride		
Malic acid, 10%		
Manganese chloride		
Methyl chloride	160	71
Methyl ethyl ketone	X	X
Methyl isobutyl ketone	160	71
Muriatic acid	300	149
Nitric acid, 5%	X	X
Nitric acid, 20%	X	X

(continued)

TABLE 3.16   *Continued*

Chemical	Maximum Temperature	
	°F	°C
Nitric acid, 70%	X	X
Nitric acid, anhydrous	X	X
Nitrous acid, conc.		
Oleum		
Perchloric acid, 10%		
Perchloric acid, 70%		
Phenol	X	X
Phosphoric acid, 50–80%	212	100
Picric acid		
Potassium bromide, 30%		
Salicylic acid		
Silver bromide, 10%		
Sodium carbonate		
Sodium chloride	300	149
Sodium hydroxide, 10%	X	X
Sodium hydroxide, 50%	X	X
Sodium hydroxide, conc.	X	X
Sodium hypochlorite, 15%	X	X
Sodium hypochlorite, conc.	X	X
Sodium sulfide to 50%		
Stannic chloride		
Stannous chloride		
Sulfuric acid, 10%	250	121
Sulfuric acid, 50%	250	121
Sulfuric acid, 70%	200	93
Sulfuric acid, 90%	70	21
Sulfuric acid, 98%	X	X
Sulfuric acid, 100%	X	X
Sulfuric acid, fuming		
Sulfurous acid	80	27
Thionyl chloride	200	93
Toluene		
Trichloroacetic acid, 30%		
White liquor		
Zinc chloride	300	149

The chemicals listed are in the pure state or in a saturated solution unless otherwise indicated. Compatibility is shown to the maximum allowable temperature for which data are available. Incompatibility is shown by an X. A blank space indicates that the data are unavailable.

Source: From P.A. Schweitzer. 2004. *Corrosion Resistance Tables*, Vols. 1–4, 5th ed., New York: Marcel Dekker.

the bonding agent for converting organic and inorganic fibers into acoustical and thermal insulation pads, baths, or cushioning for home, industrial, and automotive applications. Decorative or electrical laminates are produced by impregnating paper with phenolic resins.

The excellent water resistance of phenolics makes them particularly suitable for marine applications. They also find applications as gears, wheels, and pulleys because of their wear and abrasion resistance. Because of their suitability in a variety of environments, they are used in printed circuits and terminal blocks.

Their use is limited because of two disadvantages. The laminates are only available in dark colors, usually brown or black, because of the nature of the resin. Second, they have somewhat poor resistance to electric arcs, and even though high filler loading can improve the low-power arc resistance, moisture, dirt, or high voltage usually result in complete arcing breakdown. Although they maintain their properties in the presence of water, their water absorption is high, reaching 14% in some paper-based grades of laminates.

Haveg, a division of Ametek, Inc. produces a phenolic piping system using silica filaments and filler. It is sold under the tradename of Haveg SP. Refer to Reference [2] for more details.

3.10   Phenol-Formaldehyde

As the name implies, these resins are derived from phenol and formaldehyde. The structural formula is shown in Figure 3.7. Phenol-formaldehyde is cross-linked phenolic resin and, in general, has approximately the same basic physical and mechanical properties as other phenolic resins. However, it does not have the impact resistance of the polyesters or epoxies.

Phenol-formaldehyde laminates are generally used with mineral acids, salts, and chlorinated aromatic hydrocarbons. When graphite is used as a filler, the laminate is suitable for use with hydrofluoric acid and certain fluoride salts. Refer to Table 3.17 for the compatibility of phenol-formaldehyde with selected corrodents. Reference [1] provides a more comprehensive listing.

3.10.1   Typical Applications

Applications include gears, wheels, and pulleys, printed circuits, terminal blocks, and piping. Reference [2] provides details on phenol-formaldehyde

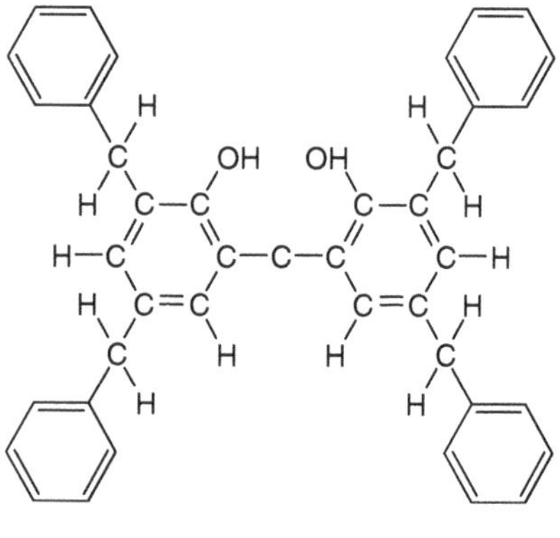

FIGURE 3.7
Structural formula for phenol-formaldehyde resins.

TABLE 3.17

Compatibility of Phenol-Formaldehyde with Selected
Corrodents

Chemical	Maximum Temperature	
	°F	°C
Acetaldehyde	X	X
Acetamide		
Acetic acid, 10%	212	100
Acetic acid, 50%	160	71
Acetic acid, 80%	120	49
Acetic acid, glacial	120	49
Acetone	X	X
Acetyl chloride	X	X
Acrylic acid, 90%	80	27
Acrylonitrile	X	X
Aluminum acetate		
Aluminum chloride, aqueous	300	149
Aluminum chloride, dry	300	149
Aluminum sulfate	300	149
Ammonium hydroxide, 25%	X	X
Ammonium hydroxide, sat.	X	X
Amyl alcohol	160	71
Aniline	X	X
Aqua regia 3:1	X	X
Benzene	160	71
Benzenesulfonic acid, 10%	160	71
Benzyl chloride	160	71
Boric acid	300	149
Bromine, liquid, 3% max	300	149
Butyric acid	260	127
Calcium chloride	300	149
Calcium hydroxide, 10%	X	X
Calcium hydroxide, sat.	X	X
Calcium hypochlorite	X	X
Carbon bisulfide	160	71
Carbon tetrachloride	212	100
Chlorine gas, dry	160	71
Chlorine gas, wet	160	71
Chlorobenzene		
Chloroform	160	71
Chlorosulfonic acid	80	27
Chromic acid, 10%	X	X
Chromic acid, 50%	X	X
Copper sulfate	300	149
Cresol		
Cupric chloride, 5%	300	149
Cupric chloride, 50%	300	149
Ethylene glycol	80	27
Ferric chloride, 50% in water	300	149
Ferric nitrate, 10–50%	300	149
Ferrous chloride, 40%	300	149
Hydrobromic acid, dil.	212	100

(continued)

TABLE 3.17 *Continued*

Chemical	Maximum Temperature	
	°F	°C
Hydrobromic acid, 20%	212	100
Hydrochloric acid, 20%	300	149
Hydrochloric acid, 38%	300	149
Hydrocyanic acid, 10%	160	71
Hydrofluoric acid, 30%	X	X
Hydrofluoric acid, 70%	X	X
Hydrofluoric acid, 100%	X	X
Hypochlorous acid		
Iodine solution, 10%	X	X
Lactic acid, 25%	160	71
Lactic acid, conc.	160	71
Methyl chloride	300	149
Methyl ethyl ketone	X	X
Methyl isobutyl ketone	X	X
Muriatic acid	300	149
Nitric acid, 5%	X	X
Nitric acid, 20%	X	X
Nitric acid, 70%	X	X
Nitric acid, anhydrous	X	X
Phenol	X	X
Phosphoric acid, 50%	212	100
Sodium carbonate	X	X
Sodium hydroxide, 10%	X	X
Sodium hydroxide, 50%	X	X
Sodium hydroxide, conc.	X	X
Sodium hypochlorite, 15%	X	X
Sodium hypochlorite, conc.	X	X
Sulfuric acid, 10%	300	149
Sulfuric acid, 50%	300	149
Sulfuric acid, 70%	250	121
Sulfuric acid, 90%	100	38
Sulfurous acid	160	71
Thionyl chloride	80	27
Toluene	212	100
Trichloroacetic acid, 30%	80	27
Zinc chloride	300	149

The chemicals listed are in the pure state or in a saturated solution unless otherwise indicated. Compatibility is shown to the maximum allowable temperature for which date are available. Incompatibility is shown by an X. A blank space indicates that the data are unavailable.

Source: From P.A. Schweitzer. 2004. *Corrosion Resistance Tables*, Vols. 1–4, 5th ed., New York: Marcel Dekker.

piping systems. Phenol-formaldehyde resins have high heat resistance and good char strengths. Because of this, they find application as ablative shields for re-entry vehicles, rocket nozzles, nose cones, and rocket motor chambers. They also produce less smoke and toxic by-products of combustion and are often used as interior aircraft panels.

3.11 Silicones

Silicon is in the same chemical group as carbon, but it is a more stable element. The silicones are a family of synthetic polymers that are partly organic and inorganic. They have a backbone structure of alternating silicon and oxygen atoms rather than a backbone of carbon–carbon atoms. The basic structure is:

$$
\left[
\begin{array}{cc}
CH_3 & CH_3 \\
| & | \\
Si{-}O{\sim}Si{-}O \\
| & | \\
CH_3 & CH_3
\end{array}
\right]_n
$$

Typically, the silicon atoms will have one or more organic side groups attached to them, generally phenol (CH_6H_5-), methyl (CH_3), or vinyl ($CH_2{=}CH-$) units. These groups impact properties such as solvent resistance, lubricity, and reactivity with organic chemicals and polymers. Silicone polymers may be filled or unfilled, depending upon the properties required and the application.

Silicone polymers possess several properties that distinguish them from their organic counterparts. These are

1. Chemical inertness
2. Weather resistance
3. Extreme water repellency
4. Uniform properties over a wide temperature range
5. Excellent electrical properties over a wide range of temperature and frequencies
6. Low surface tension
7. High degree of slip or lubricity
8. Excellent release properties
9. Inertness and compatibility both physiologically and in electronic applications

Silicone resins and composites produced with silicon resins exhibit outstanding long-term thermal stability at temperatures approaching 572°F (300°C) and excellent moisture resistance and electrical properties. These materials are also useful in the cryogenic temperature range.

Silicone laminates can be used in contact with dilute acids and alkalies, alcohol, animal and vegetable oils, and lubrication oils. They are also resistant to aliphatic hydrocarbons, but aromatic solvents such as benzene, toluene, gasoline, and chlorinated solvents will cause excessive swelling. Although they exhibit excellent resistance to water and weathering, they are not resistant to high-pressure, high-temperature steam.

As previously discussed, the silicon atoms may have one or more organic side groups attached. The addition of these side groups has an effect on the corrosion resistance. Therefore, it is necessary to check with the supplier as to the properties of the silicone laminate being supplied. Table 3.18 lists the compatibility of a silicone laminate with methyl-appended side groups.

TABLE 3.18

Compatibility of Methyl Appended Silicone Laminate with Selected Corrodents

| | Maximum Temperature | |
Chemical	°F	°C
Acetic acid, 10%	90	32
Acetic acid, 50%	90	32
Acetic acid, 80%	90	32
Acetic acid, glacial	90	32
Acetone	100	43
Acrylic acid, 75%	80	27
Acrylonitrile	X	X
Alum	220	104
Aluminum sulfate	410	210
Ammonium chloride, 10%	X	X
Ammonium chloride, 50%	80	27
Ammonium chloride, sat.	80	27
Ammonium fluoride, 25%	80	27
Ammonium hydroxide, 25%	X	X
Ammonium nitrate	210	99
Amly acetate	80	27
Amly alcohol	X	X
Amly chloride	X	X
Aniline	X	X
Antimony trichloride	80	27
Aqua regia 3:1	X	X
Benzene	X	X
Benzyl chloride	X	X
Boric acid	390	189
Butyl alcohol	80	27
Calcium bisulfide	400	204
Calcium chloride	300	149
Calcium hydroxide, 30%	200	99
Calcium hydroxide, sat.	400	204
Carbon bisulfide	X	X
Carbon disulfide	X	X
Carbon monoxide	400	204
Carbonic acid	400	204
Chlorobenszen	X	X
Chlorosulfonic acid	X	X
Ethylene glycol	400	204
Ferric chloride	400	204
Hydrobromic acid, 50%	X	X
Hydrochloric acid, 20%	90	32

(continued)

TABLE 3.18 *Continued*

Chemical	Maximum Temperature	
	°F	°C
Hydrochloric acid, 38%	X	X
Hydrofluoric acid, 30%	X	X
Lactic acid, all conc.	80	27
Lactic acid, conc.	80	27
Magnesium chloride	400	204
Methyl alcohol	410	210
Methyl ethyl ketone	X	X
Methyl isobutyl ketone	X	X
Nitric acid, 5%	80	23
Nitric acid, 20%	X	X
Nitric acid, 70%	X	X
Nitric acid, anhydrous	X	X
Oleum	X	X
Phenol	X	X
Phosphoric acid, 50–80%	X	X
Propyl alcohol	400	204
Sodium carbonate	300	149
Sodium chloride, 10%	400	204
Sodium hydroxide, 10%	90	32
Sodium hydroxide, 50%	90	32
Sodium hydroxide, conc.	90	32
Sodium hypochlorite, 20%	X	X
Sodium sulfate	400	204
Stannic chloride	80	27
Sulfuric acid, 10%	X	X
Sulfuric acid, 50%	X	X
Sulfuric acid, 70%	X	X
Sulfuric acid, 90%	X	X
Sulfuric acid, 98%	X	X
Sulfuric acid, 100%	X	X
Sulfuric acid, fuming	X	X
Sulfurous acid	X	X
Tartaric acid	400	204
Tetrahydrofuran	X	X
Toluene	X	X
Tributyl phosphate	X	X
Turpentine	X	X
Vinegar	400	204
Water, acid mine	210	99
Water, demineralized	210	99
Water, distilled	210	99
Water, salt	210	99
Water, sea	210	99
Xylene	X	X
Zinc chloride	400	204

The chemicals listed are in the pure state or in a saturated solution unless otherwise indicated. Compatibility is shown to the maximum allowable temperature. Incompatibility is shown by an X.

Source: From P.A. Schweitzer. 2004. *Corrosion Resistance Tables*, Vols. 1–4, 5th ed., New York: Marcel Dekker.

3.11.1 Typical Applications

Silicone laminates find application as radomes, structures in electronics, heaters, rocket components, slot wedges, ablation shields, coil forms, and terminal boards.

3.12 Siloxirane

Siloxirane is the registered trademark for Tankenitics homopolymerized polymer with a cross-linked organic–inorganic (SiO) backbone with oxirane endcaps. Siloxirane is a homopolymerized polymer with an ether cross-linkage (carbon–oxygen–carbon) having a very dense, highly cross-linked molecular structure. The absence of the hydroxyl (found in epoxies) and ester groups (found in vinyl esters) eliminates the built-in failure modes of other polymers and provides superior performance.

The end products are extremely resistant to material abrasion and have an operating temperature range of −80 to +500°F (−62 to +260°C). The operating temperature will be tempered by the material being handled.

Siloxirane laminates have a wide range of chemical resistance. Following is a list of some of the chemicals with which siloxirane is compatible:

Acetamide	Ethyl acetate	Molten sulfur
Acetic acid, glacial	Ferric chloride	Monochloroacetic acid
Acetic anhydride	Formaldehyde	Nickel plating solutions
Acetone	Furan	Nitrous oxide
Aluminum chloride	Furfural alcohol	Phosphoric acid
Ammonium chloride	Gasohol	Phosphoric acid, 85%
Ammonium hydroxide	Gasoline	Sodium chloride
Aqua regia	Green liquor	Sodium dichromate
Benzene	Hydraulic oil	Sodium hydroxide
Benzenesulfonic acid	Hydrazine	Sodium hypochlorite, 17%
Black liquor	Hydrochloric acid, 1%	Sodium hypochlorite, aged
Bromine hypochlorite	Hydrochloric acid, 0–37%	Sulfate liquor (paper)
Carbon tetrachloride	Hydrofluoric acid, 40%	Sulfur trioxide
Chloric acid	Hydrofluoric acid, 52%	Sulfuric acid, 1–98%
Chlorine water	Iodine	Sulfuric acid, fuming oleum
Chloroacetic acid	Jet fuel	Tallow
Chlorobenzene	Kerosene	Thionyl chloride
Chromic acid, 10%	Ketones	Toluene
Chromic acid, 50%	Latex	Trichloroethylene
Dibutyl phthalate	Methanol	Tricresyl phosphate
Dichlorobenzene	Methyl ethyl ketone	Water, deionized
Dimethylformamide	Methyl isobutyl ketone	Water, salt
Ethanol	Methylene chloride	White liquor (paper)

3.12.1 Typical Applications

Siloxirane laminates are used in piping, ductwork, storage tanks, and tank liners. Vessels produced from Siloxirane have been approved to receive ASME Section X Class 11 certification and stamping as code vessels. Class 11 vessels may be a maximum of 144 inches in diameter, and the product of pressure (psig) and the diameter in inches may not exceed 7200.

3.13 Polyurethanes

Polyurethanes are reaction products of isocyanates, polyols, and curing agents. Because of the hazards involved in handling free isocyanate, prepolymers of the isocyanate and the polyol are generally used in casting. Polyurethanes can be formulated to produce a range of materials from elastomers as soft as Shore A of 5 to tough solids with a Shore D of 90. Polyurethane thermosets can be rigid or flexible, depending on the formulation.

Polyurethane resin is resistant to most mineral and vegetable oils. Some 1–4, butane-diol-cured polyurethanes have been approved by the FDA for use in applications that will come into contact with dry, aqueous, and fatty foods. They are also resistant to greases, fuels, aliphatic, and chlorinated hydrocarbons. This makes these materials particularly useful for service in contact with lubricating oils and automotive fuels.

Aromatic hydrocarbons, polar solvents, esters, and ketones will attack urethane.

3.13.1 Typical Applications

Applications include large automotive parts and building components. Foam materials blown with halocarbons have the lowest thermal conductivity of any commercially available insulation. They are used in refrigerators, picnic boxes, and building construction. Flexible foam has also been used in furniture, packaging, and shock and vibration mounts.

3.14 Melamines

Melamine is a polymer formed by a condensation reaction between formaldehyde and amino compounds containing NH_2 groups. Consequently, they are also referred to as melamine formaldehydes. Their structural formula is shown in Figure 3.8.

Melamine resin can be combined with a variety of reinforcing fibers. However, the best properties are obtained when glass cloth is used as the

FIGURE 3.8
Structural formula for melamine-formaldehyde resins.

reinforcement material. Low temperatures have relatively little effect on the properties of the melamine.

The clarity of melamine resins permits products to be fabricated in virtually any color. Finished melamine products exhibit excellent resistance to moisture, greases, oils, and solvents, are tasteless and odorless, are self-extinguishing, resist scratching and marring, and offer excellent electrical properties.

Melamine formulations filled with cellulose are capable of producing an unlimited range of light-stable colors with a high degree of translucency. The basic material properties are not affected by the addition of color. Prolonged exposure at high temperature will affect the color and cause a loss of some strength characteristics.

Melamines exposed outdoors suffer little degradation in electrical or physical properties, but some color changes may take place.

Melamine laminates are compatible with the following chemicals:

Acetone	Detergent	Soaps
Alcohol	Fly spray	Shoe polish
Ammonium hydroxide, 10%	Gasoline	Sodium bisulfite
Amyl acetate	Moth spray	Trisodium phosphate
Carbon tetrachloride	Mustard	Urine
Citric acid	Naphtha	Water
Coffee	Olive oil	Wax and crayons

There are certain chemicals and household preparations that tend to stain the melamine laminate. However, the stain can be removed by buffing with a mild abrasive. Included are such materials as

Beet juice	Ink	Phenol (lysol)
Bluing	Iodine solution	Tea
Dyes	Mercurochrome solution	Vinegar

The following chemicals and household preparations may damage melamine laminates:

Berry juices	Lye solutions	Silver nitrate
Gentian violet	Mineral acids	Silver protein (argyrol)
Hydrogen peroxide	Potassium permanganate	Sodium bisulfate
Hypochlorite bleaches		

3.14.1 Typical Applications

The heat resistance and water-white color of melamine polymers make them ideally suited for use in the home as high-quality molded dinnerware; melamine-surfaced decorative laminates find wide application in the home. The decorative pattern may be a solid color, a wood grain, or any design.

3.15 Alkyds

Alkyds are saturated resins produced from the reaction of organic alcohols with organic acids. The ability to use any of the many suitable polyfunctional alcohols and acids permits selection of a large variation of repeating units. Formulating can provide resins that demonstrate a wide variety of characteristics involving flexibility heat resistance, chemical resistance, and electrical properties.

Alkyd compounds are chemically similar to polyester compounds but make use of higher viscosity or dry monomers. Alkyd compounds often contain glass fiber filler but may contain clay, calcium carbonate, or alumina.

The greatest limitations of alkyds are in the extremes of temperature above 350°F (117°C) and in high humidity.

The alkyds exhibit poor resistance to solvents and alkalies. Their resistance to dilute acid is fair. However, they exhibit good resistance to weather. Although alkyds are used outdoors, they are not as durable to long-term exposure as the acrylics, and their color and gloss retention are inferior.

3.15.1 Typical Applications

Alkyds are used for finishing metal and wood products but not to the degree previously used. Their durability to interior exposure is good, but their durability to exterior exposure is only fair. Because of their formulating flexibility, they are used in fillers, sealers, and caulks for wood finishing. They are still used for finishing by the machine tool and other industries. Alkyd-modified acrylic latex paints are excellent architectural finishes.

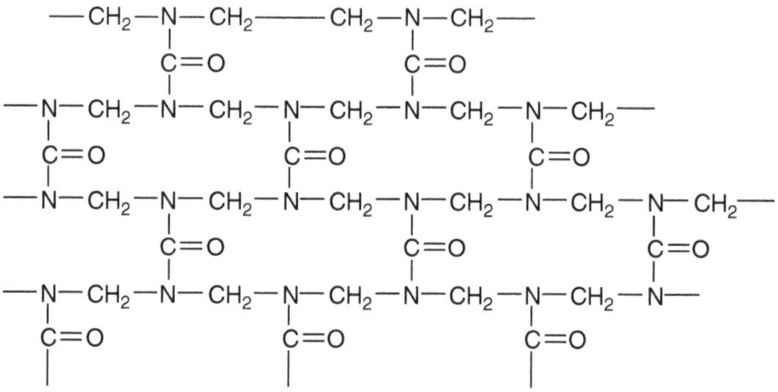

FIGURE 3.9
Chemical structure of urea-formaldehyde.

3.16 Ureas (Aminos)

Ureas (commonly referred to as *aminos*) are polymers formed by condensation reactions and do not produce by-products. They are reaction products of formaldehyde with amino compounds containing NH_2 groups. Therefore, they are often referred to as urea formaldehydes. The general chemical structure of urea formaldehyde is shown in Figure 3.9.

Amino polymers are self-extinguishing and have excellent electrical insulating properties. The most commonly used filler for the amines is alpha cellulose.

The addition of alpha cellulose produces an unlimited range of light-stable colors and high degrees of translucency. Basic material properties are unaffected by the addition of color.

When urea moldings are subjected to severe cycling between dry and wet conditions, cracks develop. Certain strength characteristics also experience a loss when amino moldings are subjected to prolonged elevated temperatures.

Finished products containing an amino resin surface exhibit excellent resistance to moisture, greases, oils, and solvents, are tasteless and odorless, and resist scratching and marring. However, the melamine resins exhibit better chemical, heat, and moisture resistance than the amines.

Amino resins are used in modifying other resins to increase hardness and accelerate cure. Aminos are unsuitable for outdoor exposure.

3.17 Allyls

The allyls, or diallyl phthalates, are produced in several variations, but the most commonly used are Diallyl phthlate (DAP) and diallyl isophthalate

TABLE 3.19

Chemical Resistance of DAP and DAIP

% Gain in Weight after 30 Days Immersion at 77°F/25°C		
Corrodent	DAP	DAIP
Acetone	1.3	−0.03
Sodium hydroxide, 1%	0.7	0.7
Sodium hydroxide, 10%	0.5	0.6
Sulfuric acid, 3%	0.8	0.7
Sulfuric acid, 30%	0.4	0.4
Water	0.9	0.8

(DAIP). The primary difference between the two is that DAIP will withstand somewhat higher temperatures than DAP.

DAP and DAIP are seldom used as cast homopolymers except for small electrical parts because of their low tensile strength and impact resistance. When physical abuse is likely, the polymer is provided with a glass or mineral filling.

DAP finds application for the impregnation of ferrous and nonferrous castings because of its low viscosity, excellent sealing properties, low resin bleed out, and ease of cleanup. It is also used to impregnate wood to reduce water absorption and increase impact, compressive, and shear strengths.

DAP and DAIP glass laminates have high-temperature electrical properties superior to those of most other structural laminates. Cure cycles are shorter and little or no post cure is required to provide usable strength up to 482°F (250°C).

Allyl carbonate (diethylene glycol) is marketed as CR-39 by PPG Industries. It is used largely in optical castings in competition with glass, or polymethyl methacrylate (PMMA) where abrasion resistance and high heat distortion or impact resistance are required.

Refer to Table 3.19 for the chemical resistance of DAP and DAIP.

3.18 Polybutadienes

Polybutadienes have essentially hydrocarbon structure as follows:

$$
\begin{bmatrix}
-CH-CH_2- \\
CH \\
CH_2
\end{bmatrix}
$$

The basic materials are monopolymers or copolymers that react through terminal hydroxyl groups, terminal carboxyl groups, vinyl groups, or a combination of these materials.

Polybutadiene polymers that have a 1,2 microstructure varying from 60 to 90% offer potential as moldings, laminating resins, coatings, and cast liquid and formed sheet products. Outstanding electrical and thermal stability results from the structure that is essentially pure hydrocarbon.

Peroxide catalysts used to cure polybutadienes produce carbon–carbon double bonds in the vinyl group. The final product is 100% hydrocarbon except where the starting polymer is the –OH or –COOH radical.

Poly BD marketed by Atochem North America, Inc. is a family of hydroxyl-terminated butadiene homopolymers and copolymers with styrene or acrylonitrile. These materials are usually reacted with isocyanates to produce polyurethanes that have excellent resistance to boiling water.

The B. F. Goodrich Chemical Company produces monopolymers and copolymers with acrylonitrile. These are known as the Hylar series. Acrylonitrile increases the viscosity and imparts oil resistance, adhesion, and compatibility with epoxy resins. Carboxy-terminated butadiene-acrylonitrile copolymer (CTBN) improves impact strength, low-temperature shear strength, and crack resistance in epoxy formulations.

Amine-terminated butadiene-acrylonitrile copolymers (ATBN) are also used to modify epoxy resins. These are formulated on the amine-hardener side of the mix.

Applications for these formulations include electrical potting compounds (with transformer oil) sealants and moisture block compounds for telephone cables.

The hydrocarbon structure is responsible for the excellent resistance shown by polybutadiene to chemicals and solvents and for the electrical properties that are good over a range of frequencies, temperatures, and humidity, and resistance to high temperature. Refer to Table 3.20 for the compatibility of polybutadiene with selected corrodents.

TABLE 3.20

Compatibility of Polybutadiene with Selected Corrodents

Chemical	Maximum Temperature	
	°F	°C
Alum	90	32
Alum ammonium	90	32
Alum ammonium sulfate	90	32
Alum chrome	90	32
Alum potassium	90	32
Aluminum chloride, aqueous	90	32
Aluminum sulfate	90	32
Ammonia gas	90	32
Ammonium chloride, 10%	90	32
Ammonium chloride, 28%	90	32

(continued)

TABLE 3.20 *Continued*

Chemical	Maximum Temperature	
	°F	°C
Ammonium chloride, 50%	90	32
Ammonium chloride, sat.	90	32
Ammonium nitrate	90	32
Ammonium sulfate, 10–40%	90	32
Calcium chloride, sat.	80	27
Calcium hypochlorite, sat.	90	32
Carbon dioxide, wet	90	32
Chlorine gas, wet	X	X
Chrome alum	90	32
Chromic acid, 10%	X	X
Chromic acid, 30%	X	X
Chromic acid, 40%	X	X
Chromic acid, 50%	X	X
Copper chloride	90	32
Copper sulfate	90	32
Fatty acids	90	32
Ferrous chloride	90	32
Ferrous sulfate	90	32
Hydrochloric acid, dil.	80	27
Hydrochloric acid, 20%	90	32
Hydrochloric acid, 35%	90	32
Hydrochloric acid, 38%	90	32
Hydrochloric acid, 50%	90	32
Hydrochloric acid fumes	90	32
Hydrogen peroxide, 90%	90	32
Hydrogen sulfide, dry	90	32
Nitric acid, 5%	80	27
Nitric acid, 10%	80	27
Nitric acid, 20%	80	27
Nitric acid, 30%	80	27
Nitric acid, 40%	X	X
Nitric acid, 50%	X	X
Nitric acid, 70%	X	X
Nitric acid, anhydrous	X	X
Nitric acid, conc.	80	27
Ozone	X	X
Phenol	80	27
Sodium bicarbonate, 20%	90	32
Sodium bisulfate	80	27
Sodium bisulfite	90	32
Sodium carbonate	90	32
Sodium chlorate	80	27
Sodium hydroxide, 10%	90	32
Sodium hydroxide, 15%	90	32
Sodium hydroxide, 30%	90	32
Sodium hydroxide, 50%	90	32
Sodium hydroxide, 70%	90	32
Sodium hydroxide, conc.	90	32

(continued)

TABLE 3.20 *Continued*

	Maximum Temperature	
Chemical	°F	°C
Sodium hypochlorite to 20%	90	32
Sodium nitrate	90	32
Sodium phosphate, acid	90	32
Sodium phosphate, alkaline	90	32
Sodium phosphate, neutral	90	32
Sodium silicate	90	32
Sodium sulfide to 50%	90	32
Sodium sulfite, 10%	90	32
Sodium dioxide, dry	X	X
Sulfur trioxide	90	32
Sulfuric acid, 10%	80	27
Sulfuric acid, 30%	80	27
Sulfuric acid, 50%	80	27
Sulfuric acid, 60%	80	27
Sulfuric acid, 70%	90	32
Toluene	X	X

The chemicals listed are in the pure state or in a saturated solution unless otherwise indicated. Compatibility is shown to the maximum allowable temperature for which data are available. Incompatibility is shown by an X. A blank space indicates that the data are unavailable.

Source: From P.A. Schweitzer. 2004. *Corrosion Resistance Tables*, Vols. 1–4, 5th ed., New York: Marcel Dekker.

3.19 Polyimides

Polyimides can be prepared as either thermoplastic or thermoset resins. There are two types of polyimides, condensation, and addition resins. Condensation polyimides are available as either thermosets or thermoplastics. The additional polyimides are available only as thermosets.

The polyimides are heterocyclic polymers having an atom of nitrogen in the inside ring as shown below:

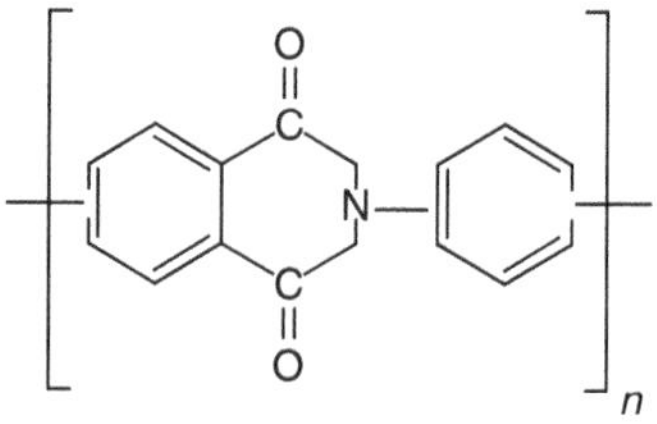

The fused rings provide stiffness that provides high-temperature strength retention. The low concentration of hydrogen provides oxidative resistance by preventing thermal fracture of the chain.

Polyimides have continuous high-temperature stability up to 500–600°F (260–315°C) and can withstand 900°F (482°C) for short-term use.

Polyimides are sensitive to alkaline chemicals and will be dissolved by hot, concentrated sodium hydroxide. They are also moisture sensitive, gaining 1% in weight after 1000 hours at 50% relative humidity and 72°F (23°C). The polyimides exhibit excellent resistance to ionizing radiation and very low outgassing in high vacuum.

3.19.1 Typical Applications

Polyimides find use in various forms, including laminates, moldings, films, coatings, and adhesives (specifically in the areas of high temperature). Coatings are used in electrical applications as insulating varnishes and magnet wire enamels at high temperatures. They also find application as a coating on cookware as an alternative to fluorocarbon coatings.

3.20 Cyanate Esters

Cyanate esters are a family of aryl dicyanate monomers that contain the reactive cyanate (–O–C=N) functional group. When heated, this cyanate functionality undergoes an exothermic cyclotrimerization reaction to form triazine ring connecting units, resulting in the formation of a thermoset polycyanate polymer.

The cyanate ester monomers are available from low-viscosity liquids to meltable solids. Ciba-Geigy produces a series of cyanate esters under the trademark AroCy L-10. Cyanate esters are used in manufacturing structural composites by filament winding, resin transfer molding, and pultrusion.

References

1. P.A. Schweitzer. 2004. *Corrosion Resistance Tables*, Vols. 1–4, 5th ed., New York: Marcel Dekker.
2. P.A. Schweitzer. 1994. *Corrosion Resistant Piping Systems*. New York: Marcel Dekker.

4

Comparative Corrosion Resistance of Thermoplastic and Thermoset Polymers

The corrosion tables on the following pages are arranged alphabetically according to the corrodent. The chemicals listed are in the pure state or in a saturated solution unless otherwise indicated. Compatibility is shown to the maximum allowable temperature for which data is available. Incompatibility is indicated by an x. A blank space indicates that data is not available.

A greater range of corrodents can be found in Reference [1].

The following polymers are identified in the table by abbreviations:

Halar–ECTFE
Noryl–PPO
Kynar–PVDF
Penton–CPE
Polyetheretherketone–PEEK
Polypropylene–PP
Polysulfone–PSF
Kel-F–CTFE
Ultem–PEI
Ryton–PPS
Polyamide imide–PAI
Polyether sulfone–PEF
Polybutadiene–PB
Polyurethane–PUR
Tefzel–ETFE
Telflon–PTFE
Saran–PVDC

Because many of the polymers may be compounded, the manufacturer should be checked to see that the material being furnished is satisfactory for the service. For example, there are three different formulations of polyamides.

Acetaldahyde

Polymer	°F/°C
ABS	X
Acrylics	X
CTFE	130/54
ECTFE	
ETFE	200/93
FEP	200/93
Polyamides	X
PAI	
PEEK	80/27
PEI	
PES	
PFA	450/232
PTFE	450/232
PVDF	150/66
EHMWPE, 40%	90/32
HMWPE	X
PET	
PPO	
PPS	230/110
PP	120/49
PVDC	150/66
PSF	X
PVC Type 1	X
PVC Type 2	X
CPVC	X
CPE	140/60
PUR	X
Acetals	80/27
Polyesters	
Bisphenol A-fumurate	X
Halogenated	X
Hydrogenated-Bisphenol A	
Isophthalic	X
PET	X
Epoxy	150/66
Vinyl esters	X
Furans	X
Phenolics	
Phenol-formaldehyde	X
Methyl appended silicone	
PB	

Acetamide

Polymer	°F/°C
ABS	
Acrylics	
CTFE	200/93
ECTFE	
ETFE	250/121
FEP	400/204
Polyamides	250/121
PAI	
PEEK	
PEI	
PES	
PFA	450/232
PTFE	450/232
PVDF	90/32
EHMWPE	
HMWPE	140/60
PET	
PPO	
PPS	240/116
PP	110/43
PVDC	80/27
PSF	
PVC Type 1	X
PVC Type 2	X
CPVC	
CPE	
PUR	X
Acetals	
Polyesters	
Bisphenol A-fumurate	
Halogenated	
Hydrogenated-Bisphenol A	
Isophthalic	
PET	X
Epoxy	90/32
Vinyl esters	
Furans	
Phenolics	
Phenol-formaldehyde	
Methyl appended silicone	
PB	

Acetic Acid, 10%

Polymer	°F/°C
ABS	100/38
Acrylics	X
CTFE	
ECTFE	250/121
ETFE	250/121
FEP	400/204
Polyamides	200/93
PAI	200/93
PEEK	80/27
PEI	80/27
PES	80/27
PFA	450/232
PTFE	450/232
PVDF	300/149
EHMWPE	140/60
HMWPE	140/60
PET	300/149
PPO	140/60
PPS	240/116
PP	220/104
PVDC	150/66
PSF	200/93
PVC Type 1	140/60
PVC Type 2	100/38
CPVC	180/82
CPE	250/121
PUR	X
Acetals	X
Polyesters	
Bisphenol A-fumurate	220/104
Halogenated	140/60
Hydrogenated-Bisphenol A	200/93
Isophthalic	180/82
PET	300/149
Epoxy	190/88
Vinyl esters	200/93
Furans	260/127
Phenolics	210/99
Phenol-formaldehyde	
Methyl appended silicone	90/32
PB	

Acetic Acid, 50%

Polymer	°F/°C
ABS	130/54
Acrylics	X
CTFE	
ECTFE	250/121
ETFE	250/121
FEP	400/204
Polyamides	X
PAI	200/93
PEEK	140/60
PEI	80/27
PES	140/60
PFA	450/232
PTFE	450/232
PVDF	300/149
EHMWPE	140/60
HMWPE	140/60
PET	300/149
PPO	140/60
PPS	250/121
PP	200/93
PVDC	130/54
PSF	200/93
PVC Type 1	140/60
PVC Type 2	90/32
CPVC	X
CPE	
PUR	
Acetals	90/32
Polyesters	
Bisphenol A-fumurate	160/71
Halogenated	90/32
Hydrogenated-Bisphenol A	160/71
Isophthalic	110/43
PET	300/149
Epoxy	110/43
Vinyl esters	
Furans	200/93
Phenolics	
Phenol-formaldehyde	160/71
Methyl appended silicone	90/32
PB	

Acetic Acid, 80%

Polymer	°F/°C
ABS	X
Acrylics	X
CTFE	
ECTFE	150/66
ETFE	230/110
FEP	400/204
Polyamides	X
PAI	200/93
PEEK	140/60
PEI	80/27
PES	200/93
PFA	450/232
PTFE	450/232
PVDF	190/88
EHMWPE	80/27
HMWPE	80/27
PET	300/149
PPO	140/60
PPS	250/121
PP	200/93
PVDC	130/54
PSF	200/93
PVC Type 1	140/60
PVC Type 2	X
CPVC	X
CPE	250/121
PUR	X
Acetals	90/32
Polyesters	
Bisphenol A-fumurate	160/71
Halogenated	90/32
Hydrogenated-Bisphenol A	
Isophthalic	X
PET	300/149
Epoxy	110/43
Vinyl esters	150/66
Furans	200/93
Phenolics	
Phenol-formaldehyde	120/49
Methyl appended silicone	90/32
PB	

Acetic Acid, Glacial

Polymer	°F/°C
ABS	X
Acrylics	X
CTFE	
ECTFE	200/93
ETFE	230/110
FEP	400/204
Polyamides	X
PAI	200/93
PEEK	140/60
PEI	80/27
PES	200/93
PFA	400/204
PTFE	460/238
PVDF	190/88
EHMWPE, 40%	100/38
HMWPE	80/27
PET	300/149
PPO	150/66
PPS	250/121
PP	190/88
PVDC	140/60
PSF	200/93
PVC Type 1	130/54
PVC Type 2	X
CPVC	X
CPE	250/121
PUR	X
Acetals	X
Polyesters	
Bisphenol A-fumurate	X
Halogenated	110/43
Hydrogenated-Bisphenol A	
Isophthalic	X
PET	300/149
Epoxy	
Vinyl esters	150/66
Furans	270/132
Phenolics	120/49
Phenol-formaldehyde	120/49
Methyl appended silicone	90/32
PB	

Acetic Anhydride

Polymer	°F/°C
ABS	X
Acrylics	X
CTFE	
ECTFE	100/38
ETFE	300/149
FEP	400/204
Polyamides	200/93
PAI	200/93
PEEK	
PEI	
PES	
PFA	450/232
PTFE	450/232
PVDF	100/38
EHMWPE	X
HMWPE	X
PET	X
PPO	X
PPS	280/138
PP	100/38
PVDC	90/32
PSF	
PVC Type 1	X
PVC Type 2	X
CPVC	X
CPE	150/66
PUR	X
Acetals	X
Polyesters	
Bisphenol A-fumurate	130/54
Halogenated	100/38
Hydrogenated-Bisphenol A	X
Isophthalic	X
PET	X
Epoxy	X
Vinyl esters	100/38
Furans	90/32
Phenolics	90/32
Phenol-formaldehyde	
Methyl appended silicone	
PB	

Acetone	
Polymer	°F/°C
ABS	X
Acrylics	X
CTFE	
ECTFE	150/66
ETFE	150/66
FEP[a]	400/204
Polyamides	80/27
PAI	80/27
PEEK	210/99
PEI	80/27
PES	X
PFA	450/232
PTFE	450/232
PVDF	X
EHMWPE	120/49
HMWPE	80/27
PET	X
PPO	X
PPS	260/127
PP	220/104
PVDC	90/32
PSF	X
PVC Type 1	X
PVC Type 2	X
CPVC	X
CPE	90/32
PUR	X
Acetals	120/49
Polyesters	
Bisphenol A-fumurate	X
Halogenated	X
Hydrogenated-Bisphenol A	X
Isophthalic	X
PET	X
Epoxy	110/43
Vinyl esters	X
Furans	200/93
Phenolics	X
Phenol-formaldehyde	
Methyl appended silicone	100/43
PB	

[a] Corrodent will be absorbed.

Acetyl Chloride

Polymer	°F/°C
ABS	X
Acrylics	
CTFE	100/38
ECTFE	150/66
ETFE	150/66
FEP	400/204
Polyamides	X
PAI DRY	120/49
PEEK	
PEI	
PES	
PFA	460/238
PTFE	460/238
PVDF	120/49
EHMWPE	
HMWPE	X
PET	80/27
PPO	X
PPS	
PP	X
PVDC	130/54
PSF	X
PVC Type 1	X
PVC Type 2	X
CPVC	X
CPE	X
PUR	
Acetals	
Polyesters	
Bisphenol A-fumurate	X
Halogenated	X
Hydrogenated-Bisphenol A	X
Isophthalic	X
PET	80/27
Epoxy	X
Vinyl esters	X
Furans	200/93
Phenolics	
Phenol-formaldehyde	X
Methyl appended silicone	
PB	

Acrylic Acid

Polymer	°F/°C
ABS	
Acrylics	
CTFE	
ECTFE	
ETFE	
FEP	200/93
Polyamides	
PAI	
PEEK	80/27
PEI	
PES	
PFA	
PTFE	460/238
PVDF	150/66
EHMWPE	
HMWPE	
PET	
PPO	
PPS, 25%	100/38
PP	X
PVDC	
PSF	
PVC Type 1	X
PVC Type 2	X
CPVC	X
CPE	
PUR	
Acetals	
Polyesters	
Bisphenol A-fumurate	100/38
Halogenated	X
Hydrogenated-Bisphenol A	
Isophthalic	X
PET	
Epoxy	X
Vinyl esters	100/38
Furans	100/38
Phenolics	
Phenol-formaldehyde, 90%	90/32
Methyl appended silicone, 75%	80/27
PB	

Acrylonitrile

Polymer	°F/°C
ABS	
Acrylics	
CTFE	
ECTFE	150/66
ETFE	150/66
FEP	400/204
Polyamides	80/27
PAI	
PEEK	80/27
PEI	
PES	
PFA	460/238
PTFE	460/238
PVDF	130/54
EHMWPE	140/60
HMWPE	150/66
PET	
PPO	
PPS	130/54
PP	90/32
PVDC	90/32
PSF	
PVC Type 1	X
PVC Type 2	X
CPVC	X
CPE	150/66
PUR	
Acetals	
Polyesters	
Bisphenol A-fumurate	X
Halogenated	X
Hydrogenated-Bisphenol A	X
Isophthalic	X
PET	80/27
Epoxy	90/32
Vinyl esters	X
Furans	260/127
Phenolics	X
Phenol-formaldehyde	
Methyl appended silicone	X
PB	

Adipic Acid

Polymer	°F/°C
ABS	140/60
Acrylics	
CTFE	
ECTFE	150/66
ETFE	280/138
FEP	400/204
Polyamides	
PAI	
PEEK	
PEI	
PES	
PFA	460/227
PTFE	460/227
PVDF	280/138
EHMWPE	140/60
HMWPE	140/60
PET	
PPO	
PPS	300/149
PP	140/60
PVDC	150/66
PSF	
PVC Type 1	140/60
PVC Type 2	140/60
CPVC	200/93
CPE	250/121
PUR	
Acetals	
Polyesters	
Bisphenol A-fumurate	220/104
Halogenated	220/104
Hydrogenated-Bisphenol A	
Isophthalic	220/104
PET	
Epoxy	250/121
Vinyl esters	180/82
Furans, 25%	280/138
Phenolics	
Phenol-formaldehyde	
Methyl appended silicone	
PB	

Allyl Alcohol	
Polymer	**°F/°C**
ABS	X
Acrylics	
CTFE	200/93
ECTFE	
ETFE	210/99
FEP	400/204
Polyamides	90/32
PAI	
PEEK	
PEI	
PES	
PFA	430/221
PTFE	430/221
PVDF	200/93
EHMWPE	140/60
HMWPE	140/60
PET	
PPO	
PPS	
PP	140/60
PVDC	80/27
PSF	
PVC Type 1	90/32
PVC Type 2	90/32
CPVC, 90%	200/93
CPE	250/121
PUR	
Acetals	
Polyesters	
Bisphenol A-fumurate	X
Halogenated	X
Hydrogenated-Bisphenol A	
Isophthalic	X
PET	
Epoxy	X
Vinyl esters	90/32
Furans	300/149
Phenolics	
Phenol-formaldehyde	
Methyl appended silicone	
PB	

Allyl Chloride

Polymer	°F/°C
ABS	X
Acrylics	
CTFE	
ECTFE	300/149
ETFE	190/88
FEP	400/204
Polyamides	
PAI	
PEEK	
PEI	
PES	
PFA	450/232
PTFE	450/232
PVDF	200/93
EHMWPE	80/27
HMWPE	110/43
PET	
PPO	X
PPS	
PP	140/60
PVDC	
PSF	
PVC Type 1	X
PVC Type 2	X
CPVC	X
CPE	90/32
PUR	
Acetals	
Polyesters	
Bisphenol A-fumurate	X
Halogenated	X
Hydrogenated-Bisphenol A	
Isophthalic	X
PET	
Epoxy	140/60
Vinyl esters	90/32
Furans	300/149
Phenolics	
Phenol-formaldehyde	
Methyl appended silicone	
PB	

Alum

Polymer	°F/°C
ABS	140/60
Acrylics	90/32
CTFE	
ECTFE	300/149
ETFE	300/149
FEP	400/204
Polyamides	X
PAI	
PEEK	
PEI	
PES	
PFA	450/232
PTFE	450/232
PVDF	210/99
EHMWPE	140/60
HMWPE	140/60
PET	150/66
PPO	120/49
PPS	300/149
PP	220/104
PVDC	180/82
PSF	
PVC Type 1	140/60
PVC Type 2	140/60
CPVC	200/93
CPE	250/121
PUR	
Acetals	
Polyesters	
Bisphenol A-fumurate	220/104
Halogenated	200/93
Hydrogenated-Bisphenol A	
Isophthalic	250/121
PET	150/66
Epoxy	300/149
Vinyl esters	240/116
Furans, 5%	140/60
Phenolics	
Phenol-formaldehyde	
Methyl appended silicone	220/104
PB	90/32

Aluminum Acetate

Polymer	°F/°C
ABS	
Acrylics	
CTFE	
ECTFE	
ETFE	
FEP	400/204
Polyamides	
PAI	
PEEK	
PEI	
PES	
PFA	200/93
PTFE	460/238
PVDF	250/121
EHMWPE	
HMWPE	
PET	
PPO	
PPS	210/99
PP	100/38
PVDC	
PSF	
PVC Type 1	100/38
PVC Type 2	100/38
CPVC	100/38
CPE	
PUR	
Acetals	
Polyesters	
Bisphenol A-fumurate	
Halogenated	
Hydrogenated-Bisphenol A	
Isophthalic	
PET	
Epoxy, 10%	250/121
Vinyl esters	210/99
Furans	
Phenolics	
Phenol-formaldehyde	
Methyl appended silicone	
PB	

Aluminum Chloride, Aqueous

Polymer	°F/°C
ABS	140/60
Acrylics	90/32
CTFE, 25%	200/93
ECTFE	300/149
ETFE	300/149
FEP	400/204
Polyamides, 10%	90/32
PAI	
PEEK	
PEI	
PES	
PFA	450/232
PTFE	450/232
PVDF	300/149
EHMWPE	140/60
HMWPE	140/60
PET	170/77
PPO	140/60
PPS	300/149
PP	200/93
PVDC	150/66
PSF, 10%	200/93
PVC Type 1	140/60
PVC Type 2	140/60
CPVC	200/93
CPE	250/121
PUR	
Acetals	
Polyesters	
Bisphenol A-fumurate	200/93
Halogenated	120/49
Hydrogenated-Bisphenol A	200/93
Isophthalic	180/82
PET	170/77
Epoxy, 1%	300/149
Vinyl esters	260/127
Furans	300/149
Phenolics	
Phenol-formaldehyde	300/149
Methyl appended silicone	
PB	90/32

Aluminum Fluoride

Polymer	°F/°C
ABS	140/60
Acrylics	
CTFE	
ECTFE	300/149
ETFE	300/149
FEP[a]	400/204
Polyamides	80/27
PAI	
PEEK	
PEI	
PES	
PFA	450/232
PTFE	450/232
PVDF	300/149
EHMWPE	140/60
HMWPE	140/60
PET	
PPO	80/27
PPS	
PP	200/93
PVDC	150/66
PSF	200/93
PVCS Type 1	140/60
PVC Type 2	140/60
CPVC	200/93
CPE	250/121
PUR	
Acetals	
Polyesters	
Bisphenol A-fumurate, 10%	90/32
Halogenated, 10%	90/32
Hydrogenated-Bisphenol A	X
Isophthalic, 10%	140/60
PET	
Epoxy	180/82
Vinyl esters	100/38
Furans	280/138
Phenolics	
Phenol-formaldehyde	
Methyl appended silicone	
PB	

[a] Corrodent will penetrate.

Aluminum Hydroxide

Polymer	°F/°C
ABS	140/60
Acrylics	90/32
CTFE	
ECTFE	300/149
ETFE	300/149
FEP	400/204
Polyamides	250/121
PAI	
PEEK	
PEI	
PES	
PFA	450/232
PTFE	450/232
PVDF	260/127
EHMWPE	140/60
HMWPE	140/60
PET	300/149
PPO	
PPS	250/121
PP	200/93
PVDC	170/77
PSF	
PVC Type 1	140/60
PVC Type 2	140/60
CPVC	200/93
CPE	250/121
PUR	
Acetals	
Polyesters	
Bisphenol A-fumurate	160/71
Halogenated	170/77
Hydrogenated-Bisphenol A	
Isophthalic	160/71
PET	300/149
Epoxy	200/93
Vinyl esters	200/93
Furans	260/127
Phenolics	
Phenol-formaldehyde	
Methyl appended silicone	
PB	

Aluminum Nitrate

Polymer	°F/°C
ABS	
Acrylics	
CTFE	
ECTFE	300/149
ETFE	300/149
FEP	400/204
Polyamides	80/27
PAI	
PEEK	
PEI	
PES	
PFA	450/232
PTFE	450/232
PVDF	300/149
EHMWPE	
HMWPE	140/60
PET	
PPO	
PPS	250/121
PP	210/99
PVDC	180/82
PSF	
PVC Type 1	140/60
PVC Type 2	140/60
CPVC	200/93
CPE	250/121
PUR	
Acetals	
Polyesters	
Bisphenol A-fumurate	200/93
Halogenated	160/77
Hydrogenated-Bisphenol A	
Isophthalic	160/77
PET	
Epoxy	250/121
Vinyl esters	200/93
Furans	
Phenolics	
Phenol-formaldehyde	
Methyl appended silicone	
PB	

Aluminum Oxychloride

Polymer	°F/°C
ABS	140/60
Acrylics	
CTFE	
ECTFE	150/66
ETFE	300/149
FEP	400/204
Polyamides	
PAI	
PEEK	
PEI	
PES	
PFA	450/232
PTFE	450/232
PVDF	280/138
EHMWPE	
HMWPE	
PET	
PPO	
PPS	460/238
PP	200/93
PVDC	140/60
PSF	150/66
PVC Type 1	140/60
PVC Type 2	
CPVC	200/93
CPE	220/104
PUR	
Acetals	
Polyesters	
Bisphenol A-fumurate	
Halogenated	
Hydrogenated-Bisphenol A	
Isophthalic	
PET	
Epoxy	
Vinyl esters	
Furans	
Phenolics	
Phenol-formaldehyde	
Methyl appended silicone	
PB	

Aluminum Sulfate

Polymer	°F/°C
ABS	140/60
Acrylics	90/32
CTFE	
ECTFE	300/149
ETFE	300/149
FEP	400/204
Polyamides	140/60
PAI, 10%	220/104
PEEK	80/27
PEI	
PES	
PFA	450/232
PTFE	450/232
PVDF	300/149
EHMWPE	140/60
HMWPE	140/60
PET	300/149
PPO	200/93
PPS	
PP	220/104
PVDC	180/82
PSF	200/93
PVC Type 1	140/60
PVC Type 2	140/60
CPVC	200/93
CPE	250/121
PUR	
Acetals	
Polyesters	
Bisphenol A-fumurate	200/93
Halogenated	250/121
Hydrogenated-Bisphenol A	200/93
Isophthalic	180/82
PET	300/149
Epoxy	300/149
Vinyl esters	250/121
Furans	260/127
Phenolics	290/143
Phenol-formaldehyde	
Methyl appended silicone	410/210
PB	90/32

Ammonia Gas

Polymer	°F/°C
ABS DRY	140/60
Acrylics	90/32
CTFE	
ECTFE	300/149
ETFE	
FEP[a]	400/204
Polyamides	200/93
PAI	
PEEK	210/99
PEI	
PES	80/27
PFA[a]	450/232
PTFE[a]	450/232
PVDF	270/132
EHMWPE	140/60
HMWPE	140/60
PET	150/66
PPO	80/27
PPS	250/121
PP	150/66
PVDC	X
PSF	X
PVC Type 1	140/60
PVC Type 2	140/60
CPVC DRY	200/93
CPE	220/104
PUR	
Acetals	
Polyesters	
Bisphenol A-fumurate	200/93
Halogenated	160/71
Hydrogenated-Bisphenol A	
Isophthalic	90/32
PET	150/66
Epoxy dry	210/99
Vinyl esters	150/66
Furans	
Phenolics	90/32
Phenol-formaldehyde	
Methyl appended silicone	
PB	100/38

[a] Corrodent will permeate.

Ammonium Bifluoride

Polymer	°F/°C
ABS	140/60
Acrylics	
CTFE	
ECTFE	300/149
ETFE	300/149
FEP[a]	400/204
Polyamides	
PAI	
PEEK	
PEI	
PES	
PFA[b]	450/232
PTFE[b]	450/232
PVDF	260/127
EHMWPE	
HMWPE	140/60
PET	
PPO	
PPS	
PP	200/93
PVDC	140/60
PSF	
PVC Type 1	90/32
PVC Type 2	90/32
CPVC	200/93
CPE	230/110
PUR	
Acetals	
Polyesters	
Bisphenol A-fumurate	
Halogenated	
Hydrogenated-Bisphenol A	
Isophthalic	
PET	
Epoxy	90/32
Vinyl esters	150/66
Furans	
Phenolics	
Phenol-formaldehyde	
Methyl appended silicone	
PB	

[a] Corrodent will be absorbed.
[b] Corrodent will permeate.

Ammonium Carbonate

Polymer	°F/°C
ABS	140/60
Acrylics	90/32
CTFE	
ECTFE	300/149
ETFE	300/149
FEP	400/204
Polyamides	250/121
PAI	
PEEK	
PEI	
PES	
PFA	430/221
PTFE	430/221
PVDF	280/138
EHMWPE	140/60
HMWPE	140/60
PET	300/149
PPO	140/60
PPS	100/38
PP	200/93
PVDC	180/82
PSF	200/93
PVC Type 1	140/60
PVC Type 2	140/60
CPVC	200/93
CPE	250/121
PUR	90/32
Acetals	
Polyesters	
Bisphenol A-fumurate	90/32
Halogenated	140/60
Hydrogenated-Bisphenol A	
Isophthalic	X
PET	300/149
Epoxy	160/71
Vinyl esters	150/66
Furans	240/116
Phenolics	90/32
Phenol-formaldehyde	
Methyl appended silicone	
PB	

Ammonium Chloride, 10%

Polymer	°F/°C
ABS	140/60
Acrylics	80/27
CTFE	
ECTFE	290/143
ETFE	300/149
FEP	400/204
Polyamides	200/93
PAI	200/93
PEEK	
PEI	
PES	
PFA	410/210
PTFE	450/232
PVDF	280/138
EHMWPE	140/60
HMWPE	140/60
PET	170/77
PPO	
PPS	300/149
PP	180/82
PVDC	
PSF	200/93
PVC Type 1	140/60
PVC Type 2	140/60
CPVC	200/93
CPE	
PUR	80/27
Acetals	
Polyesters	
Bisphenol A-fumurate	200/93
Halogenated	200/93
Hydrogenated-Bisphenol A	
Isophthalic	160/71
PET	170/77
Epoxy	200/93
Vinyl esters	200/93
Furans	220/104
Phenolics	90/32
Phenol-formaldehyde	
Methyl appended silicone	X
PB	100/38

Ammonium Chloride, 50%

Polymer	°F/°C
ABS	
Acrylics	
CTFE	
ECTFE	300/149
ETFE	300/149
FEP	400/204
Polyamides, 37%	200/93
PAI	
PEEK	
PEI	
PES	
PFA	430/221
PTFE	460/238
PVDF	280/138
EHMWPE	140/60
HMWPE	140/60
PET	170/77
PPO	
PPS	300/149
PP	180/82
PVDC	
PSF	200/93
PVC Type 1	140/60
PVC Type 2	
CPVC	180/82
CPE	
PUR	90/32
Acetals	
Polyesters	
Bisphenol A-fumurate	220/104
Halogenated	200/93
Hydrogenated-Bisphenol A	
Isophthalic	160/71
PET	170/77
Epoxy	270/132
Vinyl esters	200/93
Furans	220/104
Phenolics	
Phenol-formaldehyde	
Methyl appended silicone	80/27
PB	90/32

Ammonium Chloride, Sat.

Polymer	°F/°C
ABS	140/60
Acrylics	90/32
CTFE	
ECTFE	300/149
ETFE	300/149
FEP	400/204
Polyamides	X
PAI	
PEEK	
PEI	
PES	
PFA	410/210
PTFE	450/232
PVDF	280/138
EHMWPE	140/60
HMWPE	140/60
PET	170/77
PPO	140/60
PPS	300/149
PP	200/93
PVDC	160/71
PSF	200/93
PVC Type 1	140/60
PVC Type 2	140/60
CPVC	190/88
CPE	250/121
PUR	90/32
Acetals	
Polyesters	
Bisphenol A-fumurate	220/104
Halogenated	200/93
Hydrogenated-Bisphenol A	200/93
Isophthalic	180/82
PET	170/77
Epoxy	260/127
Vinyl esters	200/93
Furans	260/127
Phenolics	80/27
Phenol-formaldehyde	
Methyl appended silicone	80/27
PB	90/32

Ammonium Fluoride, 10%

Polymer	°F/°C
ABS	X
Acrylics	
CTFE	
ECTFE	300/149
ETFE	300/149
FEP[a]	400/204
Polyamides	80/27
PAI	
PEEK	
PEI	
PES	
PFA[a]	430/221
PTFE	450/232
PVDF	280/138
EHMWPE	140/60
HMWPE	140/60
PET	
PPO	
PPS	
PP	210/99
PVDC	90/32
PSF	
PVC Type 1	140/60
PVC Type 2	90/32
CPVC	200/93
CPE	250/121
PUR	
Acetals	
Polyesters	
Bisphenol A-fumurate[b]	180/82
Halogenated[b]	140/60
Hydrogenated-Bisphenol A	
Isophthalic[b]	90/32
PET	
Epoxy	
Vinyl esters	200/93
Furans	280/177
Phenolics	
Phenol-formaldehyde	
Methyl appended silicone	
PB	

[a] Corrodent will permeate.
[b] Synthetic veil or surfacing mat should be used.

Ammonium Fluoride, 25%

Polymer	°F/°C
ABS	X
Acrylics	
CTFE	
ECTFE	300/149
ETFE	300/149
FEP[a]	400/204
Polyamides	80/27
PAI	
PEEK	
PEI	
PES	
PFA[a]	420/216
PTFE	450/232
PVDF	280/138
EHMWPE	140/60
HMWPE	140/60
PET	
PPO	140/60
PPS	
PP	200/93
PVDC	90/32
PSF	
PVC Type 1	140/60
PVC Type 2	90/32
CPVC	200/93
CPE	250/121
PUR	
Acetals	
Polyesters	
Bisphenol A-fumurate[b]	120/49
Halogenated[b]	140/60
Hydrogenated-Bisphenol A	
Isophthalic[b]	90/32
PET	
Epoxy	150/66
Vinyl esters	220/104
Furans	260/127
Phenolics	
Phenol-formaldehyde	
Methyl appended silicone	80/27
PB	

[a] Corrodent will permeate.
[b] Synthetic veil or surfacing mat should be used.

Ammonium Hydroxide, 25%

Polymer	°F/°C
ABS	90/32
Acrylics	80/27
CTFE	
ECTFE	300/149
ETFE	300/149
FEP	400/204
Polyamides	250/121
PAI	200/93
PEEK	
PEI	80/27
PES	
PFA	390/199
PTFE	450/232
PVDF	280/138
EHMWPE	140/60
HMWPE	140/60
PET	300/149
PPO	140/60
PPS	250/121
PP	200/93
PVDC	X
PSF	200/93
PVC Type 1	140/60
PVC Type 2	140/60
CPVC	X
CPE	250/121
PUR	90/32
Acetals	X
Polyesters	
Bisphenol A-fumurate	100/38
Halogenated	90/32
Hydrogenated-Bisphenol A	
Isophthalic	X
PET	300/149
Epoxy	140/60
Vinyl esters	150/66
Furans	260/127
Phenolics	90/32
Phenol-formaldehyde	
Methyl appended silicone	80/27
PB	

Ammonium Hydroxide, Sat.

Polymer	°F/°C
ABS	80/27
Acrylics	80/27
CTFE	
ECTFE	300/149
ETFE	300/149
FEP	400/204
Polyamides	250/121
PAI	200/93
PEEK	80/27
PEI	80/27
PES	
PFA	370/188
PTFE	450/232
PVDF	280/138
EHMWPE	140/60
HMWPE	140/60
PET	300/149
PPO	140/60
PPS	250/121
PP	200/93
PVDC	X
PSF	200/93
PVC Type 1	140/60
PVC Type 2	140/60
CPVC	X
CPE	250/121
PUR	80/27
Acetals	X
Polyesters	
Bisphenol A-fumurate	
Halogenated	90/32
Hydrogenated-Bisphenol A	
Isophthalic	X
PET	300/149
Epoxy	160/71
Vinyl esters[a]	130/54
Furans	200/93
Phenolics	X
Phenol-formaldehyde	
Methyl appended silicone	
PB	

[a] Synthetic veil or surfacing mat should be used.

Ammonium Nitrate

Polymer	°F/°C
ABS	140/60
Acrylics	80/27
CTFE	
ECTFE	300/149
ETFE	230/110
FEP	400/204
Polyamides	200/93
PAI, 10%	200/93
PEEK	
PEI	
PES	
PFA	350/177
PTFE	450/232
PVDF	280/138
EHMWPE	140/60
HMWPE	140/60
PET	140/60
PPO	140/60
PPS	250/121
PP	200/93
PVDC	120/49
PSF	200/93
PVC Type 1	140/60
PVC Type 2	140/60
CPVC	200/93
CPE	240/116
PUR	
Acetals	
Polyesters	
Bisphenol A-fumurate	220/104
Halogenated	220/93
Hydrogenated-Bisphenol A	200/93
Isophthalic	160/77
PET	140/66
Epoxy, 25%	250/121
Vinyl esters	250/121
Furans	260/127
Phenolics	200/93
Phenol-formaldehyde	
Methyl appended silicone	210/99
PB	90/32

Ammonium Persulfate

Polymer	°F/°C
ABS	140/60
Acrylics	80/27
CTFE	
ECTFE	150/66
ETFE	300/149
FEP	400/204
Polyamides	X
PAI	
PEEK	
PEI	
PES	
PFA	450/232
PTFE	460/238
PVDF	280/138
EHMWPE	140/60
HMWPE, 5%	150/66
PET	180/82
PPO	140/60
PPS	
PP	220/104
PVDC	90/32
PSF	
PVC Type 1	140/60
PVC Type 2	140/60
CPVC	200/93
CPE	180/82
PUR	X
Acetals	
Polyesters	
Bisphenol A-fumurate	180/82
Halogenated	140/60
Hydrogenated-Bisphenol A	200/93
Isophthalic	160/71
PET	180/82
Epoxy	250/121
Vinyl esters	180/82
Furans	260/127
Phenolics	
Phenol-formaldehyde	
Methyl appended silicone	
PB	

Ammonium Phosphate

Polymer	°F/°C
ABS	140/60
Acrylics	
CTFE	
ECTFE	300/149
ETFE	300/149
FEP	400/204
Polyamides	80/27
PAI	
PEEK	
PEI	
PES	
PFA	300/149
PTFE	450/232
PVDF	280/138
EHMWPE	
HMWPE	80/27
PET	140/60
PPO	
PPS, 65%	300/149
PP	200/93
PVDC	150/66
PSF	200/93
PVC Type 1	140/60
PVC Type 2	140/60
CPVC	90/32
CPE	250/121
PUR	
Acetals	
Polyesters	
Bisphenol A-fumurate	80/27
Halogenated	150/66
Hydrogenated-Bisphenol A	
Isophthalic	160/71
PET	140/60
Epoxy	200/93
Vinyl esters	200/93
Furans	260/127
Phenolics	
Phenol-formaldehyde	
Methyl appended silicone	
PB	

Ammonium Sulfate, 10–40%

Polymer	°F/°C
ABS	140/60
Acrylics	80/27
CTFE	
ECTFE	300/149
ETFE	300/149
FEP	400/204
Polyamides	80/27
PAI, 10%	200/93
PEEK	
PEI	
PES	
PFA	300/149
PTFE	450/232
PVDF	280/138
EHMWPE	140/60
HMWPE	140/60
PET, sat.	170/77
PPO	140/60
PPS	300/149
PP	200/93
PVDC	120/49
PSF	200/93
PVC Type 1	140/60
PVC Type 2	140/60
CPVC	200/93
CPE	250/121
PUR	
Acetals	
Polyesters	
Bisphenol A-fumurate, sat.	220/104
Halogenated	200/93
Hydrogenated-Bisphenol A	
Isophthalic, 10%	180/82
PET, sat.	170/77
Epoxy	300/149
Vinyl esters, sat.	220/104
Furans, sat.	260/127
Phenolics	300/149
Phenol-formaldehyde	
Methyl appended silicone	
PB	90/32

Ammonium Sulfide

Polymer	°F/°C
ABS	140/60
Acrylics	
CTFE	
ECTFE	300/149
ETFE	300/149
FEP	400/204
Polyamides	
PAI	
PEEK	
PEI	
PES	
PFA	300/149
PTFE	450/232
PVDF	280/138
EHMWPE	140/60
HMWPE	140/60
PET	
PPO	
PPS	
PP	220/104
PVDC	80/27
PSF	
PVC Type 1	140/60
PVC Type 2	140/60
CPVC	200/93
CPE	250/121
PUR	
Acetals	
Polyesters	
Bisphenol A-fumurate	110/43
Halogenated	120/49
Hydrogenated-Bisphenol A	100/38
Isophthalic	X
PET	
Epoxy	
Vinyl esters	120/49
Furans	260/127
Phenolics	
Phenol-formaldehyde	
Methyl appended silicone	
PB	

Amyl Acetate

Polymer	°F/°C
ABS	X
Acrylics	X
CTFE	100/38
ECTFE	160/71
ETFE	250/121
FEP	400/204
Polyamides	150/66
PAI	200/93
PEEK	
PEI	
PES	
PFA	200/93
PTFE	450/232
PVDF	190/88
EHMWPE	140/60
HMWPE	80/27
PET	80/27
PPO	X
PPS	300/149
PP	X
PVDC	120/49
PSF	X
PVC Type 1	X
PVC Type 2	X
CPVC	X
CPE	180/82
PUR	X
Acetals	
Polyesters	
Bisphenol A-fumurate	80/27
Halogenated	190/88
Hydrogenated-Bisphenol A	X
Isophthalic	X
PET	80/27
Epoxy	80/27
Vinyl esters	110/43
Furans	260/127
Phenolics	
Phenol-formaldehyde	
Methyl appended silicone	80/27
PB	

Amyl Alcohol

Polymer	°F/°C
ABS	80/27
Acrylics	80/27
CTFE	
ECTFE	300/149
ETFE	300/149
FEP	400/204
Polyamides	200/93
PAI	
PEEK	
PEI	
PES	
PFA	300/149
PTFE	450/232
PVDF	280/138
EHMWPE	140/60
HMWPE	140/60
PET	250/121
PPO	100/49
PPS	210/99
PP	200/93
PVDC	150/66
PSF	200/93
PVC Type 1	140/60
PVC Type 2	X
CPVC	130/54
CPE	220/104
PUR	X
Acetals	
Polyesters	
Bisphenol A-fumurate	200/93
Halogenated	200/93
Hydrogenated-Bisphenol A	200/93
Isophthalic	160/71
PET	250/121
Epoxy	200/93
Vinyl esters	210/99
Furans	260/127
Phenolics	190/88
Phenol-formaldehyde	160/71
Methyl appended silicone	X
PB	

Amyl Chloride

Polymer	°F/°C
ABS	X
Acrylics	
CTFE	
ECTFE	300/149
ETFE	300/149
FEP	400/204
Polyamides	X
PAI	
PEEK	
PEI	
PES	
PFA	200/93
PTFE	450/232
PVDF	280/138
EHMWPE	X
HMWPE	X
PET	
PPO	X
PPS	200/93
PP	X
PVDC	80/27
PSF	X
PVC Type 1	X
PVC Type 2	X
CPVC	X
CPE	220/104
PUR	
Acetals	
Polyesters	
Bisphenol A-fumurate	X
Halogenated	X
Hydrogenated-Bisphenol A	90/32
Isophthalic	X
PET	
Epoxy	80/27
Vinyl esters	120/49
Furans	X
Phenolics	
Phenol-formaldehyde	
Methyl appended silicone	X
PB	

Aniline

Polymer	°F/°C
ABS	X
Acrylics	X
CTFE	
ECTFE	90/32
ETFE	230/110
FEP	400/204
Polyamides	X
PAI	200/93
PEEK	200/93
PEI	
PES	X
PFA	200/93
PTFE	450/232
PVDF	200/93
EHMWPE	130/54
HMWPE	130/54
PET	X
PPO	X
PPS	300/149
PP	180/82
PVDC	X
PSF	X
PVC Type 1	X
PVC Type 2	X
CPVC	X
CPE	150/66
PUR	X
Acetals	80/27
Polyesters	
Bisphenol A-fumurate	X
Halogenated	120/49
Hydrogenated-Bisphenol A	X
Isophthalic	X
PET	X
Epoxy	150/66
Vinyl esters	X
Furans	260/127
Phenolics	X
Phenol-formaldehyde	
Methyl appended silicone	X
PB	

Antimony Trichloride

Polymer	°F/°C
ABS	140/60
Acrylics	
CTFE	
ECTFE	100/38
ETFE	220/104
FEP	400/204
Polyamides	X
PAI	
PEEK	
PEI	
PES	
PFA	300/149
PTFE	460/238
PVDF	150/66
EHMWPE	140/60
HMWPE	140/60
PET	250/121
PPO	140/60
PPS	
PP	180/82
PVDC	160/71
PSF	
PVC Type 1	140/60
PVC Type 2	140/60
CPVC	200/93
CPE	220/104
PUR	
Acetals	
Polyesters	
Bisphenol A-fumurate	220/104
Halogenated, 50%	200/93
Hydrogenated-Bisphenol A	80/27
Isophthalic	160/71
PET	250/121
Epoxy	220/104
Vinyl esters	220/104
Furans	250/121
Phenolics	
Phenol-formaldehyde	
Methyl appended silicone	80/27
PB	

Aqua Regia 3:1

Polymer	°F/°C
ABS	X
Acrylics	X
CTFE	200/93
ECTFE	250/121
ETFE	210/99
FEP	400/204
Polyamides	X
PAI	
PEEK	X
PEI	
PES	
PFA	240/116
PTFE	450/232
PVDF	170/77
EHMWPE	130/54
HMWPE	130/54
PET	80/27
PPO	X
PPS	
PP	X
PVDC	120/49
PSF	X
PVC Type 1	X
PVC Type 2	X
CPVC	80/27
CPE	80/27
PUR	X
Acetals	X
Polyesters	
Bisphenol A-fumurate	X
Halogenated	X
Hydrogenated-Bisphenol A	X
Isophthalic	X
PET	80/27
Epoxy	X
Vinyl esters	X
Furans	X
Phenolics	X
Phenol-formaldehyde	
Methyl appended silicone	X
PB	

Arsenic Acid

Polymer	°F/°C
ABS, 80%	100/38
Acrylics	
CTFE	
ECTFE	300/149
ETFE	300/149
FEP	400/204
Polyamides	80/27
PAI	
PEEK	
PEI	
PES	
PFA	200/93
PTFE	450/232
PVDF	280/183
EHMWPE	140/60
HMWPE	140/60
PET	
PPO	60/15
PPS	
PP, 80%	210/99
PVDC	180/82
PSF	200/93
PVC Type 1, 80%	140/60
PVC Type 2, 80%	140/60
CPVC, 80%	200/93
CPE	250/121
PUR	
Acetals	
Polyesters	
Bisphenol A-fumurate	90/32
Halogenated, 19° Be	180/82
Hydrogenated-Bisphenol A	
Isophthalic	180/82
PET	
Epoxy	180/82
Vinyl esters	180/82
Furans	80/27
Phenolics	
Phenol-formaldehyde	
Methyl appended silicone	
PB	

Barium Carbonate

Polymer	°F/°C
ABS	140/60
Acrylics	80/27
CTFE	
ECTFE	300/149
ETFE	300/149
FEP	400/204
Polyamides	80/27
PAI	
PEEK	
PEI	
PES	
PFA	400/204
PTFE	450/232
PVDF	260/183
EHMWPE	140/60
HMWPE	140/60
PET	250/121
PPO	140/60
PPS	200/93
PP	200/93
PVDC	180/82
PSF	200/93
PVC Type 1	140/60
PVC Type 2	140/60
CPVC	200/93
CPE	250/121
PUR	
Acetals	
Polyesters	
Bisphenol A-fumurate	200/93
Halogenated	250/121
Hydrogenated-Bisphenol A	180/82
Isophthalic	160/71
PET	250/121
Epoxy	240/116
Vinyl esters	260/127
Furans	240/116
Phenolics	
Phenol-formaldehyde	
Methyl appended silicone	
PB	

Barium Chloride

Polymer	°F/°C
ABS	140/60
Acrylics	80/27
CTFE	
ECTFE	300/149
ETFE	300/149
FEP	400/204
Polyamides	250/121
PAI, 10%	200/93
PEEK	
PEI	
PES	
PFA	200/93
PTFE	450/232
PVDF	280/138
EHMWPE	140/60
HMWPE	140/60
PET	250/121
PPO, 25%	140/60
PPS	200/93
PP	220/104
PVDC	180/82
PSF, 30%	200/93
PVC Type 1	140/60
PVC Type 2	140/60
CPVC	180/82
CPE	250/121
PUR	90/32
Acetals	
Polyesters	
Bisphenol A-fumurate	220/104
Halogenated	250/121
Hydrogenated-Bisphenol A	200/93
Isophthalic	140/60
PET	250/121
Epoxy	250/121
Vinyl esters	200/93
Furans	260/127
Phenolics	
Phenol-formaldehyde	
Methyl appended silicone	
PB	

Barium Hydroxide

Polymer	°F/°C
ABS	140/60
Acrylics	80/27
CTFE	
ECTFE	300/149
ETFE	300/149
FEP	400/204
Polyamides	80/27
PAI	
PEEK	
PEI	
PES	
PFA	350/177
PTFE	450/232
PVDF	280/138
EHMWPE	140/60
HMWPE	140/60
PET	X
PPO	140/60
PPS, 10%	200/93
PP	200/93
PVDC	180/82
PSF	200/43
PVC Type 1	140/60
PVC Type 2	140/60
CPVC	200/93
CPE	250/121
PUR	90/32
Acetals	
Polyesters	
Bisphenol A-fumurate	150/66
Halogenated	X
Hydrogenated-Bisphenol A	
Isophthalic to 10%	X
PET	X
Epoxy, 10%	200/93
Vinyl esters	150/66
Furans	260/127
Phenolics	
Phenol-formaldehyde	
Methyl appended silicone	
PB	

Barium Nitrate

Polymer	°F/°C
ABS	
Acrylics	
CTFE	
ECTFE	80/27
ETFE	
FEP	400/204
Polyamides	80/27
PAI	
PEEK	
PEI	
PES	
PFA	
PTFE	450/232
PVDF	270/132
EHMWPE	
HMWPE	140/60
PET	170/77
PPO	
PPS	
PP	210/99
PVDC	
PSF	
PVC Type 1	140/60
PVC Type 2	140/60
CPVC	200/93
CPE	220/104
PUR	
Acetals	
Polyesters	
Bisphenol A-fumurate	
Halogenated	
Hydrogenated-Bisphenol A	
Isophthalic	
PET	170/77
Epoxy	200/93
Vinyl esters	150/66
Furans	
Phenolics	
Phenol-formaldehyde	
Methyl appended silicone	
PB	

Barium Sulfate

Polymer	°F/°C
ABS	140/60
Acrylics	80/27
CTFE	300/149
ECTFE	300/149
ETFE	300/149
FEP	400/204
Polyamides	80/27
PAI	
PEEK	
PEI	
PES	
PFA	330/166
PTFE	450/232
PVDF	280/138
EHMWPE	140/60
HMWPE	140/60
PET	X
PPO	140/60
PPS	200/93
PP	200/93
PVDC	180/82
PSF	200/93
PVC Type 1	140/60
PVC Type 2	140/60
CPVC	210/99
CPE	250/121
PUR	
Acetals	
Polyesters	
Bisphenol A-fumurate	220/104
Halogenated	180/82
Hydrogenated-Bisphenol A	
Isophthalic	160/71
PET	X
Epoxy	250/121
Vinyl esters	200/93
Furans	
Phenolics	
Phenol-formaldehyde	
Methyl appended silicone	
PB	

Barium Sulfide

Polymer	°F/°C
ABS	140/60
Acrylics	
CTFE	
ECTFE	300/149
ETFE	300/149
FEP	400/204
Polyamides	80/27
PAI	
PEEK	
PEI	
PES	
PFA	360/182
PTFE	450/232
PVDF	260/127
EHMWPE	140/60
HMWPE	140/60
PET	
PPO	140/60
PPS	220/104
PP	200/93
PVDC	150/66
PSF	
PVC Type 1	140/60
PVC Type 2	140/60
CPVC	200/93
CPE	220/104
PUR	90/32
Acetals	
Polyesters	
Bisphenol A-fumurate	140/60
Halogenated	X
Hydrogenated-Bisphenol A	
Isophthalic	90/32
PET	
Epoxy	300/149
Vinyl esters	180/82
Furans	260/127
Phenolics	
Phenol-formaldehyde	
Methyl appended silicone	
PB	

Benzaldehyde

Polymer	°F/°C
ABS	X
Acrylics	X
CTFE	200/93
ECTFE	150/66
ETFE	210/99
FEP[a]	400/204
Polyamides	150/66
PAI	200/93
PEEK	80/27
PEI	
PES	
PFA[b]	300/149
PTFE	450/232
PVDF	120/40
EHMWPE	X
HMWPE	X
PET	X
PPO	X
PPS	250/121
PP	80/27
PVDC	X
PSF	X
PVC Type 1	X
PVC Type 2	X
CPVC	X
CPE	80/27
PUR	X
Acetals	
Polyesters	
Bisphenol A-fumurate	X
Halogenated	X
Hydrogenated-Bisphenol A	X
Isophthalic	X
PET	X
Epoxy	X
Vinyl esters	X
Furans	260/127
Phenolics	60/15
Phenol-formaldehyde	
Methyl appended silicone	
PB	

[a] Corrodent will be absorbed.
[b] Material is subject to stress cracking.

Benzene	
Polymer	**°F/°C**
ABS	X
Acrylics	X
CTFE	
ECTFE	150/66
ETFE	210/99
FEP[a,b]	400/204
Polyamides	250/121
PAI	
PEEK	80/27
PEI	80/27
PES	80/27
PFA[b]	200/93
PTFE[b]	450/232
PVDF	150/66
EHMWPE	X
HMWPE	X
PET	X
PPO	X
PPS	300/149
PP	140/60
PVDC	80/27
PSF	X
PVC Type 1	X
PVC Type 2	X
CPVC	X
CPE	150/66
PUR	X
Acetals	90/32
Polyesters	
Bisphenol A-fumurate	X
Halogenated	90/32
Hydrogenated-Bisphenol A	X
Isophthalic	X
PET	X
Epoxy	180/82
Vinyl esters	X
Furans	260/127
Phenolics	160/77
Phenol-formaldehyde	
Methyl appended silicone	X
PB	

[a] Corrodent will be absorbed.
[b] Corrodent will permeate.

Benzene Sulfonic Acid, 10%

Polymer	°F/°C
ABS	80/27
Acrylics	
CTFE	
ECTFE	150/66
ETFE	210/99
FEP	400/204
Polyamides	X
PAI	X
PEEK	
PEI	X
PES	90/32
PFA	380/139
PTFE	450/232
PVDF	100/38
EHMWPE	140/60
HMWPE	90/32
PET	250/121
PPO	200/93
PPS	250/121
PP	180/82
PVDC	120/49
PSF	X
PVC Type 1	140/60
PVC Type 2	140/60
CPVC	180/82
CPE	220/104
PUR	X
Acetals	
Polyesters	
Bisphenol A-fumurate	200/93
Halogenated	120/49
Hydrogenated-Bisphenol A	
Isophthalic, 30%	180/82
PET	250/121
Epoxy	220/104
Vinyl esters	200/93
Furans	266/127
Phenolics	
Phenol-formaldehyde	160/71
Methyl appended silicone	
PB	

Benzoic Acid

Polymer	°F/°C
ABS	140/60
Acrylics	X
CTFE	
ECTFE	250/121
ETFE	200/93
FEP	400/204
Polyamides, 10%	200/93
PAI	
PEEK	170/77
PEI, 10%	90/32
PES	80/27
PFA	400/204
PTFE	450/121
PVDF	250/121
EHMWPE	140/60
HMWPE	140/60
PET	250/121
PPO	90/32
PPS	230/110
PP	190/88
PVDC	150/66
PSF	X
PVC Type 1	140/60
PVC Type 2	140/60
CPVC	90/32
CPE	250/121
PUR	X
Acetals	
Polyesters	
Bisphenol A-fumurate	200/93
Halogenated	250/121
Hydrogenated-Bisphenol A	210/99
Isophthalic	180/82
PET	250/121
Epoxy	200/93
Vinyl esters	180/82
Furans	260/127
Phenolics	
Phenol-formaldehyde	
Methyl appended silicone	
PB	90/32

Benzyl Alcohol

Polymer	°F/°C
ABS	X
Acrylics	
CTFE	200/93
ECTFE	300/149
ETFE	300/149
FEP	400/204
Polyamides	200/93
PAI	X
PEEK	
PEI	
PES	
PFA	300/149
PTFE	450/232
PVDF	280/138
EHMWPE	170/77
HMWPE	X
PET	80/27
PPO	X
PPS	200/93
PP	140/60
PVDC	
PSF	
PVC Type 1	X
PVC Type 2	X
CPVC	X
CPE	220/104
PUR	X
Acetals	
Polyesters	
Bisphenol A-fumurate	90/32
Halogenated	X
Hydrogenated-Bisphenol A	X
Isophthalic	X
PET	80/27
Epoxy	X
Vinyl esters	100/38
Furans	260/177
Phenolics	60/15
Phenol-formaldehyde	
Methyl appended silicone	
PB	

Benzyl Chloride

Polymer	°F/°C
ABS	X
Acrylics	
CTFE	110/48
ECTFE	300/149
ETFE	300/149
FEP	400/204
Polyamides	250/121
PAI	120/49
PEEK	
PEI	
PES	
PFA	300/149
PTFE	450/232
PVDF	280/138
EHMWPE	
HMWPE	
PET	250/121
PPO	X
PPS	300/149
PP	80/27
PVDC	80/27
PSF	X
PVC Type 1	X
PVC Type 2	
CPVC	X
CPE	80/27
PUR	X
Acetals	
Polyesters	
Bisphenol A-fumurate	X
Halogenated	X
Hydrogenated-Bisphenol A	X
Isophthalic	X
PET	250/121
Epoxy	80/27
Vinyl esters	90/32
Furans	260/127
Phenolics	
Phenol-formaldehyde	170/77
Methyl appended silicone	X
PB	

Black Liquor

Polymer	°F/°C
ABS	100/38
Acrylics	
CTFE	
ECTFE	300/149
ETFE	300/149
FEP	400/204
Polyamides	
PAI	
PEEK	
PEI	
PES	
PFA	400/204
PTFE	450/232
PVDF	260/127
EHMWPE	
HMWPE	
PET	
PPO	
PPS	
PP	140/60
PVDC	150/66
PSF	
PVC Type 1	140/60
PVC Type 2	140/60
CPVC	200/93
CPE	230/110
PUR	X
Acetals	
Polyesters	
Bisphenol A-fumurate, pH7	220/104
Halogenated	
Hydrogenated-Bisphenol A	
Isophthalic	X
PET	
Epoxy	210/99
Vinyl esters	180/82
Furans	X
Phenolics	
Phenol-formaldehyde	
Methyl appended silicone	
PB	

Borax

Polymer	°F/°C
ABS	140/60
Acrylics	90/32
CTFE	
ECTFE	300/149
ETFE	300/149
FEP	400/204
Polyamides	200/93
PAI	
PEEK	
PEI	
PES	
PFA	200/93
PTFE	450/232
PVDF	280/138
EHMWPE	140/60
HMWPE	140/60
PET	
PPO	140/60
PPS	210/99
PP	210/99
PVDC	150/66
PSF	200/93
PVC Type 1	140/60
PVC Type 2	140/60
CPVC	200/93
CPE	250/121
PUR	90/32
Acetals	
Polyesters	
Bisphenol A-fumurate	220/104
Halogenated	170/77
Hydrogenated-Bisphenol A	
Isophthalic	140/60
PET	
Epoxy	250/121
Vinyl esters	210/99
Furans	140/60
Phenolics	
Phenol-formaldehyde	
Methyl appended silicone	
PB	

Boric Acid

Polymer	°F/°C
ABS	140/60
Acrylics	80/27
CTFE	
ECTFE	300/149
ETFE	300/149
FEP	400/204
Polyamides	X
PAI	
PEEK	80/27
PEI	
PES	
PFA	300/149
PTFE	450/232
PVDF	280/138
EHMWPE	140/60
HMWPE	140/60
PET	200/93
PPO	140/60
PPS	210/99
PP	220/104
PVDC	170/77
PSF, 10%	200/93
PVC Type 1	140/60
PVC Type 2	140/60
CPVC	210/99
CPE	250/121
PUR	90/32
Acetals	
Polyesters	
Bisphenol A-fumurate	220/104
Halogenated	180/82
Hydrogenated-Bisphenol A	210/99
Isophthalic	180/82
PET	200/93
Epoxy	220/104
Vinyl esters	200/93
Furans	260/127
Phenolics	300/149
Phenol-formaldehyde	
Methyl appended silicone	390/189
PB	

Bromine Gas, Dry

Polymer	°F/°C
ABS	
Acrylics	
CTFE	100/38
ECTFE	X
ETFE	150/66
FEP[a]	200/93
Polyamides	
PAI	
PEEK	X
PEI	
PES	
PFA	
PTFE[a]	210/99
PVDF	210/99
EHMWPE	
HMWPE	X
PET	
PPO	
PPS	X
PP	X
PVDC	
PSF	
PVC Type 1	X
PVC Type 2	X
CPVC	X
CPE	X
PUR	
Acetals	
Polyesters	
Bisphenol A-fumurate	90/32
Halogenated	100/38
Hydrogenated-Bisphenol A	
Isophthalic	X
PET	
Epoxy	X
Vinyl esters	100/38
Furans	
Phenolics	
Phenol-formaldehyde	
Methyl appended silicone	
PB	

[a] Corrodent will penetrate.

Bromine Gas, Moist

Polymer	°F/°C
ABS	
Acrylics	
CTFE	210/99
ECTFE	
ETFE	
FEP	200/93
Polyamides	X
PAI	120/49
PEEK	X
PEI	
PES	
PFA	200/93
PTFE	250/121
PVDF	210/99
EHMWPE	
HMWPE	X
PET	
PPO	
PPS	X
PP	X
PVDC	
PSF	200/93
PVC Type 1	X
PVC Type 2	X
CPVC	X
CPE	X
PUR	
Acetals	
Polyesters	
Bisphenol A-fumurate	100/38
Halogenated	100/38
Hydrogenated-Bisphenol A	
Isophthalic	X
PET	
Epoxy	X
Vinyl esters	100/38
Furans	X
Phenolics	
Phenol-formaldehyde	
Methyl appended silicone	
PB	

Bromine, Liquid	
Polymer	**°F/°C**
ABS	X
Acrylics	X
CTFE	
ECTFE	150/66
ETFE	
FEP[a,b]	400/204
Polyamides	X
PAI	
PEEK	
PEI	
PES	
PFA[a,b]	300/149
PTFE[b]	450/232
PVDF	140/60
EHMWPE	X
HMWPE	X
PET	80/27
PPO	
PPS	X
PP	X
PVDC	X
PSF	
PVC Type 1	X
PVC Type 2	X
CPVC	X
CPE	X
PUR	X
Acetals	
Polyesters	
Bisphenol A-fumurate	X
Halogenated	X
Hydrogenated-Bisphenol A	X
Isophthalic	X
PET	80/27
Epoxy	X
Vinyl esters	
Furans	60/15
Phenolics	
Phenol-formaldehyde 3%	300/149
Methyl appended silicone	
PB	

[a] Corrodent will be absorbed.
[b] Corrodent will penetrate.

Bromine Water, Sat.

Polymer	°F/°C
ABS	X
Acrylics	
CTFE	
ECTFE	250/121
ETFE	
FEP	400/204
Polyamides	210/99
PAI	
PEEK	
PEI	
PES	
PFA	300/149
PTFE	450/232
PVDF	210/99
EHMWPE	X
HMWPE	
PET	X
PPO	
PPS	
PP	X
PVDC	X
PSF	
PVC Type 1	140/60
PVC Type 2	X
CPVC	
CPE	X
PUR	X
Acetals	
Polyesters	
Bisphenol A-fumurate	180/82
Halogenated	
Hydrogenated-Bisphenol A	
Isophthalic	
PET	X
Epoxy	100/38
Vinyl esters	
Furans	260/127
Phenolics	
Phenol-formaldehyde	
Methyl appended silicone	
PB	

Butadiene

Polymer	°F/°C
ABS	X
Acrylics	
CTFE	200/93
ECTFE	250/121
ETFE	250/121
FEP[a]	400/204
Polyamides	80/27
PAI	
PEEK	
PEI	
PES	
PFA[b]	400/204
PTFE[b]	450/232
PVDF	280/138
EHMWPE	X
HMWPE	X
PET	
PPO	X
PPS	100/38
PP	X
PVDC	150/66
PSF	X
PVC Type 1	140/60
PVC Type 2	X
CPVC	150/66
CPE	250/121
PUR	X
Acetals	
Polyesters	
Bisphenol A-fumurate	
Halogenated	
Hydrogenated-Bisphenol A	
Isophthalic	80/27
PET	
Epoxy	150/66
Vinyl esters	110/43
Furans	X
Phenolics	
Phenol-formaldehyde	
Methyl appended silicone	
PB	

[a] Corrodent will be absorbed.
[b] Corrodent will permeate.

Butyl Acetate

Polymer	°F/°C
ABS	X
Acrylics	X
CTFE	
ECTFE	150/66
ETFE	230/110
FEP	400/204
Polyamides	250/121
PAI	200/93
PEEK	
PEI	
PES	
PFA	400/204
PTFE	450/232
PVDF	140/60
EHMWPE	90/32
HMWPE	90/32
PET	250/121
PPO	X
PPS	250/121
PP	X
PVDC	120/49
PSF	X
PVC Type 1	X
PVC Type 2	X
CPVC	X
CPE	140/60
PUR	X
Acetals	
Polyesters	
Bisphenol A-fumurate	80/27
Halogenated	80/27
Hydrogenated-Bisphenol A	X
Isophthalic	X
PET	250/121
Epoxy	170/77
Vinyl esters	80/27
Furans	260/127
Phenolics	X
Phenol-formaldehyde	
Methyl appended silicone	
PB	

Butyl Alcohol

Polymer	°F/°C
ABS	X
Acrylics	80/27
CTFE	
ECTFE	300/149
ETFE	300/149
FEP	400/204
Polyamides	200/93
PAI	200/93
PEEK	
PEI	
PES	
PFA	400/204
PTFE	450/232
PVDF	280/138
EHMWPE	140/60
HMWPE	140/60
PET	100/38
PPO	
PPS	200/93
PP	200/93
PVDC	150/66
PSF	200/93
PVC Type 1	X
PVC Type 2	X
CPVC	140/60
CPE	220/104
PUR	X
Acetals	
Polyesters	
Bisphenol A-fumurate	80/27
Halogenated	100/38
Hydrogenated-Bisphenol A	
Isophthalic	80/27
PET	100/38
Epoxy	140/60
Vinyl esters	120/49
Furans	210/99
Phenolics	
Phenol-formaldehyde	
Methyl appended silicone	80/27
PB	

n-Butylamine

Polymer	°F/°C
ABS	
Acrylics	
CTFE	X
ECTFE	
ETFE	120/49
FEP[a]	400/204
Polyamides	200/93
PAI	200/93
PEEK	
PEI	
PES	
PFA[a]	400/204
PTFE	450/232
PVDF	X
EHMWPE	X
HMWPE	X
PET	
PPO	X
PPS	200/93
PP	90/32
PVDC	
PSF	X
PVC Type 1	X
PVC Type 2	X
CPVC	X
CPE	
PUR	
Acetals	X
Polyesters	
Bisphenol A-fumurate	X
Halogenated	X
Hydrogenated-Bisphenol A	X
Isophthalic	X
PET	
Epoxy	X
Vinyl esters	
Furans	X
Phenolics	
Phenol-formaldehyde	
Methyl appended silicone	
PB	

[a] Corrodent will be absorbed.

Butyl Ether

Polymer	°F/°C
ABS	
Acrylics	
CTFE	100/38
ECTFE	200/93
ETFE	200/93
FEP	400/204
Polyamides	210/99
PAI	200/93
PEEK	
PEI	
PES	
PFA	400/204
PTFE	450/232
PVDF	200/93
EHMWPE	
HMWPE	
PET	
PPO	X
PPS	210/99
PP	X
PVDC	
PSF	200/93
PVC Type 1	140/60
PVC Type 2	
CPVC	X
CPE	
PUR	
Acetals	
Polyesters	
Bisphenol A-fumurate	150/66
Halogenated	80/27
Hydrogenated-Bisphenol A	
Isophthalic	X
PET	
Epoxy	80/27
Vinyl esters	210/99
Furans	210/99
Phenolics	210/99
Phenol-formaldehyde	
Methyl appended silicone	
PB	

Butyric Acid

Polymer	°F/°C
ABS	X
Acrylics	X
CTFE	
ECTFE	250/121
ETFE	250/121
FEP	400/204
Polyamides	X
PAI	
PEEK	
PEI	
PES	
PFA	400/204
PTFE	450/232
PVDF	210/99
EHMWPE	130/54
HMWPE	X
PET	250/121
PPO	
PPS	250/121
PP	180/82
PVDC	80/27
PSF	130/54
PVC Type 1	X
PVC Type 2	X
CPVC <1%	140/60
CPE	230/110
PUR	X
Acetals	
Polyesters	
Bisphenol A-fumurate, 50%	220/104
Halogenated, 20%	200/93
Hydrogenated-Bisphenol A	X
Isophthalic, 20%	120/49
PET	250/121
Epoxy	210/99
Vinyl esters	130/54
Furans	260/127
Phenolics	
Phenol-formaldehyde	260/127
Methyl appended silicone	
PB	

Calcium Bisulfide

Polymer	°F/°C
ABS	
Acrylics	
CTFE	
ECTFE	300/149
ETFE	300/149
FEP	400/204
Polyamides	
PAI	
PEEK	
PEI	
PES	
PFA	400/204
PTFE	450/232
PVDF	260/138
EHMWPE	140/60
HMWPE	140/60
PET	
PPO	
PPS	
PP	210/99
PVDC	
PSF	
PVC Type 1	140/60
PVC Type 2	140/60
CPVC	200/93
CPE	250/121
PUR	
Acetals	
Polyesters	
Bisphenol A-fumurate	
Halogenated	X
Hydrogenated-Bisphenol A	120/49
Isophthalic	160/71
PET	
Epoxy	
Vinyl esters	
Furans	
Phenolics	
Phenol-formaldehyde	
Methyl appended silicone	400/204
PB	

Calcium Bisulfite

Polymer	°F/°C
ABS	140/60
Acrylics	90/32
CTFE	
ECTFE	300/149
ETFE	
FEP	400/204
Polyamides	140/60
PAI	
PEEK	
PEI	
PES	
PFA	400/204
PTFE	450/232
PVDF	260/138
EHMWPE	80/27
HMWPE	140/60
PET	X
PPO	80/27
PPS	200/93
PP	210/99
PVDC	80/27
PSF	200/93
PVC Type 1	140/60
PVC Type 2	140/60
CPVC	210/99
CPE	250/121
PUR	90/32
Acetals	X
Polyesters	
Bisphenol A-fumurate	180/82
Halogenated	150/66
Hydrogenated-Bisphenol A	
Isophthalic	150/66
PET	X
Epoxy	270/132
Vinyl esters	260/127
Furans	260/127
Phenolics	
Phenol-formaldehyde	
Methyl appended silicone	
PB	

Calcium Carbonate

Polymer	°F/°C
ABS	100/38
Acrylics	80/27
CTFE	
ECTFE	300/149
ETFE	300/149
FEP	400/204
Polyamides	250/121
PAI	
PEEK	80/27
PEI	
PES	
PFA	390/199
PTFE	450/232
PVDF	280/138
EHMWPE	140/60
HMWPE	140/60
PET	250/121
PPO	140/60
PPS	300/149
PP	250/121
PVDC	180/82
PSF	
PVC Type 1	140/60
PVC Type 2	140/60
CPVC	210/99
CPE	250/121
PUR	
Acetals	
Polyesters	
Bisphenol A-fumurate	210/99
Halogenated	210/99
Hydrogenated-Bisphenol A	
Isophthalic	160/71
PET	250/121
Epoxy	300/149
Vinyl esters	180/82
Furans	
Phenolics	
Phenol-formaldehyde	
Methyl appended silicone	
PB	

Calcium Chlorate

Polymer	°F/°C
ABS	140/60
Acrylics	80/27
CTFE	
ECTFE	300/149
ETFE	300/149
FEP	400/204
Polyamides	
PAI	
PEEK	
PEI	
PES	
PFA	380/193
PTFE	450/232
PVDF	280/132
EHMWPE	140/60
HMWPE	140/60
PET	
PPO	
PPS	
PP	220/104
PVDC	160/71
PSF	
PVC Type 1	140/60
PVC Type 2	140/60
CPVC	200/93
CPE	150/66
PUR	
Acetals, 5%	90/32
Polyesters	
Bisphenol A-fumurate	200/93
Halogenated	250/121
Hydrogenated-Bisphenol A	210/99
Isophthalic	160/71
PET	
Epoxy	200/93
Vinyl esters	260/127
Furans	
Phenolics	300/149
Phenol-formaldehyde	
Methyl appended silicone	
PB	

Calcium Chloride

Polymer	°F/°C
ABS	140/60
Acrylics	80/27
CTFE	300/149
ECTFE	300/149
ETFE	300/149
FEP	400/204
Polyamides	230/110
PAI, 10%	200/93
PEEK	80/27
PEI	
PES	
PFA	400/204
PTFE	450/232
PVDF	280/138
EHMWPE	140/60
HMWPE	140/60
PET	250/121
PPO	140/60
PPS	300/149
PP	220/104
PVDC	180/82
PSF	200/93
PVC Type 1	140/60
PVC Type 2	140/60
CPVC	180/82
CPE	150/66
PUR	80/27
Acetals	X
Polyesters	
Bisphenol A-fumurate	220/104
Halogenated	250/121
Hydrogenated-Bisphenol A	210/99
Isophthalic	180/82
PET	250/121
Epoxy, 37.5%	190/88
Vinyl esters	180/82
Furans	260/127
Phenolics	200/93
Phenol-formaldehyde	
Methyl appended silicone	300/149
PB	80/27

Calcium Hydroxide, 10%

Polymer	°F/°C
ABS	
Acrylics	
CTFE	
ECTFE	300/149
ETFE	300/149
FEP	400/204
Polyamides	150/66
PAI	
PEEK	90/32
PEI	
PES	
PFA	400/204
PTFE	450/232
PVDF	270/132
EHMWPE	140/60
HMWPE	140/60
PET	250/121
PPO	
PPS	300/149
PP	220/104
PVDC	160/71
PSF	100/38
PVC Type 1	140/60
PVC Type 2	140/60
CPVC	200/93
CPE	
PUR	90/32
Acetals	80/27
Polyesters	
Bisphenol A-fumurate	180/82
Halogenated	170/77
Hydrogenated-Bisphenol A	
Isophthalic	160/71
PET	250/121
Epoxy	210/99
Vinyl esters	180/82
Furans	260/127
Phenolics	X
Phenol-formaldehyde	
Methyl appended silicone, 30%	200/99
PB	

Calcium Hydroxide, Sat.

Polymer	°F/°C
ABS	140/60
Acrylics	80/27
CTFE	
ECTFE	300/149
ETFE	300/149
FEP	400/204
Polyamides	150/66
PAI	
PEEK	90/32
PEI	
PES	
PFA	400/204
PTFE	450/232
PVDF	280/138
EHMWPE	140/60
HMWPE	140/60
PET	250/121
PPO	140/60
PPS	300/149
PP	220/104
PVDC	180/82
PSF	100/38
PVC Type 1	140/60
PVC Type 2	140/60
CPVC	200/93
CPE	250/121
PUR	90/32
Acetals	
Polyesters	
Bisphenol A-fumurate	160/71
Halogenated	X
Hydrogenated-Bisphenol A	
Isophthalic	160/71
PET	250/121
Epoxy	180/82
Vinyl esters, 25%	180/82
Furans	260/127
Phenolics	X
Phenol-formaldehyde	
Methyl appended silicone	400/204
PB	

Calcium Hypochlorite, 30%

Polymer	°F/°C
ABS	
Acrylics	80/27
CTFE	
ECTFE	300/149
ETFE	300/149
FEP	400/204
Polyamides	X
PAI	X
PEEK	
PEI	
PES	
PFA	400/204
PTFE	450/232
PVDF	200/93
EHMWPE	140/60
HMWPE	100/38
PET	
PPO	
PPS	
PP	210/99
PVDC	120/49
PSF	200/93
PVC Type 1	140/60
PVC Type 2	140/60
CPVC	200/93
CPE	
PUR	X
Acetals	X
Polyesters	
Bisphenol A-fumurate	210/99
Halogenated	170/97
Hydrogenated-Bisphenol A, 10%	180/82
Isophthalic	120/49
PET	
Epoxy, 20%	150/66
Vinyl esters, 20%	160/71
Furans	X
Phenolics	X
Phenol-formaldehyde	
Methyl appended silicone	
PB	

Calcium Hypochlorite, Sat.

Polymer	°F/°C
ABS	140/60
Acrylics	80/27
CTFE	
ECTFE	300/149
ETFE	300/149
FEP	400/204
Polyamides	80/27
PAI	X
PEEK	
PEI	
PES	
PFA	400/204
PTFE	450/232
PVDF	280/138
EHMWPE	140/60
HMWPE	140/60
PET	250/121
PPO, 20%	140/60
PPS	
PP	210/99
PVDC	120/49
PSF	200/93
PVC Type 1	140/60
PVC Type 2	140/60
CPVC	200/93
CPE	200/93
PUR	X
Acetals	X
Polyesters	
Bisphenol A-fumurate, 20%	80/27
Halogenated, 20%	80/27
Hydrogenated-Bisphenol A, 10%	180/82
Isophthalic, 10%	120/49
PET	250/121
Epoxy, 70%	150/66
Vinyl esters	180/82
Furans, to 20%	240/116
Phenolics	
Phenol-formaldehyde	
Methyl appended silicone	
PB	90/32

Calcium Nitrate

Polymer	°F/°C
ABS	140/60
Acrylics	
CTFE	
ECTFE	300/149
ETFE	300/149
FEP	400/204
Polyamides	X
PAI	
PEEK	
PEI	
PES	
PFA	400/204
PTFE	450/232
PVDF	260/138
EHMWPE	140/60
HMWPE	140/60
PET	170/77
PPO	140/60
PPS	
PP	210/99
PVDC	150/66
PSF	200/93
PVC Type 1	140/60
PVC Type 2	140/60
CPVC	200/93
CPE	250/121
PUR	90/32
Acetals	X
Polyesters	
Bisphenol A-fumurate	220/104
Halogenated	220/104
Hydrogenated-Bisphenol A	
Isophthalic	160/71
PET	170/77
Epoxy	250/121
Vinyl esters	210/99
Furans	260/127
Phenolics	
Phenol-formaldehyde	
Methyl appended silicone	
PB	

Calcium Oxide

Polymer	°F/°C
ABS	140/60
Acrylics	
CTFE	
ECTFE	300/149
ETFE	270/132
FEP	400/204
Polyamides	80/27
PAI	
PEEK	
PEI	
PES	
PFA	400/204
PTFE	450/232
PVDF	250/121
EHMWPE	140/60
HMWPE	140/60
PET	80/27
PPO	
PPS	
PP	220/104
PVDC	180/82
PSF	
PVC Type 1	140/60
PVC Type 2	140/60
CPVC	200/93
CPE	250/121
PUR	
Acetals	
Polyesters	
Bisphenol A-fumurate	
Halogenated	150/66
Hydrogenated-Bisphenol A	
Isophthalic	150/66
PET	80/27
Epoxy	
Vinyl esters	160/71
Furans	
Phenolics	
Phenol-formaldehyde	
Methyl appended silicone	
PB	

Calcium Sulfate

Polymer	°F/°C
ABS, 25%	140/60
Acrylics	80/27
CTFE	300/149
ECTFE	300/149
ETFE	300/149
FEP	400/204
Polyamides	80/27
PAI	
PEEK	
PEI	
PES	
PFA	400/204
PTFE	450/232
PVDF	280/138
EHMWPE	140/60
HMWPE	140/60
PET	170/77
PPO	140/60
PPS	
PP	220/104
PVDC	180/82
PSF	200/93
PVC Type 1	140/60
PVC Type 2	140/60
CPVC	200/93
CPE	250/121
PUR	90/32
Acetals	X
Polyesters	
Bisphenol A-fumurate	220/104
Halogenated	250/121
Hydrogenated-Bisphenol A	
Isophthalic	160/71
PET	170/77
Epoxy	250/121
Vinyl esters	250/121
Furans	260/127
Phenolics	
Phenol-formaldehyde	
Methyl appended silicone	
PB	

Caprylic Acid

Polymer	°F/°C
ABS	
Acrylics	
CTFE	
ECTFE	220/104
ETFE	210/99
FEP	400/204
Polyamides	250/121
PAI	
PEEK	
PEI	
PES	
PFA	390/199
PTFE	450/232
PVDF	220/104
EHMWPE	
HMWPE	
PET	250/121
PPO	
PPS	
PP	140/60
PVDC	90/32
PSF	
PVC Type 1	120/49
PVC Type 2	
CPVC	180/82
CPE	220/104
PUR	
Acetals	
Polyesters	
Bisphenol A-fumurate	160/71
Halogenated	140/60
Hydrogenated-Bisphenol A	
Isophthalic	160/71
PET	250/121
Epoxy	X
Vinyl esters	220/104
Furans	250/121
Phenolics	
Phenol-formaldehyde	
Methyl appended silicone	
PB	

Carbon Dioxide, Dry

Polymer	°F/°C
ABS	90/32
Acrylics	
CTFE	300/149
ECTFE	300/149
ETFE	300/149
FEP	400/204
Polyamides	80/27
PAI	
PEEK	80/27
PEI	
PES	
PFA	400/204
PTFE	450/232
PVDF	280/138
EHMWPE	140/60
HMWPE	140/60
PET	
PPO	80/27
PPS	200/93
PP	220/104
PVDC	180/82
PSF	
PVC Type 1	140/60
PVC Type 2	140/60
CPVC	210/99
CPE	250/121
PUR	80/27
Acetals	
Polyesters	
Bisphenol A-fumurate	350/177
Halogenated	250/121
Hydrogenated-Bisphenol A	
Isophthalic	160/71
PET	
Epoxy	200/93
Vinyl esters	200/93
Furans	90/32
Phenolics	300/149
Phenol-formaldehyde	
Methyl appended silicone	
PB	80/27

Carbon Dioxide, Wet

Polymer	°F/°C
ABS	140/60
Acrylics	90/32
CTFE	300/149
ECTFE	300/149
ETFE	300/149
FEP	400/204
Polyamides	80/27
PAI	
PEEK	
PEI	
PES	
PFA	400/204
PTFE	450/232
PVDF	280/138
EHMWPE	140/60
HMWPE	140/60
PET	
PPO	140/60
PPS	
PP	220/104
PVDC	170/77
PSF	
PVC Type 1	140/60
PVC Type 2	140/60
CPVC	160/71
CPE	250/121
PUR	80/27
Acetals	
Polyesters	
Bisphenol A-fumurate	210/99
Halogenated	250/121
Hydrogenated-Bisphenol A	
Isophthalic	150/71
PET	
Epoxy	250/121
Vinyl esters	220/104
Furans	80/27
Phenolics	300/149
Phenol-formaldehyde	
Methyl appended silicone	
PB	90/32

Carbon Disulfide

Polymer	°F/°C
ABS	X
Acrylics	80/27
CTFE	
ECTFE	80/27
ETFE	150/66
FEP[a]	400/204
Polyamides	80/27
PAI	
PEEK	
PEI	
PES	
PFA[a,b]	390/199
PTFE[b]	450/232
PVDF	80/27
EHMWPE	X
HMWPE	X
PET	80/27
PPO	60/15
PPS	200/93
PP	X
PVDC	90/32
PSF	X
PVC Type 1	X
PVC Type 2	X
CPVC	X
CPE	X
PUR	
Acetals	80/27
Polyesters	
Bisphenol A-fumurate	X
Halogenated	X
Hydrogenated-Bisphenol A	X
Isophthalic	X
PET	80/27
Epoxy	100/38
Vinyl esters	X
Furans	260/127
Phenolics	180/82
Phenol-formaldehyde	
Methyl appended silicone	X
PB	

[a] Corrodent will permeate.
[b] Corrodent will be absorbed.

Carbon Monoxide

Polymer	°F/°C
ABS	140/60
Acrylics	
CTFE	
ECTFE	150/66
ETFE	300/149
FEP	400/204
Polyamides	80/27
PAI	
PEEK	
PEI	
PES	
PFA	400/204
PTFE	450/232
PVDF	280/138
EHMWPE	140/60
HMWPE	140/60
PET	210/99
PPO	140/60
PPS	
PP	220/104
PVDC	180/82
PSF	
PVC Type 1	140/60
PVC Type 2	140/60
CPVC	210/99
CPE	250/121
PUR	
Acetals	
Polyesters	
Bisphenol A-fumurate	350/177
Halogenated	170/77
Hydrogenated-Bisphenol A	
Isophthalic	160/71
PET	210/99
Epoxy	200/93
Vinyl esters	350/177
Furans	
Phenolics	
Phenol-formaldehyde	
Methyl appended silicone	400/204
PB	

Carbon Tetrachloride

Polymer	°F/°C
ABS	X
Acrylics	X
CTFE	
ECTFE	300/149
ETFE	270/132
FEP[a,b,c]	400/204
Polyamides	250/121
PAI	
PEEK	80/27
PEI	80/27
PES	80/27
PFA[a,b,c]	380/193
PTFE[a]	450/232
PVDF	280/138
EHMWPE	X
HMWPE	X
PET	250/121
PPO	X
PPS	120/49
PP	X
PVDC	140/60
PSF	X
PVC Type 1	X
PVC Type 2	X
CPVC	X
CPE	200/93
PUR	X
Acetals	80/27
Polyesters	
Bisphenol A-fumurate	110/43
Halogenated	120/49
Hydrogenated-Bisphenol A	X
Isophthalic	X
PET	250/121
Epoxy	
Vinyl esters	170/77
Furans	260/127
Phenolics	200/93
Phenol-formaldehyde	
Methyl appended silicone	
PB	

[a] Material is subject to stress cracking.
[b] Corrodent will be absorbed.
[c] Corrodent will permeate.

Carbonic Acid

Polymer	°F/°C
ABS	140/60
Acrylics	80/27
CTFE	
ECTFE	300/149
ETFE	300/149
FEP	400/204
Polyamides	100/38
PAI	
PEEK	80/27
PEI	
PES	
PFA	390/199
PTFE	450/232
PVDF	280/138
EHMWPE	140/60
HMWPE	140/60
PET	210/99
PPO	140/60
PPS	
PP	220/104
PVDC	180/82
PSF	200/93
PVC Type 1	140/60
PVC Type 2	140/60
CPVC	200/93
CPE	250/121
PUR	
Acetals	80/27
Polyesters	
Bisphenol A-fumurate	90/32
Halogenated	160/71
Hydrogenated-Bisphenol A	
Isophthalic	160/71
PET	210/99
Epoxy	200/93
Vinyl esters	120/49
Furans	
Phenolics	200/93
Phenol-formaldehyde	
Methyl appended silicone	400/204
PB	

Caustic Potash (Potassium Hydroxide)

Polymer	°F/°C
ABS	140/60
Acrylics	
CTFE	
ECTFE	300/149
ETFE, 10–50%	210/99
FEP	400/204
Polyamides, to 50%	120/49
PAI	
PEEK	
PEI, 10%	80/27
PES	
PFA	390/199
PTFE	450/232
PVDF	180/82
EHMWPE	140/60
HMWPE	140/60
PET, 50%	X
PPO, 50%	90/32
PPS	
PP	170/77
PVDC, 50%	80/27
PSF, 35%	130/54
PVC Type 1	140/60
PVC Type 2	140/60
CPVC, 50%	180/82
CPE	250/121
PUR, to 50%	80/27
Acetals	
Polyesters	
Bisphenol A-fumurate, 10–50%	170/77
Halogenated, 10–50%	120/49
Hydrogenated-Bisphenol A	
Isophthalic	X
PET, 50%	X
Epoxy	200/93
Vinyl esters, 10–50%	150/66
Furans	260/127
Phenolics	
Phenol-formaldehyde	
Methyl appended silicone	
PB	

Cellosolve

Polymer	°F/°C
ABS	X
Acrylics	80/27
CTFE	200/93
ECTFE	300/149
ETFE	300/149
FEP	400/204
Polyamides	250/121
PAI	200/93
PEEK	
PEI	
PES	
PFA	380/193
PTFE	450/232
PVDF	280/138
EHMWPE	X
HMWPE	80/27
PET	X
PPO	X
PPS	200/93
PP	200/93
PVDC	80/27
PSF	X
PVC Type 1	X
PVC Type 2	X
CPVC	X
CPE	220/104
PUR	
Acetals	80/27
Polyesters	
Bisphenol A-fumurate	140/60
Halogenated	80/27
Hydrogenated-Bisphenol A	
Isophthalic	X
PET	X
Epoxy	140/60
Vinyl esters	140/60
Furans	240/116
Phenolics	
Phenol-formaldehyde	
Methyl appended silicone	
PB	

Chlorine Gas, Dry

Polymer	°F/°C
ABS	140/60
Acrylics	
CTFE	X
ECTFE	150/66
ETFE	210/99
FEP	X
Polyamides	X
PAI	
PEEK	80/27
PEI	
PES	
PFA	X
PTFE	
PVDF	240/116
EHMWPE	80/27
HMWPE, 10%	80/27
PET	80/27
PPO	X
PPS	X
PP	X
PVDC	80/27
PSF	X
PVC Type 1	140/60
PVC Type 2	140/60
CPVC	X
CPE	100/38
PUR	X
Acetals	X
Polyesters	
Bisphenol A-fumurate	350/177
Halogenated	200/93
Hydrogenated-Bisphenol A	210/99
Isophthalic	160/71
PET	80/27
Epoxy	150/66
Vinyl esters	240/116
Furans	260/127
Phenolics	
Phenol-formaldehyde	160/171
Methyl appended silicone	
PB	

Chlorine Gas, Wet

Polymer	°F/°C
ABS	140/60
Acrylics	
CTFE	
ECTFE	250/121
ETFE	250/121
FEP[a]	400/204
Polyamides	X
PAI	
PEEK	
PEI	
PES	
PFA[a]	400/204
PTFE[a]	450/232
PVDF, 10%	250/121
EHMWPE, 10%	120/49
HMWPE	X
PET	80/27
PPO	80/27
PPS	X
PP	X
PVDC	80/27
PSF	X
PVC Type 1	X
PVC Type 2	X
CPVC	X
CPE	80/27
PUR	X
Acetals	80/27
Polyesters	
Bisphenol A-fumurate	200/93
Halogenated	220/104
Hydrogenated-Bisphenol A	210/99
Isophthalic	160/71
PET	80/27
Epoxy	X
Vinyl esters	250/121
Furans	260/127
Phenolics	
Phenol-formaldehyde	150/66
Methyl appended silicone	
PB	X

[a] Corrodent will permeate.

Chlorine Liquid

Polymer	°F/°C
ABS	X
Acrylics	
CTFE	
ECTFE	250/121
ETFE	
FEP[a]	400/204
Polyamides	X
PAI	
PEEK	X
PEI	
PES	
PFA[a]	X
PTFE	
PVDF	210/99
EHMWPE	X
HMWPE	X
PET	
PPO	X
PPS	200/93
PP	X
PVDC	X
PSF	X
PVC Type 1	X
PVC Type 2	X
CPVC	X
CPE	X
PUR	
Acetals	
Polyesters	
Bisphenol A-fumurate	X
Halogenated	X
Hydrogenated-Bisphenol A	
Isophthalic	X
PET	
Epoxy	
Vinyl esters	X
Furans	X
Phenolics	X
Phenol-formaldehyde	
Methyl appended silicone	
PB	

[a] Corrodent will be absorbed.

Chloracetic Acid

Polymer	°F/°C
ABS	X
Acrylics	X
CTFE	120/49
ECTFE	250/121
ETFE, 50%	230/110
FEP	400/204
Polyamides	X
PAI	
PEEK	
PEI	
PES	
PFA, 50%	400/204
PTFE	450/232
PVDF	200/93
EHMWPE	X
HMWPE	X
PET	80/27
PPO	
PPS	200/93
PP	180/82
PVDC	120/49
PSF	X
PVC Type 1	140/60
PVC Type 2	X
CPVC	X
CPE	220/104
PUR	X
Acetals	X
Polyesters	
Bisphenol A-fumurate	80/27
Halogenated, 25%	90/32
Hydrogenated-Bisphenol A	
Isophthalic, to 25%	150/66
PET	80/27
Epoxy	X
Vinyl esters, 25%	200/93
Furans	240/116
Phenolics	X
Phenol-formaldehyde	
Methyl appended silicone	
PB	

Chlorobenzene

Polymer	°F/°C
ABS	X
Acrylics	X
CTFE	
ECTFE	150/66
ETFE	210/99
FEP[a]	400/204
Polyamides	250/121
PAI DRY	200/93
PEEK	200/93
PEI	
PES	X
PFA[a]	390/199
PTFE[a]	450/232
PVDF	220/104
EHMWPE	X
HMWPE	X
PET	X
PPO	X
PPS	200/93
PP	X
PVDC	80/27
PSF	X
PVC Type 1	X
PVC Type 2	X
CPVC	X
CPE	150/66
PUR	X
Acetals, dry	80/27
Polyesters	
Bisphenol A-fumurate	X
Halogenated	X
Hydrogenated-Bisphenol A	X
Isophthalic	X
PET	X
Epoxy	190/88
Vinyl esters	110/43
Furans	260/127
Phenolics	250/121
Phenol-formaldehyde	
Methyl appended silicone	X
PB	

[a] Corrodent will permeate.

Chloroform

Polymer	°F/°C
ABS	X
Acrylics	X
CTFE	
ECTFE	250/121
ETFE	230/110
FEP[a]	400/204
Polyamides	200/93
PAI	120/49
PEEK	80/27
PEI	
PES	
PFA[a]	390/199
PTFE[a]	450/232
PVDF	250/121
EHMWPE	80/27
HMWPE	X
PET	230/110
PPO	X
PPS	150/66
PP	X
PVDC	X
PSF	X
PVC Type 1	X
PVC Type 2	X
CPVC	X
CPE	80/27
PUR	X
Acetals	X
Polyesters	
Bisphenol A-fumurate	X
Halogenated	X
Hydrogenated-Bisphenol A	X
Isophthalic	X
PET	230/110
Epoxy	110/43
Vinyl esters	X
Furans	240/116
Phenolics	160/171
Phenol-formaldehyde	
Methyl appended silicone	
PB	

[a] Corrodent will permeate.

Chlorosulfonic Acid

Polymer	°F/°C
ABS	X
Acrylics	X
CTFE	
ECTFE	80/27
ETFE	80/27
FEP[a]	400/204
Polyamides	X
PAI	
PEEK	80/27
PEI	
PES	X
PFA[a]	390/199
PTFE[a]	450/232
PVDF	110/43
EHMWPE	X
HMWPE	X
PET	X
PPO	X
PPS	X
PP	X
PVDC	X
PSF	X
PVC Type 1	X
PVC Type 2	X
CPVC	X
CPE	X
PUR	X
Acetals	X
Polyesters	
Bisphenol A-fumurate	X
Halogenated	X
Hydrogenated-Bisphenol A	
Isophthalic	X
PET	X
Epoxy	X
Vinyl esters	X
Furans	260/127
Phenolics	130/54
Phenol-formaldehyde	
Methyl appended silicone	X
PB	

[a] Corrodent will be absorbed.

Chromic Acid, 10%

Polymer	°F/°C
ABS	90/32
Acrylics	X
CTFE	
ECTFE	250/121
ETFE	150/66
FEP	400/204
Polyamides	X
PAI	200/93
PEEK	90/32
PEI	
PES	X
PFA	380/193
PTFE	450/232
PVDF	220/104
EHMWPE	140/60
HMWPE	140/60
PET	250/121
PPO	X
PPS	200/93
PP	140/60
PVDC	180/82
PSF, 12%	140/60
PVC Type 1	140/60
PVC Type 2	140/60
CPVC	210/99
CPE	250/121
PUR	X
Acetals	80/27
Polyesters	
Bisphenol A-fumurate	X
Halogenated	180/82
Hydrogenated-Bisphenol A	
Isophthalic	X
PET	250/121
Epoxy	110/43
Vinyl esters	150/66
Furans	260/127
Phenolics	X
Phenol-formaldehyde	
Methyl appended silicone	
PB	X

Chromic Acid, 50%

Polymer	°F/°C
ABS	X
Acrylics	X
CTFE	
ECTFE	250/121
ETFE	150/66
FEP[a]	400/204
Polyamides	X
PAI	
PEEK	200/93
PEI	
PES	X
PFA[a]	380/193
PTFE	450/232
PVDF	250/121
EHMWPE	90/32
HMWPE	90/32
PET	250/121
PPO	X
PPS	200/93
PP	150/66
PVDC	180/82
PSF	X
PVC Type 1	X
PVC Type 2	X
CPVC	210/99
CPE	200/93
PUR	X
Acetals	80/27
Polyesters	
Bisphenol A-fumurate	X
Halogenated	140/60
Hydrogenated-Bisphenol A	X
Isophthalic	X
PET	250/121
Epoxy	X
Vinyl esters	X
Furans	X
Phenolics	X
Phenol-formaldehyde	
Methyl appended silicone	
PB	X

[a] Corrodent will be absorbed.

Citric Acid, 15%

Polymer	°F/°C
ABS	140/60
Acrylics	80/27
CTFE	
ECTFE	300/149
ETFE	120/49
FEP	400/204
Polyamides	200/93
PAI	
PEEK	170/77
PEI	
PES	
PFA	390/199
PTFE	450/232
PVDF	250/121
EHMWPE	140/60
HMWPE	140/60
PET	250/121
PPO	120/49
PPS	250/121
PP	220/104
PVDC	180/82
PSF	100/38
PVC Type 1	140/60
PVC Type 2	140/60
CPVC	180/82
CPE	250/121
PUR	
Acetals	80/27
Polyesters	
Bisphenol A-fumurate	220/104
Halogenated	250/121
Hydrogenated-Bisphenol A	210/99
Isophthalic	160/71
PET	250/121
Epoxy	190/88
Vinyl esters	210/99
Furans	250/121
Phenolics	160/71
Phenol-formaldehyde	
Methyl appended silicone	
PB	

Citric Acid, Conc.

Polymer	°F/°C
ABS, 25%	140/60
Acrylics	80/27
CTFE	
ECTFE	300/149
ETFE	
FEP	400/204
Polyamides	200/93
PAI	
PEEK	170/77
PEI	250/121
PES	80/27
PFA	370/188
PTFE	450/232
PVDF	250/121
EHMWPE	140/60
HMWPE	140/60
PET	170/77
PPO	90/32
PPS	250/121
PP	220/104
PVDC	180/82
PSF, 40%	80/27
PVC Type 1	140/60
PVC Type 2	140/60
CPVC	180/82
CPE	250/121
PUR	
Acetals	80/27
Polyesters	
Bisphenol A-fumurate	220/104
Halogenated	250/121
Hydrogenated-Bisphenol A	210/99
Isophthalic	200/93
PET	170/77
Epoxy, 32%	190/88
Vinyl esters	200/93
Furans	250/121
Phenolics	160/71
Phenol-formaldehyde	
Methyl appended silicone	
PB	

Copper Acetate

Polymer	°F/°C
ABS	
Acrylics	
CTFE	
ECTFE	
ETFE	
FEP	400/204
Polyamides	
PAI	
PEEK	
PEI	
PES	
PFA	290/143
PTFE	450/232
PVDF	250/121
EHMWPE	
HMWPE	
PET	
PPO	
PPS	300/149
PP	80/27
PVDC	
PSF	
PVC Type 1	80/27
PVC Type 2	
CPVC	200/93
CPE	
PUR	
Acetals	
Polyesters	
Bisphenol A-fumurate	180/82
Halogenated	210/99
Hydrogenated-Bisphenol A	210/99
Isophthalic	180/71
PET	
Epoxy	200/93
Vinyl esters	210/99
Furans	260/127
Phenolics	160/71
Phenol-formaldehyde	
Methyl appended silicone	
PB	

Copper Carbonate

Polymer	°F/°C
ABS	
Acrylics	
CTFE	
ECTFE	150/66
ETFE	
FEP	400/204
Polyamides	80/27
PAI	
PEEK	
PEI	
PES	
PFA	400/204
PTFE	450/232
PVDF	250/121
EHMWPE	
HMWPE	
PET	
PPO	
PPS	
PP	200/93
PVDC	180/82
PSF	
PVC Type 1	140/60
PVC Type 2	140/60
CPVC	200/93
CPE	250/121
PUR	
Acetals	
Polyesters	
Bisphenol A-fumurate	
Halogenated	
Hydrogenated-Bisphenol A	
Isophthalic	160/83
PET	
Epoxy	150/66
Vinyl esters	
Furans	
Phenolics	
Phenol-formaldehyde	
Methyl appended silicone	
PB	

Copper Chloride

Polymer	°F/°C
ABS	140/60
Acrylics	80/27
CTFE	
ECTFE	300/149
ETFE	300/149
FEP	400/204
Polyamides	X
PAI	
PEEK	
PEI	
PES	
PFA	400/204
PTFE	450/232
PVDF	280/138
EHMWPE	140/60
HMWPE	140/60
PET	170/77
PPO	140/60
PPS	220/104
PP	200/93
PVDC	180/82
PSF	
PVC Type 1	140/60
PVC Type 2	140/60
CPVC	210/99
CPE	250/121
PUR	90/32
Acetals	
Polyesters	
Bisphenol A-fumurate	220/104
Halogenated	250/121
Hydrogenated-Bisphenol A	210/99
Isophthalic	180/82
PET	170/77
Epoxy	250/121
Vinyl esters	220/104
Furans	260/127
Phenolics	
Phenol-formaldehyde	
Methyl appended silicone	90/32
PB	

Copper Cyanide

Polymer	°F/°C
ABS	140/60
Acrylics	90/32
CTFE	
ECTFE	300/149
ETFE	300/149
FEP	400/204
Polyamides	80/27
PAI	
PEEK	
PEI	
PES	
PFA	390/199
PTFE	450/232
PVDF	280/138
EHMWPE	140/60
HMWPE	140/60
PET	
PPO	140/60
PPS	210/99
PP	200/93
PVDC	130/54
PSF	200/93
PVC Type 1	140/60
PVC Type 2	140/60
CPVC	200/93
CPE, 10%	250/121
PUR	90/32
Acetals	
Polyesters	
Bisphenol A-fumurate	220/104
Halogenated	250/121
Hydrogenated-Bisphenol A	210/99
Isophthalic	160/71
PET	
Epoxy	150/66
Vinyl esters	210/99
Furans	240/116
Phenolics	
Phenol-formaldehyde	
Methyl appended silicone	
PB	

Copper Sulfate

Polymer	°F/°C
ABS	140/60
Acrylics	X
CTFE	
ECTFE	300/149
ETFE	300/149
FEP	400/204
Polyamides	140/60
PAI	
PEEK	
PEI	
PES	
PFA	400/204
PTFE	450/232
PVDF	280/138
EHMWPE	140/60
HMWPE	140/60
PET	170/77
PPO	140/60
PPS	250/121
PP	200/93
PVDC	180/82
PSF	200/93
PVC Type 1	140/60
PVC Type 2	140/60
CPVC	210/99
CPE	250/121
PUR	90/32
Acetals	X
Polyesters	
Bisphenol A-fumurate	220/104
Halogenated	250/121
Hydrogenated-Bisphenol A	210/99
Isophthalic	200/93
PET	170/77
Epoxy, 17%	210/99
Vinyl esters	240/116
Furans	260/127
Phenolics	300/149
Phenol-formaldehyde	
Methyl appended silicone	
PB	90/32

Cresol

Polymer	°F/°C
ABS	X
Acrylics	X
CTFE	
ECTFE	300/149
ETFE	270/132
FEP	400/204
Polyamides	X
PAI	
PEEK	
PEI	
PES	
PFA	390/199
PTFE	450/232
PVDF	210/99
EHMWPE	140/60
HMWPE	X
PET	X
PPO	X
PPS	200/93
PP	X
PVDC	150/66
PSF	X
PVC Type 1	130/54
PVC Type 2	X
CPVC	X
CPE	140/60
PUR	X
Acetals	X
Polyesters	
Bisphenol A-fumurate	X
Halogenated	X
Hydrogenated-Bisphenol A	X
Isophthalic	X
PET	X
Epoxy	100/38
Vinyl esters	X
Furans	260/127
Phenolics	300/149
Phenol-formaldehyde	
Methyl appended silicone	
PB	

Cupric Chloride, 5%

Polymer	°F/°C
ABS	
Acrylics	
CTFE	300/149
ECTFE	300/149
ETFE	300/149
FEP	400/204
Polyamides	X
PAI	
PEEK	
PEI	
PES	
PFA	400/204
PTFE	450/232
PVDF	270/232
EHMWPE	
HMWPE	140/60
PET	250/121
PPO	150/66
PPS	300/149
PP	140/60
PVDC	160/71
PSF	200/93
PVC Type 1	150/66
PVC Type 2	
CPVC	
CPE	
PUR	
Acetals	X
Polyesters	
Bisphenol A-fumurate	
Halogenated	
Hydrogenated-Bisphenol A	
Isophthalic	170/77
PET	250/121
Epoxy	80/27
Vinyl esters	270/132
Furans	300/149
Phenolics	300/149
Phenol-formaldehyde	
Methyl appended silicone	
PB	

Cupric Chloride, 50%

Polymer	°F/°C
ABS	
Acrylics	100/38
CTFE	
ECTFE	300/149
ETFE, 5%	300/149
FEP	400/204
Polyamides	X
PAI	
PEEK	
PEI	
PES	
PFA	400/204
PTFE	450/232
PVDF, 5%	270/132
EHMWPE, 10%	80/27
HMWPE	140/60
PET	250/121
PPO	170/77
PPS, 5%	300/149
PP	140/60
PVDC	170/77
PSF, sat.	200/93
PVC Type 1	150/66
PVC Type 2	
CPVC	180/82
CPE	200/93
PUR	
Acetals	X
Polyesters	
Bisphenol A-fumurate	
Halogenated	
Hydrogenated-Bisphenol A	
Isophthalic	140/60
PET	250/121
Epoxy	80/27
Vinyl esters	220/104
Furans, Sat.	300/149
Phenolics	
Phenol-formaldehyde	300/149
Methyl appended silicone	
PB	

Cupric Cyanide

Polymer	°F/°C
ABS	140/60
Acrylics	80/27
CTFE	
ECTFE	300/149
ETFE	300/149
FEP	400/204
Polyamides	80/27
PAI	
PEEK	
PEI	
PES	
PFA	400/204
PTFE	450/232
PVDF, 10%	270/132
EHMWPE	140/60
HMWPE	140/60
PET	250/121
PPO	140/60
PPS	210/99
PP, 10%	200/93
PVDC, 10%	130/54
PSF	200/93
PVC Type 1	140/60
PVC Type 2	140/60
CPVC	180/82
CPE, 10%	240/116
PUR	90/32
Acetals	
Polyesters	
Bisphenol A-fumurate	220/104
Halogenated	240/116
Hydrogenated-Bisphenol A	210/99
Isophthalic	160/71
PET	250/121
Epoxy	150/66
Vinyl esters	210/99
Furans	240/116
Phenolics	
Phenol-formaldehyde	
Methyl appended silicone	
PB	

Cupric Fluoride

Polymer	°F/°C
ABS, 25%	140/60
Acrylics	
CTFE	
ECTFE	300/149
ETFE	300/149
FEP	400/204
Polyamides	
PAI	
PEEK	
PEI	
PES	
PFA	200/93
PTFE	250/121
PVDF	280/138
EHMWPE	140/60
HMWPE	140/60
PET	
PPO	140/60
PPS	
PP	200/93
PVDC	160/71
PSF	
PVC Type 1	140/60
PVC Type 2	140/60
CPVC	182/82
CPE	250/121
PUR	
Acetals	
Polyesters	
Bisphenol A-fumurate	150/66
Halogenated	120/49
Hydrogenated-Bisphenol A	
Isophthalic	X
PET	
Epoxy	250/121
Vinyl esters	200/93
Furans	
Phenolics	
Phenol-formaldehyde	
Methyl appended silicone	
PB	

Cupric Nitrate

Polymer	°F/°C
ABS	140/60
Acrylics	
CTFE	
ECTFE	300/149
ETFE	300/149
FEP	400/204
Polyamides	X
PAI	
PEEK	
PEI	
PES	
PFA	380/193
PTFE	450/232
PVDF	270/132
EHMWPE	140/60
HMWPE	140/60
PET	170/77
PPO	140/60
PPS	
PP	200/93
PVDC	170/77
PSF	200/93
PVC Type 1	140/60
PVC Type 2	140/60
CPVC	200/93
CPE	240/116
PUR	
Acetals	X
Polyesters	
Bisphenol A-fumurate	220/104
Halogenated	140/60
Hydrogenated-Bisphenol A	210/99
Isophthalic	160/71
PET	170/77
Epoxy	240/116
Vinyl esters	210/99
Furans	280/183
Phenolics	
Phenol-formaldehyde	
Methyl appended silicone	
PB	

Cupric Sulfate

Polymer	°F/°C
ABS	100/38
Acrylics	X
CTFE	
ECTFE	300/149
ETFE	300/149
FEP	390/199
Polyamides	140/60
PAI	
PEEK	
PEI	
PES	
PFA	390/199
PTFE	450/232
PVDF	280/132
EHMWPE	140/60
HMWPE	140/60
PET	170/77
PPO	140/60
PPS	250/121
PP	200/93
PVDC	180/82
PSF	200/93
PVC Type 1	140/60
PVC Type 2	140/60
CPVC	210/99
CPE	250/121
PUR	90/32
Acetals	X
Polyesters	
Bisphenol A-fumurate	220/104
Halogenated	250/121
Hydrogenated-Bisphenol A	230/110
Isophthalic	200/93
PET	170/77
Epoxy	250/121
Vinyl esters	200/93
Furans	260/127
Phenolics	300/149
Phenol-formaldehyde	
Methyl appended silicone	
PB	

Cyclohexanol	
Polymer	**°F/°C**
ABS	80/27
Acrylics	
CTFE	
ECTFE	300/149
ETFE	250/121
FEP	400/204
Polyamides	250/121
PAI	200/93
PEEK	
PEI	
PES	
PFA	390/199
PTFE	450/232
PVDF	210/99
EHMWPE	170/77
HMWPE	80/27
PET	
PPO	190/88
PPS	250/121
PP	150/66
PVDC	90/32
PSF	200/93
PVC Type 1	X
PVC Type 2	X
CPVC	X
CPE	220/104
PUR	
Acetals	80/27
Polyesters	
Bisphenol A-fumurate	
Halogenated	
Hydrogenated-Bisphenol A	
Isophthalic	120/49
PET	
Epoxy	150/66
Vinyl esters	150/66
Furans	
Phenolics	
Phenol-formaldehyde	
Methyl appended silicone	
PB	

Cyclohexanone

Polymer	°F/°C
ABS	X
Acrylics	
CTFE	100/38
ECTFE	300/149
ETFE	300/149
FEP	400/204
Polyamides	250/121
PAI	200/93
PEEK	
PEI	
PES	
PFA	400/204
PTFE	450/232
PVDF	110/43
EHMWPE	140/60
HMWPE	X
PET	250/121
PPO	X
PPS	200/93
PP	X
PVDC	X
PSF	X
PVC Type 1	X
PVC Type 2	X
CPVC	X
CPE	80/27
PUR	
Acetals	80/27
Polyesters	
Bisphenol A-fumurate	X
Halogenated	100/38
Hydrogenated-Bisphenol A	X
Isophthalic	X
PET	250/121
Epoxy	X
Vinyl esters	120/49
Furans	260/127
Phenolics	
Phenol-formaldehyde	
Methyl appended silicone	
PB	

Detergents

Polymer	°F/°C
ABS	
Acrylics	90/32
CTFE	
ECTFE	300/149
ETFE	300/149
FEP	400/204
Polyamides	110/43
PAI	
PEEK	
PEI	
PES	
PFA	390/199
PTFE	450/232
PVDF	
EHMWPE	140/60
HMWPE	140/60
PET	140/60
PPO	90/32
PPS	290/143
PP	220/104
PVDC	
PSF	80/27
PVC Type 1	140/60
PVC Type 2	140/60
CPVC	210/99
CPE	250/121
PUR	
Acetals	80/27
Polyesters	
Bisphenol A-fumurate	90/32
Halogenated	90/32
Hydrogenated-Bisphenol A	
Isophthalic	90/32
PET	140/60
Epoxy	250/121
Vinyl esters	100/38
Furans	100/38
Phenolics	
Phenol-formaldehyde	
Methyl appended silicone	
PB	

Detergent Solution, Heavy Duty	
Polymer	**°F/°C**
ABS	
Acrylics	
CTFE	
ECTFE	300/149
ETFE	300/149
FEP	400/204
Polyamides	100/38
PAI	
PEEK	
PEI	
PES	
PFA	390/199
PTFE	450/232
PVDF	
EHMWPE	140/60
HMWPE	140/60
PET	120/49
PPO	
PPS	300/149
PP	150/66
PVDC	
PSF	
PVC Type 1	140/60
PVC Type 2	140/60
CPVC	160/71
CPE	
PUR	
Acetals	80/27
Polyesters	
Bisphenol A-fumurate	
Halogenated	
Hydrogenated-Bisphenol A	
Isophthalic	80/27
PET	120/49
Epoxy	250/121
Vinyl esters	120/49
Furans	
Phenolics	
Phenol-formaldehyde	
Methyl appended silicone	
PB	

Dextrin

Polymer	°F/°C
ABS	140/60
Acrylics	
CTFE	
ECTFE	250/121
ETFE	300/149
FEP	400/204
Polyamides	
PAI	
PEEK	
PEI	
PES	
PFA	390/199
PTFE	450/232
PVDF	230/110
EHMWPE	140/60
HMWPE	140/60
PET	
PPO	
PPS	
PP	160/71
PVDC	
PSF	
PVC Type 1	140/60
PVC Type 2	140/60
CPVC	200/93
CPE	250/121
PUR	
Acetals	
Polyesters	
Bisphenol A-fumurate	
Halogenated	
Hydrogenated-Bisphenol A	
Isophthalic	
PET	
Epoxy	
Vinyl esters	
Furans	
Phenolics	
Phenol-formaldehyde	
Methyl appended silicone	
PB	

| Dextrose | |
Polymer	°F/°C
ABS, 30%	140/60
Acrylics	80/27
CTFE	
ECTFE	250/121
ETFE	
FEP	400/204
Polyamides	80/27
PAI	
PEEK	
PEI	
PES	
PFA	390/199
PTFE	450/232
PVDF	280/138
EHMWPE	140/60
HMWPE	140/60
PET	80/27
PPO, 20%	140/60
PPS	300/149
PP	220/104
PVDC	180/82
PSF	
PVC Type 1	140/60
PVC Type 2	140/60
CPVC	210/99
CPE	250/121
PUR	X
Acetals	
Polyesters	
Bisphenol A-fumurate	220/104
Halogenated	220/104
Hydrogenated-Bisphenol A	
Isophthalic	160/71
PET	80/27
Epoxy	200/93
Vinyl esters	240/116
Furans	260/127
Phenolics	
Phenol-formaldehyde	
Methyl appended silicone	
PB	

Dibutyl Ether

Polymer	°F/°C
ABS	
Acrylics	
CTFE	100/38
ECTFE	200/93
ETFE	200/93
FEP	400/204
Polyamides	200/93
PAI	200/93
PEEK	
PEI	
PES	
PFA	400/204
PTFE	350/177
PVDF	200/93
EHMWPE	
HMWPE	
PET	
PPO	X
PPS	210/99
PP	X
PVDC	
PSF	200/93
PVC Type 1	140/60
PVC Type 2	
CPVC	X
CPE	
PUR	
Acetals	
Polyesters	
Bisphenol A-fumurate	150/66
Halogenated	80/27
Hydrogenated-Bisphenol A	
Isophthalic	X
PET	
Epoxy	80/27
Vinyl esters	200/93
Furans	210/99
Phenolics	210/99
Phenol-formaldehyde	
Methyl appended silicone	
PB	

Dibutyl Phthalate

Polymer	°F/°C
ABS	X
Acrylics	
CTFE	
ECTFE	
ETFE	150/66
FEP	400/204
Polyamides	80/27
PAI	200/93
PEEK	
PEI	
PES	
PFA	380/193
PTFE	450/232
PVDF	X
EHMWPE	140/60
HMWPE	X
PET	80/27
PPO	120/49
PPS	300/149
PP	110/43
PVDC	120/49
PSF	180/82
PVC Type 1	X
PVC Type 2	X
CPVC	X
CPE	
PUR	X
Acetals	
Polyesters	
Bisphenol A-fumurate	180/82
Halogenated	100/38
Hydrogenated-Bisphenol A	230/121
Isophthalic	160/71
PET	80/27
Epoxy	250/121
Vinyl esters	200/93
Furans	230/110
Phenolics	210/99
Phenol-formaldehyde	
Methyl appended silicone	
PB	

Dichloroacetic Acid

Polymer	°F/°C
ABS	X
Acrylics	
CTFE	
ECTFE	
ETFE	150/66
FEP	400/204
Polyamides	X
PAI	
PEEK	
PEI	
PES	
PFA	400/204
PTFE	450/232
PVDF	120/49
EHMWPE	80/27
HMWPE	
PET	100/38
PPO	
PPS	
PP	100/38
PVDC	120/49
PSF	
PVC Type 1	100/38
PVC Type 2	100/38
CPVC, 20%	100/38
CPE	
PUR	
Acetals	
Polyesters	
Bisphenol A-fumurate	100/38
Halogenated	100/38
Hydrogenated-Bisphenol A	
Isophthalic	X
PET	100/38
Epoxy	X
Vinyl esters	100/38
Furans	X
Phenolics	
Phenol-formaldehyde	
Methyl appended silicone	
PB	

Dichlorobenzene

Polymer	°F/°C
ABS	X
Acrylics	
CTFE	
ECTFE	X
ETFE	150/66
FEP	400/204
Polyamides	250/121
PAI	
PEEK	
PEI	
PES	
PFA	380/193
PTFE	450/232
PVDF	120/49
EHMWPE	X
HMWPE	X
PET	250/121
PPO	
PPS	200/93
PP	150/66
PVDC	X
PSF	X
PVC Type 1	X
PVC Type 2	X
CPVC	X
CPE	
PUR	
Acetals	
Polyesters	
Bisphenol A-fumurate	110/43
Halogenated	X
Hydrogenated-Bisphenol A	X
Isophthalic	X
PET	250/121
Epoxy	190/88
Vinyl esters	110/43
Furans	260/127
Phenolics	200/93
Phenol-formaldehyde	
Methyl appended silicone	
PB	

Dichloroethane (Ethylene Dichloride)

Polymer	°F/°C
ABS	X
Acrylics	X
CTFE	100/38
ECTFE	300/149
ETFE	300/149
FEP[a]	400/204
Polyamides	250/121
PAI	
PEEK	
PEI	
PES	
PFA[a]	340/171
PTFE[a]	450/232
PVDF	180/138
EHMWPE	X
HMWPE	X
PET	X
PPO	X
PPS	
PP	X
PVDC	80/27
PSF	X
PVC Type 1	X
PVC Type 2	X
CPVC	X
CPE	120/49
PUR	
Acetals	90/32
Polyesters	
Bisphenol A-fumurate	X
Halogenated	X
Hydrogenated-Bisphenol A	X
Isophthalic	X
PET	X
Epoxy	X
Vinyl esters	110/43
Furans	250/121
Phenolics	200/93
Phenol-formaldehyde	
Methyl appended silicone	
PB	

[a] Corrodent will permeate.

Ethyl Alcohol

Polymer	°F/°C
ABS	140/60
Acrylics	X
CTFE	
ECTFE	300/149
ETFE	300/149
FEP	400/204
Polyamides	250/121
PAI	
PEEK	80/27
PEI	90/32
PES	
PFA	390/199
PTFE	450/232
PVDF	280/138
EHMWPE	140/60
HMWPE	140/60
PET	250/121
PPO	90/38
PPS	
PP	180/82
PVDC	150/66
PSF	100/38
PVC Type 1	140/60
PVC Type 2	140/60
CPVC	210/99
CPE	220/104
PUR	X
Acetals	
Polyesters	
Bisphenol A-fumurate	90/32
Halogenated	150/66
Hydrogenated-Bisphenol A	90/32
Isophthalic	80/27
PET	250/121
Epoxy	150/66
Vinyl esters	100/38
Furans	260/127
Phenolics	150/66
Phenol-formaldehyde	
Methyl appended silicone	
PB	

Ethyl Chloride

Polymer	°F/°C
ABS	X
Acrylics	X
CTFE	
ECTFE	300/149
ETFE	300/149
FEP	400/204
Polyamides	250/127
PAI	200/93
PEEK	
PEI	
PES	
PFA	390/199
PTFE	450/232
PVDF	260/138
EHMWPE	X
HMWPE	X
PET	250/121
PPO	X
PPS	200/93
PP	X
PVDC	X
PSF	X
PVC Type 1	X
PVC Type 2	X
CPVC	X
CPE	220/104
PUR	X
Acetals	80/27
Polyesters	
Bisphenol A-fumurate	
Halogenated	80/27
Hydrogenated-Bisphenol A	
Isophthalic	X
PET	250/121
Epoxy	X
Vinyl esters	X
Furans	260/127
Phenolics	160/71
Phenol-formaldehyde	
Methyl appended silicone	
PB	

Ethyl Ether

Polymer	°F/°C
ABS	X
Acrylics	X
CTFE	
ECTFE	150/66
ETFE	210/99
FEP	400/204
Polyamides	200/93
PAI	
PEEK	
PEI	80/27
PES	
PFA	370/188
PTFE	450/232
PVDF	150/66
EHMWPE	X
HMWPE	X
PET	250/121
PPO	X
PPS	150/66
PP	90/32
PVDC	X
PSF	X
PVC Type 1	X
PVC Type 2	X
CPVC	X
CPE	220/104
PUR	X
Acetals	80/27
Polyesters	
Bisphenol A-fumurate	X
Halogenated	X
Hydrogenated-Bisphenol A	X
Isophthalic	X
PET	250/121
Epoxy	X
Vinyl esters	X
Furans	260/127
Phenolics	
Phenol-formaldehyde	
Methyl appended silicone	
PB	

Ethylene Glycol

Polymer	°F/°C
ABS	140/60
Acrylics	80/27
CTFE	
ECTFE	300/149
ETFE	300/149
FEP	400/204
Polyamides	205/121
PAI	200/93
PEEK	160/71
PEI	80/27
PES	90/32
PFA	390/199
PTFE	450/232
PVDF	280/138
EHMWPE	140/60
HMWPE	140/60
PET	200/93
PPO	140/60
PPS	300/149
PP	230/110
PVDC	180/82
PSF	200/93
PVC Type 1	140/60
PVC Type 2	140/60
CPVC	210/99
CPE	220/104
PUR	90/32
Acetals	180/82
Polyesters	
Bisphenol A-fumurate	220/104
Halogenated	250/121
Hydrogenated-Bisphenol A	
Isophthalic	250/121
PET	200/93
Epoxy	300/149
Vinyl esters	210/99
Furans	260/177
Phenolics	100/38
Phenol-formaldehyde	
Methyl appended silicone	400/204
PB	

Fatty Acids

Polymer	°F/°C
ABS	140/60
Acrylics	90/32
CTFE	
ECTFE	300/149
ETFE	300/149
FEP	400/204
Polyamides	80/27
PAI	
PEEK	
PEI	
PES	
PFA	390/199
PTFE	450/232
PVDF	280/138
EHMWPE	140/60
HMWPE	140/60
PET	200/93
PPO	60/15
PPS	250/121
PP	140/60
PVDC	160/71
PSF	
PVC Type 1	140/60
PVC Type 2	140/60
CPVC	180/82
CPE	150/66
PUR	
Acetals	100/38
Polyesters	
Bisphenol A-fumurate	200/93
Halogenated	250/121
Hydrogenated-Bisphenol A	220/104
Isophthalic	180/82
PET	200/93
Epoxy	200/93
Vinyl esters	250/121
Furans	260/127
Phenolics	80/27
Phenol-formaldehyde	
Methyl appended silicone	
PB	90/32

Ferric Chloride

Polymer	°F/°C
ABS	140/60
Acrylics	80/27
CTFE	
ECTFE	300/149
ETFE, 50%	300/149
FEP	400/204
Polyamides	X
PAI	
PEEK	
PEI	
PES	
PFA	390/199
PTFE	450/232
PVDF	280/138
EHMWPE	140/60
HMWPE	140/60
PET	250/121
PPO	140/60
PPS	210/99
PP	210/99
PVDC	140/60
PSF	200/93
PVC Type 1	140/60
PVC Type 2	140/60
CPVC	210/99
CPE	250/121
PUR	90/32
Acetals	X
Polyesters	
Bisphenol A-fumurate	220/104
Halogenated	250/121
Hydrogenated-Bisphenol A	210/99
Isophthalic	180/82
PET	250/121
Epoxy	300/149
Vinyl esters	210/99
Furans	260/127
Phenolics	300/149
Phenol-formaldehyde	
Methyl appended silicone	400/204
PB	

Ferric Nitrate, 10–50%

Polymer	°F/°C
ABS	140/60
Acrylics	80/27
CTFE	
ECTFE	300/149
ETFE	300/149
FEP	400/204
Polyamides	X
PAI	
PEEK	
PEI	
PES	
PFA	400/204
PTFE	450/232
PVDF	280/138
EHMWPE	140/60
HMWPE	140/60
PET	170/77
PPO	140/60
PPS	210/99
PP	200/93
PVDC	130/54
PSF	200/93
PVC Type 1	140/60
PVC Type 2	140/60
CPVC	200/93
CPE	250/121
PUR	
Acetals	X
Polyesters	
Bisphenol A-fumurate	220/104
Halogenated	250/121
Hydrogenated-Bisphenol A	200/93
Isophthalic	180/82
PET	170/77
Epoxy	250/121
Vinyl esters	200/93
Furans	260/127
Phenolics	
Phenol-formaldehyde	300/149
Methyl appended silicone	
PB	

Ferrous Chloride

Polymer	°F/°C
ABS	140/60
Acrylics	80/27
CTFE	
ECTFE	300/149
ETFE	300/149
FEP	400/204
Polyamides	X
PAI	
PEEK	200/93
PEI	
PES	90/32
PFA	390/199
PTFE	450/232
PVDF	280/138
EHMWPE	140/60
HMWPE	140/60
PET	250/121
PPO	140/60
PPS	210/99
PP	210/99
PVDC	130/54
PSF	200/93
PVC Type 1	140/60
PVC Type 2	140/60
CPVC	210/99
CPE	250/121
PUR	
Acetals	80/27
Polyesters	
Bisphenol A-fumurate	220/104
Halogenated	250/121
Hydrogenated-Bisphenol A	210/99
Isophthalic	180/82
PET	250/121
Epoxy	250/121
Vinyl esters	200/93
Furans	260/127
Phenolics, 40%	300/149
Phenol-formaldehyde	
Methyl appended silicone	
PB	90/32

Ferrous Nitrate

Polymer	°F/°C
ABS	
Acrylics	
CTFE	
ECTFE	300/149
ETFE	300/149
FEP	400/204
Polyamides	
PAI	
PEEK	
PEI	
PES	
PFA	390/199
PTFE	450/232
PVDF	280/138
EHMWPE	140/60
HMWPE	140/60
PET	
PPO	
PPS	210/99
PP	210/99
PVDC	80/27
PSF	
PVC Type 1	140/60
PVC Type 2	140/60
CPVC	200/93
CPE	250/121
PUR	
Acetals	X
Polyesters	
Bisphenol A-fumurate	220/104
Halogenated	160/71
Hydrogenated-Bisphenol A	210/99
Isophthalic	160/71
PET	
Epoxy	200/93
Vinyl esters	220/104
Furans	
Phenolics	
Phenol-formaldehyde	
Methyl appended silicone	
PB	

Fluorine Gas, Dry

Polymer	°F/°C
ABS	90/32
Acrylics	
CTFE	X
ECTFE	X
ETFE	100/38
FEP	200/93
Polyamides	440/227
PAI	
PEEK	X
PEI	
PES	
PFA	X
PTFE	310/154
PVDF	80/27
EHMWPE	X
HMWPE	80/27
PET	
PPO	
PPS	X
PP	X
PVDC	X
PSF	X
PVC Type 1	X
PVC Type 2	X
CPVC	X
CPE	X
PUR	
Acetals	80/27
Polyesters	
Bisphenol A-fumurate	
Halogenated	
Hydrogenated-Bisphenol A	
Isophthalic	X
PET	
Epoxy	90/32
Vinyl esters	X
Furans	X
Phenolics	
Phenol-formaldehyde	
Methyl appended silicone	
PB	

Fluorine Gas, Moist

Polymer	°F/°C
ABS	
Acrylics	
CTFE	80/27
ECTFE	80/27
ETFE	100/38
FEP	X
Polyamides	X
PAI	
PEEK	X
PEI	
PES	
PFA	X
PTFE	200/93
PVDF	250/121
EHMWPE	X
HMWPE	80/27
PET	80/27
PPO	
PPS	
PP	
PVDC	X
PSF	X
PVC Type 1	X
PVC Type 2	X
CPVC	X
CPE	X
PUR	
Acetals	
Polyesters	
Bisphenol A-fumurate	
Halogenated	
Hydrogenated-Bisphenol A	
Isophthalic	X
PET	80/27
Epoxy	
Vinyl esters	X
Furans	X
Phenolics	
Phenol-formaldehyde	
Methyl appended silicone	
PB	

Formaldehyde, Dil.

Polymer	°F/°C
ABS	100/38
Acrylics	80/27
CTFE	
ECTFE	140/60
ETFE	230/110
FEP	400/204
Polyamides	250/121
PAI	200/93
PEEK	
PEI	
PES	
PFA	200/93
PTFE	300/149
PVDF	120/49
EHMWPE	140/60
HMWPE	140/60
PET	240/116
PPO	
PPS	200/93
PP	200/93
PVDC	140/60
PSF	200/93
PVC Type 1	140/60
PVC Type 2	
CPVC	140/60
CPE	220/104
PUR	
Acetals	80/27
Polyesters	
Bisphenol A-fumurate	80/27
Halogenated	150/66
Hydrogenated-Bisphenol A	220/104
Isophthalic	140/60
PET	240/116
Epoxy	110/43
Vinyl esters	150/66
Furans	160/71
Phenolics	200/93
Phenol-formaldehyde	
Methyl appended silicone	
PB	

Formaldehyde, 35%

Polymer	°F/°C
ABS	100/38
Acrylics	80/27
CTFE	
ECTFE	150/66
ETFE	230/110
FEP	400/204
Polyamides	250/121
PAI	200/93
PEEK	
PEI	
PES	
PFA	200/93
PTFE	450/232
PVDF	140/60
EHMWPE	140/60
HMWPE	140/60
PET	250/121
PPO	
PPS	200/93
PP	200/93
PVDC	120/49
PSF	200/93
PVC Type 1	140/60
PVC Type 2	100/38
CPVC	140/60
CPE	220/104
PUR	
Acetals	80/27
Polyesters	
Bisphenol A-fumurate	80/27
Halogenated	150/66
Hydrogenated-Bisphenol A	200/93
Isophthalic	140/60
PET	250/121
Epoxy	110/43
Vinyl esters	140/60
Furans	160/71
Phenolics	200/93
Phenol-formaldehyde	
Methyl appended silicone	
PB	

Formaldehyde, 37%

Polymer	°F/°C
ABS	100/38
Acrylics	80/27
CTFE	
ECTFE	150/66
ETFE	230/110
FEP	400/204
Polyamides	250/121
PAI	200/93
PEEK	
PEI	
PES	
PFA	200/93
PTFE	450/232
PVDF	120/49
EHMWPE	140/60
HMWPE	140/60
PET	250/121
PPO	200/93
PPS	200/93
PP	210/99
PVDC	140/60
PSF	200/93
PVC Type 1	140/60
PVC Type 2	100/38
CPVC	140/60
CPE	220/104
PUR	
Acetals	80/27
Polyesters	
Bisphenol A-fumurate	80/27
Halogenated	150/66
Hydrogenated-Bisphenol A	210/99
Isophthalic	140/60
PET	250/121
Epoxy	150/66
Vinyl esters	200/93
Furans	260/127
Phenolics	200/93
Phenol-formaldehyde	
Methyl appended silicone	
PB	

Formaldehyde, 50%

Polymer	°F/°C
ABS	100/38
Acrylics	80/27
CTFE	
ECTFE	80/27
ETFE	
FEP	400/204
Polyamides	250/166
PAI	
PEEK	
PEI	
PES	
PFA	
PTFE	450/232
PVDF	280/138
EHMWPE	
HMWPE	
PET	250/121
PPO	
PPS	200/93
PP	200/93
PVDC	130/54
PSF	
PVC Type 1	140/60
PVC Type 2	120/49
CPVC	150/66
CPE	220/104
PUR	
Acetals	
Polyesters	
Bisphenol A-fumurate	80/27
Halogenated	120/49
Hydrogenated-Bisphenol A	
Isophthalic	190/88
PET	250/121
Epoxy	110/43
Vinyl esters	140/60
Furans	240/116
Phenolics	
Phenol-formaldehyde	
Methyl appended silicone	
PB	

Formic Acid, 10–85%

Polymer	°F/°C
ABS	X
Acrylics	X
CTFE	
ECTFE	250/121
ETFE	270/132
FEP	400/204
Polyamides	X
PAI	
PEEK, 10%	90/32
PEI, 10%	90/32
PES	80/27
PFA	390/199
PTFE	450/121
PVDF	250/121
EHMWPE	140/60
HMWPE	140/60
PET	250/121
PPO	140/60
PPS	
PP	210/99
PVDC	150/66
PSF	200/93
PVC Type 1	110/38
PVC Type 2	90/32
CPVC, to 25%	180/82
CPE	220/104
PUR	
Acetals	X
Polyesters	
Bisphenol A-fumurate	150/66
Halogenated	150/66
Hydrogenated-Bisphenol A	90/32
Isophthalic	X
PET	250/121
Epoxy	X
Vinyl esters	100/38
Furans	260/127
Phenolics	200/93
Phenol-formaldehyde	
Methyl appended silicone	
PB	

	Furfural
Polymer	**°F/°C**
ABS	X
Acrylics	
CTFE	
ECTFE	
ETFE	210/99
FEP	400/204
Polyamides	90/32
PAI	80/27
PEEK	
PEI	
PES	
PFA	390/199
PTFE	450/232
PVDF	110/43
EHMWPE	80/27
HMWPE	130/54
PET	80/27
PPO	X
PPS	300/149
PP	X
PVDC	80/27
PSF	X
PVC Type 1	X
PVC Type 2	X
CPVC	X
CPE	140/60
PUR	X
Acetals	
Polyesters	
Bisphenol A-fumurate	X
Halogenated	X
Hydrogenated-Bisphenol A	
Isophthalic	X
PET	80/27
Epoxy	X
Vinyl esters	X
Furans	260/127
Phenolics	X
Phenol-formaldehyde	
Methyl appended silicone	
PB	

Gallic Acid

Polymer	°F/°C
ABS	140/60
Acrylics	
CTFE	
ECTFE	150/66
ETFE	210/99
FEP	400/204
Polyamides	80/27
PAI	
PEEK	
PEI	
PES	
PFA	200/93
PTFE	450/232
PVDF	160/71
EHMWPE	140/60
HMWPE	140/60
PET	250/121
PPO	
PPS	250/121
PP	180/82
PVDC	160/71
PSF	
PVC Type 1	140/60
PVC Type 2	140/60
CPVC	180/82
CPE	250/121
PUR	X
Acetals	
Polyesters	
Bisphenol A-fumurate	250/121
Halogenated	200/93
Hydrogenated-Bisphenol A	
Isophthalic	
PET	250/121
Epoxy	
Vinyl esters	120/49
Furans	260/127
Phenolics	
Phenol-formaldehyde	
Methyl appended silicone	
PB	

Gas, Natural

Polymer	°F/°C
ABS	X
Acrylics	
CTFE	
ECTFE	300/149
ETFE	300/149
FEP	400/204
Polyamides	140/60
PAI	
PEEK	80/27
PEI	
PES	
PFA	200/93
PTFE	450/232
PVDF	
EHMWPE	X
HMWPE	X
PET	
PPO	140/60
PPS	
PP	160/71
PVDC	160/71
PSF	
PVC Type 1	140/60
PVC Type 2	140/60
CPVC	180/82
CPE	250/121
PUR	
Acetals	
Polyesters	
Bisphenol A-fumurate	
Halogenated	
Hydrogenated-Bisphenol A	
Isophthalic	90/32
PET	
Epoxy	200/93
Vinyl esters	200/93
Furans	280/138
Phenolics	
Phenol-formaldehyde	
Methyl appended silicone	
PB	

Gasoline, Refined	
Polymer	**°F/°C**
ABS	X
Acrylics	80/27
CTFE	
ECTFE	
ETFE	200/93
FEP	400/204
Polyamides	250/127
PAI	120/49
PEEK	
PEI	
PES	
PFA	380/193
PTFE	450/232
PVDF	
EHMWPE	X
HMWPE	
PET	250/121
PPO	90/32
PPS	300/149
PP	X
PVDC	90/32
PSF	200/93
PVC Type 1	120/49
PVC Type 2	X
CPVC	X
CPE	
PUR	80/27
Acetals	80/27
Polyesters	
Bisphenol A-fumurate	80/27
Halogenated	200/93
Hydrogenated-Bisphenol A	90/32
Isophthalic	90/32
PET	250/121
Epoxy	150/66
Vinyl esters	180/82
Furans	260/127
Phenolics	200/93
Phenol-formaldehyde	
Methyl appended silicone	
PB	

Gasoline, Sour

Polymer	°F/°C
ABS	X
Acrylics	
CTFE	
ECTFE	300/149
ETFE	300/149
FEP	400/204
Polyamides	250/121
PAI	
PEEK	80/27
PEI	
PES	
PFA	370/160
PTFE	450/232
PVDF	280/138
EHMWPE	
HMWPE	
PET	250/121
PPO	
PPS	300/149
PP	X
PVDC	
PSF	
PVC Type 1	140/60
PVC Type 2	140/60
CPVC	X
CPE	250/121
PUR	80/27
Acetals	X
Polyesters	
Bisphenol A-fumurate	90/32
Halogenated	
Hydrogenated-Bisphenol A	
Isophthalic	80/27
PET	250/121
Epoxy	300/149
Vinyl esters	200/93
Furans	190/88
Phenolics	200/93
Phenol-formaldehyde	
Methyl appended silicone	
PB	

Gasoline, Unleaded

Polymer	°F/°C
ABS	X
Acrylics	90/32
CTFE	
ECTFE	300/149
ETFE	300/149
FEP	400/204
Polyamides	250/121
PAI	
PEEK	
PEI	
PES	
PFA	200/93
PTFE	450/232
PVDF	280/138
EHMWPE	140/60
HMWPE	X
PET	250/121
PPO	X
PPS	300/149
PP	X
PVDC	150/66
PSF	200/93
PVC Type 1	140/60
PVC Type 2	
CPVC	X
CPE	250/121
PUR	80/27
Acetals	80/27
Polyesters	
Bisphenol A-fumurate	90/32
Halogenated	200/93
Hydrogenated-Bisphenol A	90/32
Isophthalic	100/38
PET	250/121
Epoxy	250/121
Vinyl esters	100/38
Furans	
Phenolics	200/93
Phenol-formaldehyde	
Methyl appended silicone	
PB	

Glycerine (Glycerol)

Polymer	°F/°C
ABS	140/60
Acrylics	90/32
CTFE	
ECTFE	300/149
ETFE	300/149
FEP	400/204
Polyamides	250/121
PAI	200/93
PEEK	200/93
PEI	80/27
PES	80/27
PFA	200/93
PTFE	450/232
PVDF	280/138
EHMWPE	140/60
HMWPE	140/60
PET	250/121
PPO	90/32
PPS	300/149
PP	210/99
PVDC	180/82
PSF	80/27
PVC Type 1	140/60
PVC Type 2	140/60
CPVC	210/99
CPE	220/104
PUR	90/32
Acetals	80/27
Polyesters	
Bisphenol A-fumurate	220/104
Halogenated, 75%	250/121
Hydrogenated-Bisphenol A	210/99
Isophthalic	180/82
PET	250/121
Epoxy	280/138
Vinyl esters	300/149
Furans	260/127
Phenolics	160/71
Phenol-formaldehyde	
Methyl appended silicone	
PB	

Green Liquor

Polymer	°F/°C
ABS	140/60
Acrylics	
CTFE	
ECTFE	
ETFE	
FEP	400/204
Polyamides	
PAI	
PEEK	
PEI	
PES	
PFA	380/193
PTFE	450/232
PVDF	280/138
EHMWPE	
HMWPE	
PET	
PPO	
PPS	250/121
PP	140/60
PVDC	
PSF	
PVC Type 1	140/60
PVC Type 2	140/60
CPVC	180/82
CPE	250/121
PUR	
Acetals	
Polyesters	
Bisphenol A-fumurate	220/104
Halogenated	X
Hydrogenated-Bisphenol A	
Isophthalic	X
PET	
Epoxy	190/88
Vinyl esters	200/93
Furans	
Phenolics	
Phenol-formaldehyde	
Methyl appended silicone	
PB	

Hydrobromic Acid, Dil.

Polymer	°F/°C
ABS	
Acrylics	
CTFE	
ECTFE	300/149
ETFE	300/149
FEP[a,b]	400/204
Polyamides	X
PAI	
PEEK	X
PEI	
PES	
PFA[a,b]	200/93
PTFE[a,b]	450/232
PVDF	260/127
EHMWPE	140/60
HMWPE	140/60
PET	250/121
PPO	90/32
PPS	200/93
PP	230/110
PVDC	120/49
PSF	300/149
PVC Type 1	140/60
PVC Type 2	140/60
CPVC	130/54
CPE	250/121
PUR	
Acetals	
Polyesters	
Bisphenol A-fumurate	220/104
Halogenated	200/93
Hydrogenated-Bisphenol A	
Isophthalic	140/60
PET	250/121
Epoxy	180/82
Vinyl esters	180/82
Furans	250/121
Phenolics	200/93
Phenol-formaldehyde	
Methyl appended silicone	
PB	

[a] Material is subject to stress cracking.
[b] Corrodent will permeate.

Hydrobromic Acid, 20%

Polymer	°F/°C
ABS	120/49
Acrylics	
CTFE	
ECTFE	300/149
ETFE	300/149
FEP[a,b]	400/204
Polyamides	X
PAI	
PEEK	X
PEI	
PES	
PFA[a,b]	200/93
PTFE[a,b]	450/232
PVDF	280/138
EHMWPE	140/60
HMWPE	140/60
PET	250/121
PPO	90/32
PPS	200/93
PP	200/93
PVDC	120/49
PSF	180/82
PVC Type 1	140/60
PVC Type 2	140/60
CPVC	180/82
CPE	250/121
PUR	
Acetals	
Polyesters	
Bisphenol A-fumurate	220/104
Halogenated	160/71
Hydrogenated-Bisphenol A	90/32
Isophthalic	140/60
PET	250/121
Epoxy	180/82
Vinyl esters	180/82
Furans	250/121
Phenolics	200/93
Phenol-formaldehyde	
Methyl appended silicone	
PB	

[a] Material is subject to stress cracking.
[b] Corrodent will permeate.

Hydrobromic Acid, 50%

Polymer	°F/°C
ABS	
Acrylics	
CTFE	
ECTFE	300/149
ETFE	300/149
FEP[a,b]	400/204
Polyamides	X
PAI	
PEEK	X
PEI	
PES	
PFA[a,b]	200/93
PTFE[a,b]	450/232
PVDF	280/138
EHMWPE	140/60
HMWPE	140/60
PET	250/121
PPO	90/32
PPS	200/93
PP	190/88
PVDC	130/54
PSF	
PVC Type 1	140/60
PVC Type 2	140/60
CPVC	190/88
CPE	250/121
PUR	
Acetals	
Polyesters	
Bisphenol A-fumurate	160/71
Halogenated	200/93
Hydrogenated-Bisphenol A	90/32
Isophthalic	150/66
PET	250/121
Epoxy	110/43
Vinyl esters	200/93
Furans	260/127
Phenolics	200/93
Phenol-formaldehyde	
Methyl appended silicone	X
PB	X

[a] Material subject to stress cracking.
[b] Corrodent will permeate.

Hydrochloric Acid, 20%

Polymer	°F/°C
ABS	90/32
Acrylics	90/32
CTFE	
ECTFE	300/149
ETFE	300/149
FEP[a,b]	400/204
Polyamides	X
PAI	200/93
PEEK	90/32
PEI	90/32
PES	140/60
PFA[a,b]	250/121
PTFE[a,b]	450/232
PVDF	280/138
EHMWPE	140/60
HMWPE	140/60
PET	250/121
PPO	160/71
PPS	210/99
PP	220/104
PVDC	180/82
PSF	140/60
PVC Type 1	140/60
PVC Type 2	140/60
CPVC	180/82
CPE	250/121
PUR	X
Acetals	X
Polyesters	
Bisphenol A-fumurate	210/99
Halogenated	230/110
Hydrogenated-Bisphenol A	180/82
Isophthalic	160/71
PET	250/121
Epoxy	200/93
Vinyl esters	220/104
Furans	260/127
Phenolics	300/149
Phenol-formaldehyde	
Methyl appended silicone	90/32
PB	90/32

[a] Material is subject to stress cracking.
[b] Corrodent will permeate.

Hydrochloric Acid, 38%

Polymer	°F/°C
ABS	140/60
Acrylics	90/32
CTFE	
ECTFE	300/149
ETFE	300/149
FEP[a,b]	400/204
Polyamides	X
PAI	200/93
PEEK	80/27
PEI	90/32
PES	140/60
PFA[a,b]	200/93
PTFE[b]	450/232
PVDF	280/138
EHMWPE	140/60
HMWPE	140/60
PET	90/32
PPO	190/88
PPS	210/99
PP	200/93
PVDC	180/82
PSF	140/60
PVC Type 1	140/60
PVC Type 2	140/60
CPVC	170/77
CPE	250/121
PUR	
Acetals	X
Polyesters	
Bisphenol A-fumurate	X
Halogenated	180/82
Hydrogenated-Bisphenol A	170/77
Isophthalic	160/71
PET	90/32
Epoxy	140/60
Vinyl esters	180/82
Furans	250/121
Phenolics	300/149
Phenol-formaldehyde	
Methyl appended silicone	X
PB	90/32

[a] Material subject to stress cracking.
[b] Corrodent will permeate.

Hydrocyanic Acid, 10%

Polymer	°F/°C
ABS	
Acrylics	
CTFE	
ECTFE	300/149
ETFE	300/149
FEP	400/204
Polyamides	X
PAI	
PEEK	
PEI	
PES	
PFA	380/183
PTFE	450/232
PVDF	280/138
EHMWPE	140/60
HMWPE	140/60
PET	80/27
PPO	
PPS	250/121
PP	150/60
PVDC	120/49
PSF	
PVC Type 1	140/60
PVC Type 2	140/60
CPVC	80/27
CPE	250/121
PUR	
Acetals	
Polyesters	
Bisphenol A-fumurate	200/93
Halogenated	150/66
Hydrogenated-Bisphenol A	X
Isophthalic	90/32
PET	80/27
Epoxy	160/71
Vinyl esters	160/71
Furans	240/116
Phenolics	
Phenol-formaldehyde	
Methyl appended silicone	
PB	

Hydrofluoric Acid, 30%

Polymer	°F/°C
ABS	X
Acrylics	
CTFE	200/93
ECTFE	250/121
ETFE	270/132
FEP[a]	400/204
Polyamides	X
PAI	
PEEK	X
PEI	
PES	
PFA[a]	200/93
PTFE[a]	450/232
PVDF	260/127
EHMWPE	140/60
HMWPE	140/60
PET	X
PPO	90/32
PPS	200/93
PP	180/82
PVDC	160/71
PSF	100/38
PVC Type 1	130/54
PVC Type 2	130/54
CPVC	X
CPE	250/121
PUR	
Acetals	X
Polyesters	
Bisphenol A-fumurate[b]	90/32
Halogenated[b]	120/49
Hydrogenated-Bisphenol A	X
Isophthalic	X
PET	X
Epoxy	X
Vinyl esters	X
Furans	X
Phenolics	X
Phenol-formaldehyde	
Methyl appended silicone	X
PB	

[a] Corrodent will permeate.
[b] Synthetic veil surfacing mat should be used.

Hydrofluoric Acid, 70%

Polymer	°F/°C
ABS	X
Acrylics	
CTFE	
ECTFE	240/116
ETFE	250/121
FEP[a]	400/204
Polyamides	X
PAI	
PEEK	X
PEI	
PES	
PFA[a]	200/93
PTFE[a]	450/232
PVDF	210/99
EHMWPE	X
HMWPE	X
PET	
PPO	X
PPS	
PP	200/93
PVDC	
PSF	X
PVC Type 1	80/27
PVC Type 2	
CPVC	90/32
CPE	X
PUR	
Acetals	X
Polyesters	
Bisphenol A-fumurate	
Halogenated	
Hydrogenated-Bisphenol A	X
Isophthalic	X
PET	X
Epoxy	X
Vinyl esters	X
Furans	
Phenolics	X
Phenol-formaldehyde	
Methyl appended silicone	X
PB	

[a] Corrodent will permeate.

Hydrogen Chloride Gas, Dry

Polymer	°F/°C
ABS	
Acrylics	
CTFE	
ECTFE	300/149
ETFE	300/149
FEP[a]	400/204
Polyamides	
PAI	
PEEK	
PEI	
PES	
PFA[a]	200/93
PTFE[a]	450/232
PVDF	280/138
EHMWPE	140/60
HMWPE	
PET	
PPO	150/66
PPS	
PP	220/104
PVDC	180/82
PSF	
PVC Type 1	90/32
PVC Type 2	80/27
CPVC	
CPE	250/121
PUR	
Acetals	
Polyesters	
Bisphenol A-fumurate	100/38
Halogenated	250/121
Hydrogenated-Bisphenol A	
Isophthalic	230/110
PET	
Epoxy	140/60
Vinyl esters	
Furans	250/121
Phenolics	
Phenol-formaldehyde	
Methyl appended silicone	
PB	

[a] Corrodent will permeate.

Hydrogen Chloride Gas, Moist

Polymer	°F/°C
ABS	
Acrylics	
CTFE	
ECTFE	
ETFE	
FEP	400/204
Polyamides	
PAI	
PEEK	
PEI	
PES	
PFA	
PTFE	450/232
PVDF	270/132
EHMWPE	140/60
HMWPE	
PET	
PPO	
PPS	
PP	
PVDC	
PSF	
PVC Type 1	
PVC Type 2	
CPVC	80/27
CPE	
PUR	
Acetals	
Polyesters	
Bisphenol A-fumurate	210/99
Halogenated	210/99
Hydrogenated-Bisphenol A	
Isophthalic	120/49
PET	
Epoxy	
Vinyl esters	350/177
Furans	
Phenolics	
Phenol-formaldehyde	
Methyl appended silicone	
PB	

Hydrogen Peroxide, Dil.

Polymer	°F/°C
ABS, 3%	80/27
Acrylics	
CTFE	
ECTFE	270/132
ETFE	250/121
FEP	400/204
Polyamides	80/27
PAI	
PEEK	200/93
PEI	
PES	80/27
PFA	200/93
PTFE	450/232
PVDF	250/121
EHMWPE	120/49
HMWPE	120/49
PET	120/49
PPO	90/32
PPS	
PP	100/38
PVDC	120/49
PSF	90/32
PVC Type 1	140/60
PVC Type 2	
CPVC	X
CPE	150/66
PUR	
Acetals	80/27
Polyesters	
Bisphenol A-fumurate	150/66
Halogenated, 5%	200/93
Hydrogenated-Bisphenol A	80/27
Isophthalic, 5%	180/27
PET	120/49
Epoxy, 10%	150/66
Vinyl esters	140/60
Furans	X
Phenolics	
Phenol-formaldehyde	
Methyl appended silicone	
PB	

Hydrogen Peroxide, 30%

Polymer	°F/°C
ABS	X
Acrylics	
CTFE	
ECTFE	270/132
ETFE	250/121
FEP	400/204
Polyamides	80/27
PAI	
PEEK	200/93
PEI	80/27
PES	80/27
PFA	200/93
PTFE	450/232
PVDF	250/121
EHMWPE	140/60
HMWPE	140/60
PET	120/49
PPO	90/32
PPS	
PP	100/38
PVDC	120/49
PSF	180/82
PVC Type 1	140/60
PVC Type 2	X
CPVC	180/82
CPE	150/66
PUR	
Acetals	X
Polyesters	
Bisphenol A-fumurate	90/32
Halogenated	100/38
Hydrogenated-Bisphenol A	90/32
Isophthalic	X
PET	120/49
Epoxy	140/60
Vinyl esters	170/77
Furans	X
Phenolics	
Phenol-formaldehyde	
Methyl appended silicone	
PB	

Hydrogen Peroxide, 50%

Polymer	°F/°C
ABS	X
Acrylics	80/27
CTFE	
ECTFE	150/66
ETFE	150/66
FEP	400/204
Polyamides	80/27
PAI	
PEEK	200/83
PEI	90/32
PES	80/27
PFA	200/93
PTFE	450/232
PVDF	250/121
EHMWPE	140/60
HMWPE	140/60
PET	80/27
PPO	X
PPS	
PP	150/66
PVDC	130/54
PSF	80/27
PVC Type 1	140/60
PVC Type 2	100/38
CPVC	90/32
CPE	150/66
PUR	
Acetals	X
Polyesters	
Bisphenol A-fumurate	X
Halogenated	100/38
Hydrogenated-Bisphenol A	
Isophthalic	X
PET	80/27
Epoxy	X
Vinyl esters	110/43
Furans	
Phenolics	
Phenol-formaldehyde	
Methyl appended silicone	
PB	

Hydrogen Peroxide, 90%

Polymer	°F/°C
ABS	X
Acrylics	80/27
CTFE	
ECTFE	150/66
ETFE	150/66
FEP	400/204
Polyamides	80/27
PAI	
PEEK	200/23
PEI	80/27
PES	80/27
PFA	200/93
PTFE	450/232
PVDF	120/49
EHMWPE	80/27
HMWPE	140/60
PET	80/27
PPO	X
PPS	
PP	110/43
PVDC	120/49
PSF	100/38
PVC Type 1	140/60
PVC Type 2	X
CPVC	80/27
CPE	140/60
PUR	
Acetals	X
Polyesters	
Bisphenol A-fumurate	X
Halogenated	
Hydrogenated-Bisphenol A	
Isophthalic	X
PET	80/27
Epoxy	X
Vinyl esters	150/66
Furans	80/27
Phenolics	80/27
Phenol-formaldehyde	
Methyl appended silicone	
PB	90/32

Hydrogen Sulfide, Dry

Polymer	°F/°C
ABS	140/60
Acrylics	150/66
CTFE	
ECTFE	300/149
ETFE	300/149
FEP	400/204
Polyamides	X
PAI	
PEEK	200/93
PEI	
PES	80/27
PFA	200/93
PTFE	450/232
PVDF[a]	280/138
EHMWPE	140/60
HMWPE	140/60
PET	80/27
PPO	80/27
PPS	250/121
PP	170/77
PVDC	160/71
PSF	
PVC Type 1	140/60
PVC Type 2	140/60
CPVC	180/60
CPE	220/104
PUR	
Acetals	80/27
Polyesters	
Bisphenol A-fumurate	200/93
Halogenated	250/121
Hydrogenated-Bisphenol A	
Isophthalic	140/60
PET	80/27
Epoxy	250/121
Vinyl esters	250/121
Furans	250/121
Phenolics	300/149
Phenol-formaldehyde	
Methyl appended silicone	
PB	90/32

[a] Corrodent will permeate.

Hydrogen Sulfide, Wet

Polymer	°F/°C
ABS	
Acrylics	140/60
CTFE	
ECTFE	300/149
ETFE	300/149
FEP	400/232
Polyamides	X
PAI	
PEEK	200/93
PEI	
PES	80/27
PFA	200/93
PTFE[a]	450/232
PVDF	280/138
EHMWPE	140/60
HMWPE	140/60
PET	80/27
PPO	X
PPS	
PP	180/82
PVDC	160/71
PSF	
PVC Type 1	140/60
PVC Type 2	140/60
CPVC	
CPE	180/82
PUR	
Acetals	X
Polyesters	
Bisphenol A-fumurate	210/99
Halogenated	210/99
Hydrogenated-Bisphenol A	
Isophthalic	140/60
PET	80/27
Epoxy	300/149
Vinyl esters	180/82
Furans	250/121
Phenolics	300/149
Phenol-formaldehyde	
Methyl appended silicone	
PB	

[a] Corrodent will permeate.

Hypochlorous Acid

Polymer	°F/°C
ABS	140/60
Acrylics	
CTFE	
ECTFE	300/149
ETFE	300/149
FEP	400/204
Polyamides	
PAI	
PEEK	
PEI	
PES	
PFA	400/204
PTFE	450/232
PVDF	280/138
EHMWPE	
HMWPE	140/60
PET	
PPO	
PPS	
PP	140/60
PVDC	120/49
PSF	
PVC Type 1	140/60
PVC Type 2	140/60
CPVC	180/82
CPE	150/66
PUR	
Acetals	
Polyesters	
Bisphenol A-fumurate, 20%	90/32
Halogenated, 10%	100/38
Hydrogenated-Bisphenol A, 50%	210/99
Isophthalic	90/32
PET	
Epoxy	200/93
Vinyl esters, 50%	150/66
Furans, 10%	X
Phenolics	
Phenol-formaldehyde	
Methyl appended silicone	
PB	

Ketones, General	
Polymer	**°F/°C**
ABS	X
Acrylics	X
CTFE	
ECTFE	
ETFE	
FEP	400/204
Polyamides	150/66
PAI	
PEEK	80/27
PEI	
PES	
PFA	400/204
PTFE	450/232
PVDF	110/43
EHMWPE	80/27
HMWPE	80/27
PET	X
PPO	X
PPS	
PP	110/43
PVDC	90/32
PSF	X
PVC Type 1	X
PVC Type 2	X
CPVC	X
CPE	X
PUR	
Acetals	
Polyesters	
Bisphenol A-fumurate	
Halogenated	
Hydrogenated-Bisphenol A	X
Isophthalic	X
PET	
Epoxy	X
Vinyl esters	X
Furans	100/38
Phenolics	
Phenol-formaldehyde	
Methyl appended silicone	
PB	

Lactic Acid, 25%

Polymer	°F/°C
ABS	140/60
Acrylics	X
CTFE	
ECTFE	150/66
ETFE	250/121
FEP	400/204
Polyamides	200/93
PAI	200/93
PEEK	60/15
PEI	
PES	
PFA	400/204
PTFE	450/232
PVDF	130/54
EHMWPE	140/60
HMWPE	150/66
PET	250/121
PPO	190/88
PPS	250/121
PP	150/66
PVDC	120/49
PSF	200/93
PVC Type 1	140/60
PVC Type 2	140/60
CPVC	200/93
CPE	250/121
PUR	
Acetals	X
Polyesters	
Bisphenol A-fumurate	210/99
Halogenated	200/93
Hydrogenated-Bisphenol A	210/99
Isophthalic	160/71
PET	250/121
Epoxy	220/104
Vinyl esters	210/99
Furans	250/121
Phenolics	160/71
Phenol-formaldehyde	
Methyl appended silicone	80/27
PB	

Lactic Acid, Conc.

Polymer	°F/°C
ABS	
Acrylics	X
CTFE	
ECTFE	150/66
ETFE	250/121
FEP	400/204
Polyamides	200/93
PAI	200/93
PEEK	90/32
PEI	
PES	
PFA	400/204
PTFE	450/232
PVDF	110/43
EHMWPE	140/60
HMWPE	
PET	250/121
PPO	190/88
PPS	250/121
PP	150/66
PVDC	80/27
PSF	200/93
PVC Type 1	80/27
PVC Type 2	80/27
CPVC	X
CPE	150/66
PUR	
Acetals, 10%	80/27
Polyesters	
Bisphenol A-fumurate	220/104
Halogenated	200/93
Hydrogenated-Bisphenol A	210/99
Isophthalic	160/71
PET	250/121
Epoxy	200/93
Vinyl esters	200/93
Furans	260/127
Phenolics	
Phenol-formaldehyde	160/71
Methyl appended silicone	80/27
PB	

Magnesium Chloride

Polymer	°F/°C
ABS	140/60
Acrylics	80/27
CTFE	300/149
ECTFE	300/149
ETFE	300/149
FEP	400/204
Polyamides	250/121
PAI, 10%	200/93
PEEK	
PEI	
PES	
PFA	380/193
PTFE	450/232
PVDF	280/138
EHMWPE	140/60
HMWPE	140/60
PET	250/121
PPO	200/93
PPS	300/149
PP	210/99
PVDC	180/82
PSF	
PVC Type 1	140/60
PVC Type 2	140/60
CPVC	230/110
CPE	270/132
PUR	80/27
Acetals, 10%	80/27
Polyesters	
Bisphenol A-fumurate	220/104
Halogenated	250/121
Hydrogenated-Bisphenol A	210/99
Isophthalic	180/82
PET	250/121
Epoxy	190/88
Vinyl esters	260/127
Furans	260/127
Phenolics	
Phenol-formaldehyde	
Methyl appended silicone	400/204
PB	

Malic Acid

Polymer	°F/°C
ABS	140/60
Acrylics	
CTFE	
ECTFE	250/121
ETFE	270/132
FEP	400/204
Polyamides	X
PAI	
PEEK	
PEI	
PES	
PFA	390/199
PTFE	450/232
PVDF	250/121
EHMWPE	140/60
HMWPE	140/60
PET	80/27
PPO	
PPS	
PP	130/54
PVDC	80/27
PSF	
PVC Type 1	140/60
PVC Type 2	140/60
CPVC	180/82
CPE	250/121
PUR	
Acetals	
Polyesters	
Bisphenol A-fumurate	160/71
Halogenated, 10%	90/32
Hydrogenated-Bisphenol A	
Isophthalic	90/32
PET	80/27
Epoxy	
Vinyl esters	140/60
Furans	260/127
Phenolics	
Phenol-formaldehyde	
Methyl appended silicone	
PB	

Methyl Chloride

Polymer	°F/°C
ABS	X
Acrylics	X
CTFE	
ECTFE	300/149
ETFE	300/149
FEP[a]	400/204
Polyamides	210/99
PAI	
PEEK	
PEI	X
PES	
PFA[a]	390/199
PTFE[b]	450/232
PVDF	X
EHMWPE	X
HMWPE	X
PET	80/27
PPO	X
PPS	
PP	X
PVDC	80/27
PSF	X
PVC Type 1	X
PVC Type 2	X
CPVC	X
CPE	230/110
PUR	X
Acetals	
Polyesters	
Bisphenol A-fumurate	
Halogenated	80/27
Hydrogenated-Bisphenol A	
Isophthalic	
PET	80/27
Epoxy	X
Vinyl esters	X
Furans	120/49
Phenolics	
Phenol-formaldehyde	300/149
Methyl appended silicone	
PB	

[a] Material is subject to stress cracking.
[b] Corrodent will permeate.

Methyl Ethyl Ketone

Polymer	°F/°C
ABS	X
Acrylics	X
CTFE	
ECTFE	150/66
ETFE	230/110
FEP[a]	400/204
Polyamides	240/116
PAI	200/93
PEEK	370/188
PEI	X
PES	X
PFA[a]	400/204
PTFE[a]	450/232
PVDF	X
EHMWPE	X
HMWPE	X
PET	250/121
PPO	X
PPS	200/93
PP	150/66
PVDC	X
PSF	X
PVC Type 1	X
PVC Type 2	X
CPVC	X
CPE	90/32
PUR	X
Acetals	X
Polyesters	
Bisphenol A-fumurate	X
Halogenated	X
Hydrogenated-Bisphenol A	X
Isophthalic	X
PET	250/121
Epoxy	90/32
Vinyl esters	X
Furans	170/77
Phenolics	X
Phenol-formaldehyde	
Methyl appended silicone	X
PB	

[a] Corrodent will permeate.

Methyl Isobutyl Ketone

Polymer	°F/°C
ABS	X
Acrylics	
CTFE	
ECTFE	150/66
ETFE	300/149
FEP[a]	400/204
Polyamides	110/43
PAI	
PEEK	
PEI	
PES	
PFA[a]	400/204
PTFE[a,b]	450/232
PVDF	110/43
EHMWPE	80/27
HMWPE	80/27
PET	X
PPO	X
PPS	250/121
PP	80/27
PVDC	80/27
PSF	
PVC Type 1	X
PVC Type 2	X
CPVC	X
CPE	80/27
PUR	
Acetals	
Polyesters	
Bisphenol A-fumurate	X
Halogenated	80/27
Hydrogenated-Bisphenol A	X
Isophthalic	X
PET	X
Epoxy	140/60
Vinyl esters	
Furans	
Phenolics	160/71
Phenol-formaldehyde	
Methyl appended silicone	X
PB	

[a] Corrodent will permeate.
[b] Material subject to stress cracking.

Muriatic Acid

Polymer	°F/°C
ABS	140/60
Acrylics	
CTFE	
ECTFE	300/149
ETFE	300/149
FEP[a]	400/204
Polyamides	X
PAI	
PEEK	
PEI	
PES	
PFA[a]	400/204
PTFE[a]	450/232
PVDF	
EHMWPE	140/60
HMWPE	60/15
PET	80/27
PPO	
PPS	210/99
PP	200/93
PVDC	
PSF	
PVC Type 1	160/71
PVC Type 2	150/66
CPVC	180/82
CPE	
PUR	
Acetals	
Polyesters	
Bisphenol A-fumurate	130/54
Halogenated	190/88
Hydrogenated-Bisphenol A	180/82
Isophthalic	160/71
PET	80/27
Epoxy	140/60
Vinyl esters	180/82
Furans	80/27
Phenolics	300/149
Phenol-formaldehyde	
Methyl appended silicone	
PB	

[a] Corrodent will permeate.

Nitric Acid, 5%

Polymer	°F/°C
ABS	140/60
Acrylics	90/32
CTFE	
ECTFE	300/149
ETFE	150/66
FEP[a]	400/204
Polyamides	X
PAI	
PEEK	200/93
PEI	80/27
PES	80/27
PFA[a]	400/204
PTFE[a]	450/232
PVDF	200/93
EHMWPE	140/60
HMWPE	140/60
PET	150/66
PPO	140/66
PPS	150/66
PP	140/60
PVDC	90/32
PSF	X
PVC Type 1	140/60
PVC Type 2	100/38
CPVC	150/66
CPE	180/82
PUR	X
Acetals	X
Polyesters	
Bisphenol A-fumurate	160/71
Halogenated	210/99
Hydrogenated-Bisphenol A	90/32
Isophthalic	120/49
PET	150/66
Epoxy	160/71
Vinyl esters	180/82
Furans	200/93
Phenolics	X
Phenol-formaldehyde	
Methyl appended silicone	80/27
PB	80/27

[a] Corrodent will permeate.

Nitric Acid, 20%

Polymer	°F/°C
ABS	130/54
Acrylics	X
CTFE	
ECTFE	250/121
ETFE	160/71
FEP[a]	400/204
Polyamides	X
PAI	
PEEK	200/93
PEI	80/27
PES	X
PFA[a]	400/204
PTFE[a]	450/232
PVDF	180/82
EHMWPE	140/60
HMWPE	140/60
PET	80/27
PPO	90/32
PPS	100/38
PP	140/60
PVDC	150/66
PSF	X
PVC Type 1	140/60
PVC Type 2	140/60
CPVC	160/71
CPE	80/27
PUR	X
Acetals	X
Polyesters	
Bisphenol A-fumurate	100/38
Halogenated	80/27
Hydrogenated-Bisphenol A	
Isophthalic	X
PET	80/27
Epoxy	100/38
Vinyl esters	150/66
Furans	X
Phenolics	
Phenol-formaldehyde	
Methyl appended silicone	X
PB	80/27

[a] Corrodent will permeate.

Nitric Acid, 70%

Polymer	°F/°C
ABS	X
Acrylics	X
CTFE	
ECTFE	150/66
ETFE	250/121
FEP[a]	400/204
Polyamides	X
PAI	
PEEK	
PEI	
PES	
PFA[a]	400/204
PTFE[a]	450/232
PVDF	120/49
EHMWPE	X
HMWPE	X
PET	80/27
PPO	X
PPS	
PP	X
PVDC	X
PSF	X
PVC Type 1	140/60
PVC Type 2	140/60
CPVC	110/43
CPE	80/27
PUR	X
Acetals	X
Polyesters	
Bisphenol A-fumurate	
Halogenated	80/27
Hydrogenated-Bisphenol A	
Isophthalic	X
PET	80/27
Epoxy	X
Vinyl esters	X
Furans	X
Phenolics	
Phenol-formaldehyde	
Methyl appended silicone	X
PB	X

[a] Corrodent will permeate.

Nitric Acid, Anhydrous

Polymer	°F/°C
ABS	X
Acrylics	X
CTFE	
ECTFE	150/66
ETFE	X
FEP[a]	400/204
Polyamides	X
PAI	
PEEK	
PEI	
PES	
PFA[a], 90%	400/204
PTFE[a]	450/232
PVDF	150/66
EHMWPE	X
HMWPE	X
PET	
PPO	X
PPS	
PP	X
PVDC	X
PSF	X
PVC Type 1	X
PVC Type 2	X
CPVC	X
CPE	X
PUR	X
Acetals	X
Polyesters	
Bisphenol A-fumurate	
Halogenated	
Hydrogenated-Bisphenol A	
Isophthalic	X
PET	
Epoxy	X
Vinyl esters	X
Furans	X
Phenolics	80/27
Phenol-formaldehyde	
Methyl appended silicone	X
PB	X

[a] Corrodent will permeate.

Nitric Acid, 70%

Polymer	°F/°C
ABS	X
Acrylics	X
CTFE	
ECTFE	150/66
ETFE	250/121
FEP[a]	400/204
Polyamides	X
PAI	
PEEK	
PEI	
PES	
PFA[a]	400/204
PTFE[a]	450/232
PVDF	120/49
EHMWPE	X
HMWPE	X
PET	80/27
PPO	X
PPS	
PP	X
PVDC	X
PSF	X
PVC Type 1	140/60
PVC Type 2	140/60
CPVC	110/43
CPE	80/27
PUR	X
Acetals	X
Polyesters	
Bisphenol A-fumurate	
Halogenated	80/27
Hydrogenated-Bisphenol A	
Isophthalic	X
PET	80/27
Epoxy	X
Vinyl esters	X
Furans	X
Phenolics	
Phenol-formaldehyde	
Methyl appended silicone	X
PB	X

[a] Corrodent will permeate.

Oleum

Polymer	°F/°C
ABS	X
Acrylics	X
CTFE	
ECTFE	60/15
ETFE	150/66
FEP	400/204
Polyamides	X
PAI	120/49
PEEK	
PEI	
PES	
PFA	400/204
PTFE	450/232
PVDF	X
EHMWPE	X
HMWPE	X
PET	
PPO	
PPS	80/27
PP	X
PVDC	X
PSF	
PVC Type 1	X
PVC Type 2	X
CPVC	X
CPE	
PUR	X
Acetals	X
Polyesters	
Bisphenol A-fumurate	X
Halogenated	X
Hydrogenated-Bisphenol A	X
Isophthalic	X
PET	
Epoxy	X
Vinyl esters	X
Furans	190/88
Phenolics	
Phenol-formaldehyde	
Methyl appended silicone	X
PB	

Perchloric Acid, 10%

Polymer	°F/°C
ABS	X
Acrylics	
CTFE	200/93
ECTFE	150/66
ETFE	230/110
FEP	400/204
Polyamides	X
PAI	
PEEK	
PEI	
PES	
PFA	400/204
PTFE	450/232
PVDF	250/121
EHMWPE	140/60
HMWPE	140/60
PET	X
PPO	X
PPS	
PP	150/66
PVDC	130/54
PSF	X
PVC Type 1	140/60
PVC Type 2	60/15
CPVC	180/82
CPE	140/60
PUR	X
Acetals	X
Polyesters	
Bisphenol A-fumurate	
Halogenated	90/32
Hydrogenated-Bisphenol A	X
Isophthalic	X
PET	X
Epoxy	90/32
Vinyl esters	150/66
Furans	X
Phenolics	
Phenol-formaldehyde	
Methyl appended silicone	
PB	

Perchloric Acid, 70%

Polymer	°F/°C
ABS	X
Acrylics	
CTFE	200/93
ECTFE	150/66
ETFE	150/66
FEP	400/204
Polyamides	X
PAI	
PEEK	
PEI	
PES	
PFA	400/204
PTFE	450/232
PVDF	120/49
EHMWPE	X
HMWPE	X
PET	X
PPO	X
PPS	
PP	X
PVDC	120/49
PSF	
PVC Type 1	60/15
PVC Type 2	60/15
CPVC	180/82
CPE	X
PUR	X
Acetals	X
Polyesters	
Bisphenol A-fumurate	
Halogenated	90/32
Hydrogenated-Bisphenol A	X
Isophthalic	X
PET	X
Epoxy	80/27
Vinyl esters	X
Furans	260/127
Phenolics	
Phenol-formaldehyde	
Methyl appended silicone	
PB	

Phenol, 10%

Polymer	°F/°C
ABS	X
Acrylics	90/32
CTFE	
ECTFE	150/66
ETFE	230/110
FEP[a]	400/204
Polyamides	X
PAI	
PEEK	140/60
PEI	X
PES	X
PFA[a]	400/204
PTFE[a]	450/232
PVDF	210/99
EHMWPE	100/38
HMWPE	100/38
PET	X
PPO	X
PPS	290/143
PP	200/93
PVDC	80/27
PSF	
PVC Type 1	X
PVC Type 2	X
CPVC	90/32
CPE	90/32
PUR	X
Acetals	X
Polyesters	
Bisphenol A-fumurate	X
Halogenated, 5%	90/32
Hydrogenated-Bisphenol A	X
Isophthalic	X
PET	X
Epoxy	X
Vinyl esters	100/38
Furans	210/99
Phenolics	90/32
Phenol-formaldehyde	
Methyl appended silicone	X
PB	80/27

[a] Corrodent will permeate.

Phenol

Polymer	°F/°C
ABS	X
Acrylics	90/32
CTFE	
ECTFE	150/66
ETFE	210/99
FEP[a]	400/201
Polyamides	X
PAI	
PEEK	140/60
PEI	
PES	X
PFA[a]	400/202
PTFE[a]	450/232
PVDF	200/93
EHMWPE	100/38
HMWPE	100/38
PET	X
PPO	
PPS	300/149
PP	180/82
PVDC	X
PSF	
PVC Type 1	60/15
PVC Type 2	X
CPVC	140/60
CPE	90/32
PUR	X
Acetals	X
Polyesters	
Bisphenol A-fumurate	X
Halogenated, 5%	90/32
Hydrogenated-Bisphenol A	X
Isophthalic	X
PET	X
Epoxy	X
Vinyl esters	X
Furans, 85%	210/99
Phenolics	X
Phenol-formaldehyde	
Methyl appended silicone	X
PB	80/27

[a] Corrodent will permeate.

Phosphoric Acid, 50–80%

Polymer	°F/°C
ABS	130/54
Acrylics	80/27
CTFE	
ECTFE	250/121
ETFE	250/121
FEP	400/204
Polyamides	X
PAI	
PEEK	200/93
PEI	80/27
PES	200/93
PFA	200/93
PTFE	450/232
PVDF	250/121
EHMWPE	100/38
HMWPE	100/38
PET	250/121
PPO	140/60
PPS	220/104
PP	210/99
PVDC	130/54
PSF	80/27
PVC Type 1	140/60
PVC Type 2	140/60
CPVC	180/82
CPE	250/121
PUR	
Acetals	X
Polyesters	
Bisphenol A-fumurate	220/104
Halogenated	250/121
Hydrogenated-Bisphenol A	210/99
Isophthalic	180/82
PET	250/121
Epoxy	110/43
Vinyl esters	210/99
Furans	260/127
Phenolics	X
Phenol-formaldehyde	
Methyl appended silicone	X
PB	

Picric Acid

Polymer	°F/°C
ABS	X
Acrylics	90/32
CTFE	
ECTFE	80/27
ETFE	120/49
FEP	400/204
Polyamides	X
PAI	
PEEK	
PEI	
PES	
PFA	400/204
PTFE	450/232
PVDF	80/27
EHMWPE	100/30
HMWPE	X
PET	X
PPO	
PPS	
PP	140/60
PVDC	120/49
PSF	
PVC Type 1	X
PVC Type 2	X
CPVC	X
CPE	150/66
PUR	
Acetals	
Polyesters	
Bisphenol A-fumurate	110/43
Halogenated	100/38
Hydrogenated-Bisphenol A	
Isophthalic	X
PET	X
Epoxy	80/27
Vinyl esters, 10%	200/93
Furans	160/71
Phenolics	
Phenol-formaldehyde	
Methyl appended silicone	
PB	

Potassium Bromide, 30%

Polymer	°F/°C
ABS	140/60
Acrylics	80/27
CTFE	
ECTFE	300/149
ETFE	300/149
FEP	400/204
Polyamides, 10%	210/99
PAI	
PEEK	140/60
PEI	
PES	140/66
PFA	400/204
PTFE	450/232
PVDF	280/138
EHMWPE	140/60
HMWPE	140/60
PET	80/27
PPO	140/60
PPS	200/93
PP	210/99
PVDC	110/43
PSF	200/93
PVC Type 1	140/60
PVC Type 2	140/60
CPVC	200/93
CPE	250/121
PUR	90/32
Acetals	
Polyesters	
Bisphenol A-fumurate	200/93
Halogenated, 30%	230/110
Hydrogenated-Bisphenol A	
Isophthalic	160/71
PET	80/27
Epoxy	220/104
Vinyl esters	200/93
Furans	260/127
Phenolics	
Phenol-formaldehyde	
Methyl appended silicone	
PB	

Salicylic Acid

Polymer	°F/°C
ABS	
Acrylics	
CTFE	
ECTFE	250/121
ETFE	250/121
FEP	400/204
Polyamides	80/27
PAI	
PEEK	
PEI	
PES	
PFA	400/204
PTFE	450/232
PVDF	220/104
EHMWPE	
HMWPE	140/60
PET	
PPO	
PPS	
PP	130/54
PVDC	130/54
PSF	100/38
PVC Type 1	140/60
PVC Type 2	X
CPVC	X
CPE	220/104
PUR	
Acetals	
Polyesters	
Bisphenol A-fumurate	150/66
Halogenated	130/54
Hydrogenated-Bisphenol A	
Isophthalic	100/38
PET	
Epoxy	250/121
Vinyl esters	150/66
Furans	260/127
Phenolics	
Phenol-formaldehyde	
Methyl appended silicone	
PB	

Silver Bromide, 10%

Polymer	°F/°C
ABS	
Acrylics	
CTFE	
ECTFE	
ETFE	
FEP	400/204
Polyamides	
PAI	
PEEK	
PEI	
PES	
PFA	
PTFE	450/232
PVDF	250/121
EHMWPE	
HMWPE	
PET	
PPO	
PPS	
PP	170/77
PVDC	
PSF	
PVC Type 1	140/60
PVC Type 2	90/32
CPVC	170/77
CPE	
PUR	
Acetals	
Polyesters	
Bisphenol A-fumurate	
Halogenated	
Hydrogenated-Bisphenol A	
Isophthalic	180/82
PET	
Epoxy	
Vinyl esters	
Furans	
Phenolics	
Phenol-formaldehyde	
Methyl appended silicone	
PB	

Sodium Carbonate

Polymer	°F/°C
ABS, 25%	140/60
Acrylics	X
CTFE	
ECTFE	300/149
ETFE	300/149
FEP	400/204
Polyamides	240/116
PAI, 10%	200/93
PEEK	210/99
PEI	
PES	80/27
PFA	400/204
PTFE	450/232
PVDF	280/138
EHMWPE	140/60
HMWPE	140/60
PET	250/121
PPO	200/93
PPS	300/149
PP	220/104
PVDC	180/82
PSF	200/93
PVC Type 1	140/60
PVC Type 2	140/60
CPVC	210/99
CPE	250/121
PUR	
Acetals	80/27
Polyesters	
Bisphenol A-fumurate	160/71
Halogenated, 10%	190/88
Hydrogenated-Bisphenol A, 10%	100/38
Isophthalic, 20%	90/32
PET	250/121
Epoxy	300/149
Vinyl esters	180/82
Furans, 50%	260/127
Phenolics	X
Phenol-formaldehyde	
Methyl appended silicone	300/149
PB	90/32

Sodium Chloride

Polymer	°F/°C
ABS	140/60
Acrylics	80/27
CTFE	
ECTFE	300/149
ETFE	300/149
FEP	400/204
Polyamides	250/121
PAI, 10%	200/93
PEEK	
PEI	80/27
PES	80/27
PFA	400/204
PTFE	450/232
PVDF	280/138
EHMWPE	140/60
HMWPE	140/60
PET	250/121
PPO	100/38
PPS	300/149
PP	220/104
PVDC	180/82
PSF	200/93
PVC Type 1	140/60
PVC Type 2	140/60
CPVC	210/99
CPE	250/121
PUR	80/27
Acetals, 10%	180/82
Polyesters	
Bisphenol A-fumurate	220/104
Halogenated	250/121
Hydrogenated-Bisphenol A	210/99
Isophthalic	200/93
PET	250/121
Epoxy	210/99
Vinyl esters	180/82
Furans	260/127
Phenolics	300/149
Phenol-formaldehyde	
Methyl appended silicone, 10%	400/204
PB	

Sodium Hydroxide, 10%

Polymer	°F/°C
ABS	140/60
Acrylics	80/27
CTFE	
ECTFE	300/149
ETFE	230/110
FEP	400/204
Polyamides	250/121
PAI	X
PEEK	220/104
PEI	90/32
PES	80/27
PFA	400/204
PTFE	450/232
PVDF[a]	210/99
EHMWPE	170/77
HMWPE	150/66
PET	130/54
PPO	140/66
PPS	210/99
PP	220/104
PVDC	90/32
PSF	200/93
PVC Type 1	140/60
PVC Type 2	140/60
CPVC	190/88
CPE	180/82
PUR	
Acetals	80/27
Polyesters	
Bisphenol A-fumurate	130/54
Halogenated	110/43
Hydrogenated-Bisphenol A	100/38
Isophthalic	X
PET	130/54
Epoxy	190/88
Vinyl esters	170/77
Furans	X
Phenolics	X
Phenol-formaldehyde	
Methyl appended silicone	90/32
PB	90/32

[a] Material is subject to stress cracking.

Sodium Hydroxide, 50%

Polymer	°F/°C
ABS	140/60
Acrylics	80/27
CTFE	
ECTFE	250/121
ETFE	230/110
FEP	400/204
Polyamides	250/121
PAI	X
PEEK	180/82
PEI	
PES	80/27
PFA	400/204
PTFE	450/232
PVDF[a]	220/104
EHMWPE	170/77
HMWPE	150/66
PET	X
PPO	140/60
PPS	210/99
PP	220/104
PVDC	150/66
PSF	200/93
PVC Type 1	140/60
PVC Type 2	140/60
CPVC	180/82
CPE	250/121
PUR	90/32
Acetals	X
Polyesters	
Bisphenol A-fumurate	220/104
Halogenated	X
Hydrogenated-Bisphenol A	X
Isophthalic	X
PET	X
Epoxy	200/93
Vinyl esters	220/104
Furans	X
Phenolics	X
Phenol-formaldehyde	
Methyl appended silicone	90/32
PB	90/32

[a] Material is subject to stress cracking.

Sodium Hydroxide, Conc.

Polymer	°F/°C
ABS	140/60
Acrylics	90/32
CTFE	
ECTFE	150/66
ETFE	
FEP	400/204
Polyamides	
PAI	
PEEK	200/93
PEI	
PES	
PFA	
PTFE	450/232
PVDF[a]	150/66
EHMWPE	
HMWPE	
PET	
PPO	X
PPS	
PP	140/60
PVDC	X
PSF	80/27
PVC Type 1	140/60
PVC Type 2	140/60
CPVC	190/88
CPE	180/82
PUR	
Acetals	X
Polyesters	
Bisphenol A-fumurate	
Halogenated	
Hydrogenated-Bisphenol A	
Isophthalic	
PET	
Epoxy	
Vinyl esters	X
Furans	X
Phenolics	X
Phenol-formaldehyde	
Methyl appended silicone	90/32
PB	90/32

[a] Material is subject to stress cracking.

Sodium Hypochlorite, 20%

Polymer	°F/°C
ABS	140/60
Acrylics, 15%	100/38
CTFE	
ECTFE	300/149
ETFE	300/149
FEP	400/204
Polyamides	X
PAI, 10%	200/93
PEEK	80/27
PEI	80/27
PES	80/27
PFA	400/204
PTFE	450/232
PVDF	280/138
EHMWPE	140/60
HMWPE	140/60
PET	80/27
PPO, 20%	80/27
PPS, 5%	200/93
PP	120/49
PVDC, 10%	130/54
PSF	300/149
PVC Type 1	140/60
PVC Type 2	140/60
CPVC	200/93
CPE	180/82
PUR	X
Acetals	X
Polyesters	
Bisphenol A-fumurate	X
Halogenated	X
Hydrogenated-Bisphenol A, 10%	160/71
Isophthalic	X
PET	80/27
Epoxy, 5%	80/27
Vinyl esters	180/82
Furans	X
Phenolics	X
Phenol-formaldehyde	
Methyl appended silicone	X
PB	90/32

Sodium Hypochlorite, Conc.

Polymer	°F/°C
ABS	140/60
Acrylics	
CTFE	250/121
ECTFE	300/149
ETFE	300/149
FEP	400/204
Polyamides	X
PAI	X
PEEK	80/27
PEI	80/27
PES	80/27
PFA	400/204
PTFE	450/232
PVDF	280/138
EHMWPE	140/60
HMWPE	140/60
PET	80/27
PPO	80/27
PPS	250/121
PP	110/43
PVDC	120/49
PSF	300/149
PVC Type 1	140/60
PVC Type 2	140/60
CPVC	200/93
CPE	200/93
PUR	X
Acetals	X
Polyesters	
Bisphenol A-fumurate	140/60
Halogenated	X
Hydrogenated-Bisphenol A	
Isophthalic	X
PET	80/27
Epoxy	
Vinyl esters	100/38
Furans	X
Phenolics	X
Phenol-formaldehyde	
Methyl appended silicone	X
PB	

Sodium Sulfide, to 50%

Polymer	°F/°C
ABS	140/60
Acrylics	90/32
CTFE	
ECTFE	300/149
ETFE	300/149
FEP	400/204
Polyamides	250/121
PAI	X
PEEK	
PEI	
PES	
PFA	400/204
PTFE	450/232
PVDF	260/138
EHMWPE	140/60
HMWPE	140/60
PET	240/116
PPO, 25%	140/60
PPS, all	250/121
PP	190/88
PVDC	140/60
PSF	200/93
PVC Type 1	140/60
PVC Type 2	140/60
CPVC	200/93
CPE	230/110
PUR	
Acetals	
Polyesters	
Bisphenol A-fumurate, sat.	210/99
Halogenated	X
Hydrogenated-Bisphenol A	
Isophthalic, sat.	X
PET	240/116
Epoxy, to 10%	250/121
Vinyl esters, all	220/104
Furans, 10%	260/127
Phenolics	
Phenol-formaldehyde	
Methyl appended silicone	
PB	90/32

Stannic Chloride

Polymer	°F/°C
ABS	140/60
Acrylics	80/27
CTFE	
ECTFE	300/149
ETFE	300/149
FEP	400/204
Polyamides	80/27
PAI	
PEEK	
PEI	
PES	
PFA	400/204
PTFE	450/232
PVDF	280/138
EHMWPE	140/60
HMWPE	140/60
PET	170/77
PPO	90/32
PPS, all	230/110
PP	200/93
PVDC	180/82
PSF	200/93
PVC Type 1	140/60
PVC Type 2	140/60
CPVC	200/93
CPE	250/121
PUR	
Acetals	
Polyesters	
Bisphenol A-fumurate	200/93
Halogenated	80/27
Hydrogenated-Bisphenol A	
Isophthalic	180/82
PET	170/77
Epoxy	200/93
Vinyl esters, all	210/99
Furans	260/127
Phenolics	
Phenol-formaldehyde	
Methyl appended silicone	80/27
PB	

Stannous Chloride

Polymer	°F/°C
ABS	100/38
Acrylics	80/27
CTFE	210/99
ECTFE	300/149
ETFE	300/149
FEP	400/204
Polyamides, dry	450/232
PAI	
PEEK	
PEI	
PES	
PFA	400/204
PTFE	450/232
PVDF	280/138
EHMWPE	140/60
HMWPE	140/60
PET	170/77
PPO	140/60
PPS, all	210/99
PP	200/93
PVDC	180/82
PSF	
PVC Type 1	140/60
PVC Type 2	140/60
CPVC	200/93
CPE	250/121
PUR	
Acetals	
Polyesters	
Bisphenol A-fumurate	220/104
Halogenated	250/121
Hydrogenated-Bisphenol A	
Isophthalic	180/82
PET	170/77
Epoxy	220/104
Vinyl esters	200/93
Furans	240/116
Phenolics	
Phenol-formaldehyde	
Methyl appended silicone	
PB	

Sulfuric Acid, 10%

Polymer	°F/°C
ABS	140/60
Acrylics	80/27
CTFE	
ECTFE	250/121
ETFE	300/149
FEP	400/204
Polyamides	80/27
PAI	200/93
PEEK	80/27
PEI	80/27
PES	80/27
PFA	400/204
PTFE	450/232
PVDF	250/121
EHMWPE	140/60
HMWPE	140/60
PET	160/71
PPO	140/60
PPS	250/121
PP	200/93
PVDC	120/49
PSF	300/149
PVC Type 1	140/60
PVC Type 2	140/60
CPVC	180/82
CPE	250/121
PUR	X
Acetals	X
Polyesters	
Bisphenol A-fumurate	220/104
Halogenated	260/127
Hydrogenated-Bisphenol A	210/99
Isophthalic	160/71
PET	160/71
Epoxy	140/60
Vinyl esters	200/93
Furans	250/121
Phenolics	300/149
Phenol-formaldehyde	
Methyl appended silicone	X
PB	80/27

Sulfuric Acid, 50%

Polymer	°F/°C
ABS	130/54
Acrylics	90/32
CTFE	230/110
ECTFE	250/121
ETFE	300/149
FEP	400/204
Polyamides	X
PAI	
PEEK	200/93
PEI	80/27
PES	X
PFA	400/204
PTFE	450/232
PVDF	220/104
EHMWPE	140/60
HMWPE	140/60
PET	140/60
PPO	90/32
PPS	250/121
PP	200/93
PVDC	X
PSF	300/149
PVC Type 1	140/60
PVC Type 2	140/60
CPVC	180/82
CPE	250/121
PUR	X
Acetals	X
Polyesters	
Bisphenol A-fumurate	220/104
Halogenated	200/93
Hydrogenated-Bisphenol A	210/99
Isophthalic	150/66
PET	140/60
Epoxy	110/43
Vinyl esters	210/99
Furans	260/127
Phenolics	280/138
Phenol-formaldehyde	
Methyl appended silicone	X
PB	80/27

Sulfuric Acid, 70%

Polymer	°F/°C
ABS	X
Acrylics	80/27
CTFE	230/110
ECTFE	250/121
ETFE	300/149
FEP	400/204
Polyamides	X
PAI	
PEEK	X
PEI	X
PES	X
PFA	400/204
PTFE	450/232
PVDF	220/104
EHMWPE	80/27
HMWPE	80/27
PET	X
PPO	90/32
PPS	250/121
PP	180/82
PVDC	X
PSF	X
PVC Type 1	140/60
PVC Type 2	140/60
CPVC	200/93
CPE	220/104
PUR	X
Acetals	X
Polyesters	
Bisphenol A-fumurate	160/71
Halogenated	190/88
Hydrogenated-Bisphenol A	90/32
Isophthalic	X
PET	X
Epoxy	110/43
Vinyl esters	180/82
Furans	260/127
Phenolics	200/93
Phenol-formaldehyde	
Methyl appended silicone	X
PB	90/32

Sulfuric Acid, 90%

Polymer	°F/°C
ABS	X
Acrylics	X
CTFE	210/99
ECTFE	150/66
ETFE	300/149
FEP	400/204
Polyamides	X
PAI	
PEEK	X
PEI	X
PES	X
PFA	400/204
PTFE	450/232
PVDF	210/99
EHMWPE	X
HMWPE	X
PET	X
PPO	80/27
PPS	220/104
PP	180/82
PVDC	X
PSF	X
PVC Type 1	140/60
PVC Type 2	X
CPVC	120/49
CPE	140/60
PUR	X
Acetals	X
Polyesters	
Bisphenol A-fumurate	X
Halogenated	X
Hydrogenated-Bisphenol A	X
Isophthalic	X
PET	X
Epoxy	X
Vinyl esters	X
Furans	X
Phenolics	80/27
Phenol-formaldehyde	
Methyl appended silicone	X
PB	

Sulfuric Acid, 98%

Polymer	°F/°C
ABS	X
Acrylics	X
CTFE	210/99
ECTFE	150/66
ETFE	300/149
FEP	400/204
Polyamides	X
PAI	
PEEK	X
PEI	X
PES	X
PFA	400/204
PTFE	450/232
PVDF	140/60
EHMWPE	X
HMWPE	X
PET	X
PPO	180/82
PPS	
PP	120/49
PVDC	X
PSF	X
PVC Type 1	X
PVC Type 2	X
CPVC	120/149
CPE	X
PUR	X
Acetals	X
Polyesters	
Bisphenol A-fumurate	X
Halogenated	X
Hydrogenated-Bisphenol A	X
Isophthalic	X
PET	X
Epoxy	X
Vinyl esters	X
Furans	X
Phenolics	X
Phenol-formaldehyde	
Methyl appended silicone	X
PB	

Sulfuric Acid, 100%

Polymer	°F/°C
ABS	X
Acrylics	X
CTFE	210/99
ECTFE	80/27
ETFE	300/149
FEP	400/204
Polyamides	X
PAI	
PEEK	X
PEI	X
PES	X
PFA	400/204
PTFE	450/232
PVDF	X
EHMWPE	X
HMWPE	X
PET	X
PPO	
PPS	
PP	X
PVDC	X
PSF	X
PVC Type 1	X
PVC Type 2	X
CPVC	X
CPE	X
PUR	X
Acetals	X
Polyesters	
Bisphenol A-fumurate	
Halogenated	
Hydrogenated-Bisphenol A	X
Isophthalic	X
PET	X
Epoxy	X
Vinyl esters	X
Furans	X
Phenolics	X
Phenol-formaldehyde	
Methyl appended silicone	X
PB	

Sulfuric Acid, Fuming

Polymer	°F/°C
ABS	X
Acrylics	X
CTFE	200/93
ECTFE	300/149
ETFE	120/49
FEP[a]	400/204
Polyamides	X
PAI	120/40
PEEK	X
PEI	X
PES	X
PFA[a]	400/204
PTFE[a]	450/232
PVDF	X
EHMWPE	X
HMWPE	X
PET	
PPO	60/15
PPS	
PP	X
PVDC	X
PSF	X
PVC Type 1	X
PVC Type 2	X
CPVC	X
CPE	
PUR	
Acetals	X
Polyesters	
Bisphenol A-fumurate	
Halogenated	
Hydrogenated-Bisphenol A	
Isophthalic	X
PET	
Epoxy	X
Vinyl esters	X
Furans	X
Phenolics	
Phenol-formaldehyde	
Methyl appended silicone	X
PB	

[a] Corrodent will permeate.

Sulfurous Acid

Polymer	°F/°C
ABS	140/60
Acrylics	80/27
CTFE	
ECTFE	250/121
ETFE	210/99
FEP	400/204
Polyamides	X
PAI	
PEEK	
PEI	
PES	
PFA	400/204
PTFE	450/232
PVDF	250/121
EHMWPE	140/60
HMWPE	140/60
PET	
PPO	
PPS, 10%	200/93
PP	180/82
PVDC	80/27
PSF	200/93
PVC Type 1	140/60
PVC Type 2	140/60
CPVC	180/82
CPE	220/104
PUR	
Acetals	X
Polyesters	
Bisphenol A-fumurate	110/43
Halogenated, 10%	80/27
Hydrogenated-Bisphenol A, 25%	210/99
Isophthalic, 10%	X
PET	
Epoxy, to 20%	240/116
Vinyl esters, 10%	120/49
Furans	160/71
Phenolics	
Phenol-formaldehyde	160/71
Methyl appended silicone	X
PB	

Thionyl Chloride

Polymer	°F/°C
ABS	X
Acrylics	
CTFE	
ECTFE	150/66
ETFE	210/99
FEP[a]	400/204
Polyamides	X
PAI	
PEEK	
PEI	
PES	
PFA[a]	400/204
PTFE[a]	450/232
PVDF	X
EHMWPE	X
HMWPE	X
PET	
PPO	
PPS	X
PP	100/38
PVDC	X
PSF	X
PVC Type 1	X
PVC Type 2	X
CPVC	X
CPE	X
PUR	
Acetals	X
Polyesters	
Bisphenol A-fumurate	X
Halogenated	X
Hydrogenated-Bisphenol A	
Isophthalic	X
PET	
Epoxy	X
Vinyl esters	X
Furans	X
Phenolics	200/93
Phenol-formaldehyde	
Methyl appended silicone	
PB	

[a] Corrodent will permeate.

Toluene

Polymer	°F/°C
ABS	X
Acrylics	X
CTFE	
ECTFE	150/66
ETFE	250/121
FEP	400/204
Polyamides	200/93
PAI	200/93
PEEK	80/27
PEI	80/27
PES	80/27
PFA	200/93
PTFE	450/232
PVDF	210/99
EHMWPE	X
HMWPE	X
PET	250/121
PPO	X
PPS	300/149
PP	60/15
PVDC	80/27
PSF	X
PVC Type 1	X
PVC Type 2	X
CPVC	X
CPE	80/27
PUR	X
Acetals	X
Polyesters	
Bisphenol A-fumurate	X
Halogenated	110/43
Hydrogenated-Bisphenol A	90/32
Isophthalic	110/43
PET	250/121
Epoxy	150/66
Vinyl esters	120/49
Furans	260/127
Phenolics	200/93
Phenol-formaldehyde	
Methyl appended silicone	X
PB	X

Trichloroacetic Acid

Polymer	°F/°C
ABS	
Acrylics	
CTFE	
ECTFE	150/66
ETFE	210/99
FEP	400/204
Polyamides	X
PAI	
PEEK	
PEI	
PES	
PFA	400/204
PTFE	450/232
PVDF	130/54
EHMWPE, 50%	140/60
HMWPE	80/27
PET	250/121
PPO, 10%	140/60
PPS, 50%	210/99
PP	150/66
PVDC	80/27
PSF	
PVC Type 1	90/32
PVC Type 2	X
CPVC, 20%	140/60
CPE	X
PUR	
Acetals	
Polyesters	
Bisphenol A-fumurate, 50%	180/82
Halogenated, 50%	200/93
Hydrogenated-Bisphenol A	90/32
Isophthalic, 50%	170/77
PET	250/121
Epoxy	X
Vinyl esters, 50%	210/99
Furans	260/127
Phenolics	
Phenol-formaldehyde, 30%	100/38
Methyl appended silicone	
PB	

Water, Acid Mine

Polymer	°F/°C
ABS	140/60
Acrylics	80/27
CTFE	
ECTFE	300/149
ETFE	210/99
FEP	400/204
Polyamides	80/27
PAI	
PEEK	80/27
PEI	
PES	
PFA	200/93
PTFE	450/232
PVDF	220/104
EHMWPE	140/60
HMWPE	140/60
PET	
PPO	200/93
PPS	
PP	220/104
PVDC	180/82
PSF	
PVC Type 1	140/60
PVC Type 2	140/60
CPVC	180/82
CPE	250/121
PUR	
Acetals	100/38
Polyesters	
Bisphenol A-fumurate	
Halogenated	210/99
Hydrogenated-Bisphenol A	
Isophthalic	110/43
PET	
Epoxy	300/149
Vinyl esters	210/99
Furans	
Phenolics	
Phenol-formaldehyde	
Methyl appended silicone	210/99
PB	

Water, Demineralized

Polymer	°F/°C
ABS	140/60
Acrylics	
CTFE	
ECTFE	300/149
ETFE	210/99
FEP	400/204
Polyamides	80/27
PAI	
PEEK	150/66
PEI	
PES	
PFA	400/204
PTFE	450/232
PVDF	280/138
EHMWPE	140/60
HMWPE	140/60
PET	250/121
PPO	140/60
PPS	300/149
PP	220/104
PVDC	170/77
PSF	200/93
PVC Type 1	140/60
PVC Type 2	140/60
CPVC	200/93
CPE	250/121
PUR	
Acetals	
Polyesters	
Bisphenol A-fumurate	160/71
Halogenated	210/99
Hydrogenated-Bisphenol A	210/99
Isophthalic	150/66
PET	250/121
Epoxy	250/121
Vinyl esters	210/99
Furans	250/121
Phenolics	
Phenol-formaldehyde	
Methyl appended silicone	210/99
PB	

Water, Distilled

Polymer	°F/°C
ABS	140/60
Acrylics	80/27
CTFE	
ECTFE	300/149
ETFE	210/99
FEP	400/204
Polyamides	80/27
PAI	200/93
PEEK	90/32
PEI	
PES	
PFA	400/204
PTFE	450/232
PVDF	280/138
EHMWPE	140/60
HMWPE	140/60
PET	250/121
PPO	140/60
PPS	300/149
PP	220/104
PVDC	170/77
PSF	210/99
PVC Type 1	140/60
PVC Type 2	140/60
CPVC	200/93
CPE	250/121
PUR	80/27
Acetals	180/82
Polyesters	
Bisphenol A-fumurate	200/93
Halogenated	210/99
Hydrogenated-Bisphenol A	210/99
Isophthalic	180/82
PET	250/121
Epoxy	210/99
Vinyl esters	210/99
Furans	200/93
Phenolics	
Phenol-formaldehyde	
Methyl appended silicone	210/99
PB	

Water, Potable

Polymer	°F/°C
ABS	80/27
Acrylics	
CTFE	
ECTFE	300/149
ETFE	
FEP	
Polyamides	
PAI	
PEEK	
PEI	
PES	
PFA	
PTFE	450/232
PVDF	280/138
EHMWPE	
HMWPE	
PET	
PPO	
PPS	
PP	180/82
PVDC	170/77
PSF	
PVC Type 1	140/60
PVC Type 2	140/60
CPVC	210/99
CPE	
PUR	
Acetals	
Polyesters	
Bisphenol A-fumurate	210/99
Halogenated	170/77
Hydrogenated-Bisphenol A	
Isophthalic	210/99
PET	
Epoxy	210/99
Vinyl esters	210/99
Furans	
Phenolics	
Phenol-formaldehyde	
Methyl appended silicone	
PB	

Water, Salt

Polymer	°F/°C
ABS	140/60
Acrylics	90/32
CTFE	
ECTFE	300/149
ETFE	250/121
FEP	400/204
Polyamides	140/60
PAI	
PEEK	90/32
PEI	
PES	
PFA	400/204
PTFE	450/232
PVDF	280/138
EHMWPE	140/60
HMWPE	140/60
PET	250/121
PPO	
PPS	
PP	220/104
PVDC	180/82
PSF	200/93
PVC Type 1	140/60
PVC Type 2	140/60
CPVC	180/82
CPE	250/121
PUR	X
Acetals	80/27
Polyesters	
Bisphenol A-fumurate	180/82
Halogenated	
Hydrogenated-Bisphenol A	210/99
Isophthalic	160/76
PET	250/121
Epoxy, 10%	210/99
Vinyl esters	180/82
Furans	
Phenolics	
Phenol-formaldehyde	
Methyl appended silicone	210/99
PB	

Water, Sea

Polymer	°F/°C
ABS	90/32
Acrylics	
CTFE	
ECTFE	300/149
ETFE	250/121
FEP	400/204
Polyamides	140/60
PAI	
PEEK	150/66
PEI	
PES	
PFA	400/204
PTFE	450/232
PVDF	280/138
EHMWPE	140/60
HMWPE	140/60
PET	250/121
PPO	90/32
PPS	300/149
PP	220/104
PVDC	170/77
PSF	200/93
PVC Type 1	140/60
PVC Type 2	140/60
CPVC	200/93
CPE	250/121
PUR	
Acetals	
Polyesters	
Bisphenol A-fumurate	220/104
Halogenated	180/82
Hydrogenated-Bisphenol A	200/93
Isophthalic	120/49
PET	250/121
Epoxy	300/149
Vinyl esters	180/82
Furans	250/121
Phenolics	
Phenol-formaldehyde	
Methyl appended silicone	210/99
PB	

Water, Sewage

Polymer	°F/°C
ABS	80/27
Acrylics	
CTFE	
ECTFE	300/149
ETFE	270/132
FEP	400/204
Polyamides	
PAI	
PEEK	110/43
PEI	
PES	
PFA	400/204
PTFE	450/232
PVDF	250/121
EHMWPE	140/60
HMWPE	140/60
PET	
PPO	
PPS	
PP	220/104
PVDC	170/77
PSF	
PVC Type 1	140/60
PVC Type 2	140/60
CPVC	180/82
CPE	250/121
PUR	
Acetals	
Polyesters	
Bisphenol A-fumurate	
Halogenated	
Hydrogenated-Bisphenol A	
Isophthalic	
PET	
Epoxy	
Vinyl esters	
Furans	
Phenolics	
Phenol-formaldehyde	
Methyl appended silicone	
PB	

White Liquor

Polymer	°F/°C
ABS	140/60
Acrylics	
CTFE	
ECTFE	250/121
ETFE	
FEP	400/204
Polyamides	80/27
PAI	
PEEK	
PEI	
PES	
PFA	400/204
PTFE	450/232
PVDF	200/93
EHMWPE	
HMWPE	
PET	
PPO	90/32
PPS	
PP	220/104
PVDC	
PSF	
PVC Type 1	140/60
PVC Type 2	140/60
CPVC	200/93
CPE	220/104
PUR	
Acetals	
Polyesters	
Bisphenol A-fumurate	180/82
Halogenated	X
Hydrogenated-Bisphenol A	
Isophthalic	X
PET	
Epoxy	200/93
Vinyl esters	180/82
Furans	140/60
Phenolics	
Phenol-formaldehyde	
Methyl appended silicone	
PB	

Zinc Chloride

Polymer	°F/°C
ABS	140/60
Acrylics	80/27
CTFE	
ECTFE	300/149
ETFE	300/149
FEP	400/204
Polyamides	X
PAI	
PEEK	90/32
PEI	
PES	
PFA[a]	400/204
PTFE[a]	450/232
PVDF	260/127
EHMWPE	140/60
HMWPE	140/60
PET	250/121
PPO	140/60
PPS, 70%	250/121
PP	200/93
PVDC	170/77
PSF	200/93
PVC Type 1	140/60
PVC Type 2	140/60
CPVC	200/93
CPE	250/121
PUR	
Acetals	
Polyesters	
Bisphenol A-fumurate	250/121
Halogenated	200/93
Hydrogenated-Bisphenol A	200/93
Isophthalic	180/82
PET	250/121
Epoxy	250/121
Vinyl esters	180/82
Furans	260/127
Phenolics	300/149
Phenol-formaldehyde	
Methyl appended silicone	400/204
PB	

[a] Corrodent will be absorbed.

Reference

1.   P.A. Schweitzer. 2004. *Corrosion Resistance Tables*, Vols. 1–4, 5th ed., New York: Marcel Dekker.

5

Elastomers

5.1 Introduction

The technical definition of an *elastomer*, as given by ASTM, states:

> An elastomer is a polymeric material which at room temperature can be
> stretched to at least twice its original length and upon immediate release
> of the stress will return quickly to its original length.

More commonly, an elastomer is generally considered to be any material
that is elastic or resilient and, in general, resembles natural rubber in feeling
and appearance. These materials are sometimes referred to as *rubbers*.

Elastomers are primarily composed of large molecules that tend to form
spiral threads, similar to a coiled spring, that are attached to each other at
infrequent intervals. As a small stress is applied, these coils tend to stretch or
compress but exert an increasing resistance as additional stresses are
applied. This property is illustrated by the reaction of an elastic band.

The maximum utility of elastomers, either natural or synthetic, is achieved
by compounding. In the raw state, elastomers tend to be soft and sticky
when hot and hard and brittle when cold. Ingredients are added to make
elastomers stronger, tougher, or harder; to make them age better; to
color them; and in general, to impart specific properties to meet specific
application needs. Vulcanizing agents are also added because the
vulcanizing process extends the temperature range within which they are
flexible and elastic.

Depending on the application of the elastomer, certain specific properties
may be required. The following examples illustrate some of the important
properties that are required of elastomers and the typical services that
require these properties:

Resistance to abrasive wear: automobile tires, conveyor belt covers, soles and heels (shoes), cables, hose covers

Resistance to tearing: tire treads, footwear, hot-water bags, hose covers, belt covers, O-rings

Resistance to flexing: auto tires, transmission belts, V belts, mountings, and footwear

Resistance to high temperatures: auto tires, belts conveying hot materials, steam hose, steam packing, O-rings

Resistance to cold: airplane parts, automotive parts, auto tires, refrigeration hose, O-rings

Minimum heat buildup: auto tires, transmission belts, V belts, mountings

High resilience: sponge rubber, mountings, elastic bands, thread, sandblast hose, jar rings, O-rings

High rigidity: packing, soles and heels (shoes), valve cups, suction hose, battery boxes

Long life: fire hose, transmission belts, tubing

Electrical resistivity: electrician's tape, switchboard mats, electrician's gloves, wire insulation

Electrical conductivity: hospital flooring, nonstatic hose, matting

Impermeability to gases: balloons, life rafts, gasoline hose, special diaphragms, stack linings

Resistance to ozone: ignition distributor gaskets, ignition cables, windshield wipers

Resistance to sunlight: wearing apparel, hose covers, bathing caps, windshield wipers

Resistance to chemicals: tank linings, gaskets, valve diaphragms, hose for chemicals, O-rings

Resistance to oils: gasoline hose, oil suction hose, paint hose, creamery hose, packing house hose, special belts, tank linings, gaskets, O-rings, special footwear

Stickiness: cements, electrician's tape, adhesive tapes, pressure-sensitive tapes

Low specific gravity: airplane parts, forestry hose, balloons

Lack of odor or taste: milk tubing, brewery and winery hose, nipples, jar rings, gaskets, O-rings

Acceptance of color pigments: ponchos, life rafts, welding hose

These materials are sometimes referred to as *rubbers*. Natural rubber is polymerized hydrocarbon whose commercial synthesis proved to be very difficult. Synthetic rubbers now produced are similar to, but not identical to, natural rubber. Natural rubber has the hydrocarbon butadiene as its simplest unit. Butadiene, $CH_2=CH–CH=CH_2$, has two unsaturated linkages and is easily polarized. It is commercially produced by cracking petroleum and also from ethyl alcohol. Natural rubber is a polymer of methyl butadiene (isoprene):

$$CH_2 = \underset{\underset{\displaystyle CH_3}{|}}{C}{-}CH = CH_2$$

When butadiene or its derivatives become polarized, the units link together to form long chains that contain over 1000 units. In early attempts to develop a synthetic rubber, it was found that simple butadiene does not yield a good grade of rubber because the chains are too smooth and do not sufficiently interlock strongly. Better results are obtained by introducing side groups into the chain either by modifying butadiene or by making a copolymer of butadiene and some other compound.

As development work continued in the production of synthetic rubbers, other compounds were used as the parent material in place of butadiene. Two of them were isobutylene and ethylene.

5.1.1 Importance of Compounding

The rubber chemist is able to optimize selected properties or fine-tune the formulation to meet a desired balance of properties. However, this does involve trade-offs. Improvement in one property can come at the expense of another property. Therefore, it is important that these effects be taken into consideration. Table 5.1 illustrates some of these effects.

TABLE 5.1

Compounding Trade-Offs

An Improvement in	Can Be at the Expense of
Tensile strength, hardness	Extensibility
	Dynamic properties
Dynamic properties	Thermal stability
	Compression set resistance
Balanced properties	Processability
Optimized properties	Higher cost
Compression set resistance	Flex fatigue strength
	Resilience
Oil resistance	Low-temperature flexibility
Abrasion resistance	Resilience
Dampening	Resilience
	Compression set resistance

5.1.2 Similarities of Elastomers and Thermoplastic Polymers

All thermoplastic polymers and elastomers, with the exception of silicones, are carbon-based. They are made up from the linking of one or more monomers into long molecular chains. Many of the same monomers are found in both thermoplastic and elastomeric polymers. Typical examples include styrene, acrylonitrile, ethylene, propylene, and acrylic acid and its esters. An elastomer is in a thermoplastic state prior to vulcanization.

5.1.3 Differences between Elastomers and Thermoplasts

At room temperature, an uncured elastomer can be a soft, pliable gum or a leathery, flexible solid, whereas an engineering thermoplastic is a rigid solid. Molecular mobility is the fundamental property that accounts for the differences among elastomers and the differences between elastomers and engineering thermoplasts.

The distinction refers to a modulus of the material—a ratio between an applied force and the amount of resulting deformation. When the molecular resistance to motion is the least, deformation is the greatest. The factors that influence the resistance to motion include various intermolecular attractions, crystallinity, the presence of side chains, and physical entanglements of the molecular strands. The glass transition temperature of the polymer is determined by the cumulative effect of these factors. Below this temperature, a thermoplast or elastomeric polymer is a supercooled liquid that behaves in many ways like a rigid solid. Above this temperature, a cross-linked elastomer will display rubber-like properties. The behavior of a thermoplast above its glass transition temperature will depend on its level of crystallinity. A noncrystalline (amorphus) polymer will display a large decrease in modulus at the glass transition temperature. Between the glass transition temperature and the melting temperature, the modulus is relatively insensitive to temperature. A partially crystalline thermoplast will display a relatively small modulus change at the glass transition temperature followed by a steady decreasing modulus as the temperature increases.

Engineering thermoplastics have glass-transition temperatures that fall in a wide temperature range, extending well above and below room temperature. Elastomers in the uncured state have glass-transition temperature well below room temperature.

The major physical differences between a cured elastomer and an engineering thermoplastic is the presence of cross-links. The cross-links join the elastomer's molecular chains together at a cross-link density such that many molecular units exist between cross-link sites. This only slightly increases the resistance of deformation, but it increases the resilience of the elastomer, causing it to spring back to its original shape when deforming stress is removed. In principle, most amorphous thermoplastic materials could be converted into elastomers in some appropriate temperature range by the controlled addition of a limited number of cross-links.

5.1.4 Causes of Failure

Chemical deterioration occurs as the result of a chemical reaction between the elastomer and the medium or by absorption of the medium into the elastomer. This attack results in a swelling of the elastomer and a reduction in its tensile strength. The temperature and concentration of the corrodent will determine the degree of deterioration. Normally, the chemical attack is greater as the temperature and/or concentration of the corrodent increases. Unlike metals, elastomers absorb varying quantities of the material they are in contact with, especially organic liquids. This can result in swelling, cracking, and penetration to the substrate in an elastomer-lined vessel. Swelling can cause softening of the elastomer and, in a lined vessel, introduce high stresses and failure of the bond. Permeation is another factor that can cause failure of a lining. When an elastomer exhibits a high absorption, permeation usually results. However, it is not necessary for an elastomer to have a high absorption rate for permeation to occur. Some elastomers, such as the fluorocarbons, are easily permeated but have very little absorption. An approximation of the expected permeation and/or absorption of an elastomer can be based on the absorption of water, for which data are usually available.

All materials tend to be somewhat permeable to chemical molecules, but the permeability rate of some elastomers tends to be an order of magnitude greater than that of metals. Though permeation is a factor closely related to absorption, factors that influence the permeation rate are diffusion and temperature rather than concentration and temperature. Permeation can pose a serious problem in elastomer-lined equipment. When the corrodent permeates the elastomer, it comes into contact with the metal substrate that is then subject to chemical attack.

This can result in

1. Bond failure and blistering, caused by an accumulation of fluids at the bond when the substrate is less permeable than the lining or from the formation of corrosion or reaction products if the substrate is attacked by the corrodent
2. Failure of the substrate because of corrosive attack
3. Loss of contents through lining and substrate as the result of eventual failure of the substrate

The degree of permeation is affected by lining thickness. For general corrosion resistance, thicknesses of 0.010–0.020 in. are usually satisfactory, depending on the elastomeric material and the specific corrodent. Thick linings may be required when mechanical factors such as thinning because of cold flow, mechanical abuse, and permeation rates are taken into consideration.

Increasing lining thickness will normally decrease permeation by the square of the thickness. However, this is not necessarily the answer to the problem because increasing the liner thickness can introduce other problems. As the liner thickness increases, the thermal stresses on the boundary increase that can cause bond failure. Temperature changes and large differences in coefficients of thermal expansion are the most common causes of bond failure. These stresses are influenced by thickness and modulus of elasticity of the elastomer. In addition, the labor cost of installing the liner increases as the thickness increases.

Temperature and temperature gradient in the liner also affect the rate of permeation. Lowering these will reduce the permeation rate. Lined vessels, when used under ambient conditions such as storage tanks, provide the best service.

Linings can be installed either bonded or unbonded to the substrate. In unbonded linings, it is important that the space between the liner and support member be vented to the atmosphere to permit the escape of minute quantities of permeant vapors and also to prevent the expansion of entrapped air that could cause collapse of the liner.

Although elastomers can be damaged by mechanical means alone, this is usually not the case. Most mechanical damage occurs as a result of chemical deterioration of the elastomer. When the elastomer is in a deteriorated condition, the material is weakened, and consequently, it is more susceptible to mechanical damage from flowing or agitated media.

Some elastomeric materials are subject to degradation when placed in outdoor applications as a result of weathering. The action of sunlight, ozone, and oxygen can cause surface cracking, discoloration of colored stocks, serious loss of tensile strength, elongation, and other rubber-like properties. Therefore, the resistance to weathering must also be taken into account when selecting an elastomer as well as other corrosion resistance properties, when the material is to be installed where it will be subject to weathering.

Elastomers in outdoor use can be subject to degradation as a result of the action of ozone, oxygen, and sunlight. These three weathering agents can greatly affect the properties and appearance of a large number of elastomeric materials. Surface cracking, discoloration of colored stocks, and serious loss of tensile strength, elongation, and other rubber-like properties are the result of the attack. Table 5.2 shows the ozone resistance of selected elastomers. The low-temperature properties must also be taken into account for outdoor use. With many elastomers, crystallization takes place at low temperatures, when the elastomer is brittle and will easily fracture. Table 5.3 gives the relative low-temperature flexibility of the more common elastomers. Table 5.4 gives the brittle points of the common elastomers. Table 5.5 provides the operating temperatures of the common elastomers.

5.1.5 Selecting an Elastomer

When the need arises to specify an elastomer for a specific application, physical, mechanical, and chemical resistance properties must all be taken

TABLE 5.2

Ozone Resistance of Selected Elastomers

Excellent Resistance	
EPDM	Chlorosulfonated polyethylene
Epichlorhydrin	Polyacrylate
ETFE	Polysulfide
ECTFE	Polyamides
Fluorosilicones	Polyesters
Ethylene acrylate	Perfluoroelastomers
Ethylene vinyl acetate	Urethanes
Chlorinated polyethylene	Vinylidene fluoride
Good Resistance	
Polychloroprene	Butyl
Fair Resistance	
Nitrile	
Poor Resistance	
Buna S	Polybutadiene
Isoprene	SBR
Natural rubber	SBS

TABLE 5.3

Relative Low-Temperature Flexibility of the Common Elastomers

Elastomer	Relative Flexibility[a]
Natural rubber NR	G–E
Butyl rubber IIR	F
Ethylene–propylene rubber EPDM	E
Ethylene–acrylic rubber EA	P–F
Fluoroelastomer FKM	F–G
Fluorosilicone FSI	E
Chlorosulfonated polyethylene (Hypalon) CSM	F–G
Polychloroprene (Neoprene) CR	F–G
Nitrile rubber (Buna-N) NBR	F–G
Polybutadiene BR	G–E
Polyisoprene IR	G
Polysulfide T rubber	F–G
Butadiene–styrene rubber (Buna-S) SBR	G
Silicone rubber SI	E
Polyurethane rubber AU	G
Polyether–urethane rubber EU	G

[a] E, excellent; G, good; F, fair; P, poor.

TABLE 5.4

Brittle Points of Common Elastomers

Elastomer	°F	°C
NR; IR	−68	−56
CR	−40	−40
SBR	−76	−60
NBR	−32 to −40	−1 to −40
CSM	−40 to −80	−40 to −62
BR	−60	−56
EA	−75	−60
EPDM; EPT	−90	−68
SEBS	−58 to −148	−50 to −100
Polysulfide SF	−60	−51
Polysulfide FA	−30	−35
AU	−85 to −100	−65 to −73
Polyamide No. 11	−94	−70
PE	−94	−70
TPE	−67 to −70	−55 to −60
SI	−75	−60
FSI	−75	−60
HEP	−80	−62
FKM	−25 to −75	−32 to −59
ETFE	−150	−101
ECTFE	−105	−70
FPM	−9 to −58	−23 to −50

into account. The major physical and mechanical properties that may have to be considered, depending on the application are:

Abrasion resistance

Electrical properties

Compression set resistance

Tear resistance

Tensile strength

Adhesion to metals

Adhesion to fabrics

Rebound, cold and hot

Resistance to heat aging and flame

It should be remembered that these properties may be altered by compounding, but improvement of one property may result in an adverse effect on another. Because of this, it is best to provide a competent manufacturer with complete specifications and let that manufacturer provide an appropriate elastomer.

TABLE 5.5

Operating Temperature Ranges of Common Elastomers

	Temperature Range			
	°F		°C	
Elastomer	Min	Max	Min	Max
Natural rubber, NR	−59	175	−50	80
Isoprene rubber, IR	−59	175	−50	80
Neoprene rubber, CR	−13	203	−25	95
Buna-S, SBR	−66	175	−56	80
Nitrile Rubber Buna, NBR	−40	250	−40	105
Butyl rubber, IIR	−30	300	−34	149
Chlorobutyl rubber, CIIR	−30	300	−34	149
Hypalon, CSM	−20	250	−30	105
Polybutadiene rubber, BR	−150	200	−101	93
Ethylene–acrylic rubber, EA	−40	340	−40	170
Acrylate–butadiene rubber, ABR	−40	340	−40	170
Ethylene–propylene, EPDM	−65	300	−54	149
Styrene–butadiene–styrene, SBS		150		65
Styrene–ethylene–butylene–styrene, SEBS	−102	220	−75	105
Polysulfide, ST	−50	212	−45	100
Polysulfide, FA	−30	250	−35	121
Polyurethane, AU	−65	250	−54	121
Polyamides	−40	300	−40	149
Polyesters, PE	−40	302	−40	150
Thermoplastic elastomers, TPE	−40	277	−40	136
Silicone, SI	−60	450	−51	232
Fluorosilicone, FSI	−100	375	−73	190
Vinylidene fluoride, HEP	−40	450	−40	232
Fluoroelastomers, FKM	−10	400	−18	204
Ethylene–tetrafluoroethylene elastomer, ETFE	−370	300	−223	149
Ethylene–chlorotrifluoroethylene elastomer, ECTFE	−105	340	−76	171
Perfluoroelastomers, FPM	−58	600	−50	316

The primary requirement of the elastomer is that it must be compatible with the corrodent to be handled. Therefore, all temperatures and concentrations of the corrodent to which it will be exposed, must be provided.

Specifications should include any specific properties required for the application, such as resilience, hysteresis, static or dynamic shear and compression modulus, flex fatigue and cracking, creep resistance to oils and chemicals, permeability, and brittle point, all in the temperature ranges to be encountered in service.

5.1.6 Corrosion Resistance

The corrodent's concentration and temperature is a determining factor in the capacity of the elastomer to resist attack by the corrodent. Another important factor is the elastomer's composition. It is a common practice in the manufacture of elastomers to incorporate additives into the formulation to improve certain of the physical and/or mechanical properties. These additives may have an adverse effect on the corrosion resistance of the base elastomer, particularly at elevated temperatures. Conversely, some manufacturers compound their elastomer to improve their corrosion resistance at the expense of physical and/or mechanical properties. Because of this, it is important to know whether or not any additives have been used as the corrosion resistance charts are applicable only for the pure elastomer.

Keep in mind that there are several manufacturers of each generic compound. Because each may compound slightly differently, the corrosion resistance may be affected. When a generic compound is listed as being compatible with a specific corrodent, it indicates that at least one of the trade name materials is resistant to the corrodent, but not necessarily all. The manufacturers must be checked.

5.1.7 Applications

Elastomeric or rubber materials find a wide range of applications. One of the major areas of application is that of vessels' linings. Both natural and synthetic materials are used for this purpose. These linings have provided many years of service in the protection of steel vessels from corrosion. They are sheet-applied and bonded to a steel substrate.

These materials are also used extensively as membranes in acid-brick-lined vessels to protect the steel shell from corrosive attack. The acid-brick lining, in turn, protects the elastomer from abrasion and excessive temperature. Another major use is as an impermeable lining for settling ponds and basins. These materials are employed to prevent pond contaminants from seeping into the soil and causing pollution of groundwater and contamination of the soil.

Natural rubber and most of the synthetic elastomers are unsaturated compounds that oxidize and rapidly deteriorate when exposed to air in thin films. These materials can be saturated by reacting with chlorine under the proper conditions, producing compounds that are clear, odorless, nontoxic, and noninflammable. They may be dissolved and blended with varnishes to impart high resistance to moisture and to the action of alkalies. This makes these products particularly useful in paints for concrete where the combination of moisture and alkali causes the disintegration of ordinary paints and varnishes. These materials also resist mildew and are used to impart flame resistance and waterproofing properties to canvas. Application of these paints to steel will provide a high degree of protection against corrosion.

Large quantities of elastomeric materials are used to produce a myriad of products such as hoses, cable insulation, O-rings, seals and gaskets, belting, vibration mounts, flexible couplings, expansion joints, automotive and airplane parts, electrical parts and accessories. With such a wide variety of applications requiring very diverse properties, it is essential that an understanding of the properties of each elastomer be acquired so that proper choices can be made.

5.1.8 Elastomer Designations

Many of the elastomers have common names or trade names as well as the chemical name. For example, *polychloroprene* also goes under the name of *neoprene* (DuPont's tradename) or *Bayprene* (Mobay Corp's tradename). In addition, they also have an ASTM designation that for polychloroprene is CR.

5.2 Natural Rubber

Natural rubber of the best quality is prepared by coagulating the latex of the *Hevea brasiliensis* tree that is primarily cultivated in the Far East. However, there are other sources such as the wild rubbers of the same tree growing in Central America, guyayule rubber coming from shrubs grown mostly in Mexico, and balata. Balata is a resinous material and cannot be tapped like the *Hevea* tree sap. The balata tree must be cut down and boiled to extract balata that cures to a hard, tough product used as golf ball covers.

Another source of rubber is the planation leaf gutta-percha. This material is produced from the leaves of trees grown in bush formation. The leaves are picked and the rubber is boiled out as with the balata. Gutta-percha has been used successfully for submarine-cable insulation for more than 40 years.

Chemically, natural rubber is a polymer of methyl butadiene (isoprene):

$$\overset{\displaystyle CH_3}{\underset{\displaystyle CH_2=C-CH=CH_2}{|}}$$

When polymerized, the units link together, forming long chains that each contain over 1000 units. Simple butadiene does not yield a good grade of rubber, apparently because the chains are too smooth and do not form a strong enough interlock. Synthetic rubbers are produced by introducing side groups into the chain either by modifying butadiene or by making a copolymer of butadiene and some other compound.

Purified raw rubber becomes sticky in hot weather and brittle in cold weather. Its valuable properties become apparent after vulcanization.

Depending upon the degree of curing, natural rubber is classified as soft, semihard, or hard rubber. Only soft rubber meets the ASTM definition of an elastomer, and therefore, the information that follows pertain only to soft rubber. The properties of semihard and hard rubber differ somewhat, particularly in the area of corrosion resistance.

Most rubber is made to combine with sulfur or sulfur-bearing organic compounds or with other cross-linking chemical agents in a process known as vulcanization that was invented by Charles Goodyear in 1839 and forms the basis of all later developments in the rubber industry.

When properly carried out, vulcanization improves mechanical properties, eliminates tackiness, renders the rubber less susceptible to temperature changes, and makes it insoluble in all known solvents. Other materials are added for various purposes as follows:

Carbon blacks, precipitated pigments, and organic vulcanization accelerators are added to increase tensile strength and resistance to abrasion.

Whiting, barite, talc, silica, silicates, clays, and fibrous materials are added to cheapen and stiffen.

Bituminous substances, coal tar and its products, vegetable and mineral oils, paraffin, petrolatum, petroleum, oils, and asphalt are added to soften (for purposes of processing or for final properties).

Condensation amines and waxes are added as protective agents against natural aging, sunlight, heat, and flexing.

Pigments are added to provide coloration.

Vulcanization has the greatest effect on the mechanical properties of natural rubber. Unvulcanized rubber can be stretched to approximately ten times its length and, at this point, will bear a load of 10 tons/in.^{2}. It can be compressed to one-third of its thickness thousands of times without injury. When most types of vulcanized rubbers are stretched, their resistance increases in greater proportion than their extension. Even when stretched just short of the point of rupture, they recover almost all of their original dimensions on being released and then gradually recover a portion of the residual distortion. The outstanding property of natural rubber in comparison to the synthetic rubbers is its resilience. It has excellent rebound properties, either hot or cold.

5.2.1 Resistance to Sun, Weather, and Ozone

Cold water preserves natural rubber, but if exposed to the air, particularly in sunlight, rubber tends to become hard and brittle. It has only fair resistance to ozone. Unlike the synthetic elastomers, natural rubber softens and reverts with aging to sunlight. In general, it has relatively poor weathering and aging properties.

5.2.2 Chemical Resistance

Natural rubber offers excellent resistance to most inorganic salt solutions, alkalies, and nonoxidizing acids. Hydrochloric acid will react with soft rubber to form rubber hydrochloride, and therefore, it is not recommended that natural rubber be used for items that will come into contact with that acid. Strong oxidizing media such as nitric acid, concentrated sulfuric acid, permanganates, dichromates, chlorine dioxide, and sodium hypochlorite will severely attack rubber. Mineral and vegetable oils, gasoline, benzene, toluene, and chlorinated hydrocarbons also affect rubber. Cold water tends to preserve natural rubber. Natural rubber offers good resistance to radiation and alcohols.

Unvulcanized rubber is soluble in gasoline, naphtha, carbon bisulfide, benzene, petroleum ether, turpentine, and other liquids.

Refer to Table 5.6 for the compatibility of natural rubber with selected corrodents.

TABLE 5.6

Compatibility of Natural Rubber with Selected Corrodents

Chemical	Maximum Temperature	
	°F	°C
Acetaldehyde	X	X
Acetamide	X	X
Acetic acid, 10%	150	66
Acetic acid, 50%	X	X
Acetic acid, 80%	X	X
Acetic acid, glacial	X	X
Acetic anhydride	X	X
Acetone	140	60
Acetyl chloride	X	X
Acrylic acid		
Acrylonitrile		
Adipic acid		
Allyl alcohol		
Allyl chloride		
Alum	140	60
Aluminum acetate		
Aluminum chloride, aqueous	140	60
Aluminum chloride, dry	160	71
Aluminum fluoride	X	X
Aluminum hydroxide		
Aluminum nitrate	X	X
Aluminum oxychloride		
Aluminum sulfate	140	60
Ammonia gas		
Ammonium bifluoride		
Ammonium carbonate	140	60
Ammonium chloride, 10%	140	60
Ammonium chloride, 50%	140	60
Ammonium chloride, sat.	140	60
Ammonium fluoride, 10%	X	X
Ammonium fluoride, 25%	X	X
Ammonium hydroxide, 25%	140	60

(continued)

TABLE 5.6   *Continued*

Chemical	Maximum Temperature	
	°F	°C
Ammonium hydroxide, sat.	140	60
Ammonium nitrate	140	60
Ammonium persulfate		
Ammonium phosphate	140	60
Ammonium sulfate, 10–40%	140	60
Ammonium sulfide	140	60
Ammonium sulfite		
Amyl acetate	X	X
Amyl alcohol	140	60
Amyl chloride	X	X
Aniline	X	X
Antimony trichloride		
Aqua regia, 3:1	X	X
Barium carbonate	140	60
Barium chloride	140	60
Barium hydroxide	140	60
Barium sulfate	140	60
Barium sulfide	140	60
Benzaldehyde	X	X
Benzene	X	X
Benzenesulfonic acid, 10%	X	X
Benzoic acid	140	60
Benzyl alcohol	X	X
Benzyl chloride	X	X
Borax	140	60
Boric acid	140	60
Bromine gas, dry		
Bromine gas, moist		
Bromine, liquid		
Butadiene		
Butyl acetate	X	X
Butyl alcohol *n*-Butylamine	140	60
Butyric acid	X	X
Calcium bisulfide		
Calcium bisulfite	140	60
Calcium carbonate	140	60
Calcium chlorate	140	60
Calcium chloride	140	60
Calcium hydroxide, 10%	140	60
Calcium hydroxide, sat.	140	60
Calcium hypochlorite	X	X
Calcium nitrate	X	X
Calcium oxide	140	60
Calcium sulfate	140	60
Caprylic acid		
Carbon bisulfide	X	X

(continued)

TABLE 5.6 *Continued*

Chemical	Maximum Temperature	
	°F	°C
Carbon dioxide, dry		
Carbon dioxide, wet		
Carbon disulfide	X	X
Carbon monoxide	X	X
Carbon tetrachloride	X	X
Carbonic acid	140	60
Cellosolve	X	X
Chloroacetic acid, 50% water	X	X
Chloroacetic acid	X	X
Chlorine gas, dry	X	X
Chlorine gas, wet	X	X
Chlorine, liquid	X	X
Chlorobenzene	X	X
Chloroform	X	X
Chlorosulfonic acid	X	X
Chromic acid, 10%	X	X
Chromic acid, 50%	X	X
Chromyl chloride		
Citric acid, 15%	140	60
Citric acid, conc.	X	X
Copper acetate		
Copper carbonate	X	X
Copper chloride	X	X
Copper cyanide	140	60
Copper sulfate	140	60
Cresol	X	X
Cupric chloride, 5%	X	X
Cupric chloride, 50%	X	X
Cyclohexane	X	X
Cyclohexanol		
Dibutyl phthalate		
Dichloroacetic acid		
Dichloroethane (ethylene dichlor-ide)	X	X
Ethylene glycol	140	60
Ferric chloride	140	60
Ferric chloride, 50% in water	140	60
Ferric nitrate, 10–50%	X	X
Ferrous chloride	140	60
Ferrous nitrate	X	X
Fluorine gas, dry	X	X
Fluorine gas, moist		
Hydrobromic acid, dil.	140	60
Hydrobromic acid, 20%	140	60
Hydrobromic acid, 50%	140	60
Hydrochloric acid, 20%	X	X

(continued)

TABLE 5.6 *Continued*

Chemical	Maximum Temperature	
	°F	°C
Hydrochloric acid, 38%	140	60
Hydrocyanic acid, 10%		
Hydrofluoric acid, 30%	X	X
Hydrofluoric acid, 70%	X	X
Hydrofluoric acid, 100%	X	X
Hypochlorous acid		
Iodine solution, 10%		
Ketones, general		
Lactic acid, 25%	X	X
Lactic acid, conc.	X	X
Magnesium chloride	140	60
Malic acid	X	X
Manganese chloride		
Methyl chloride	X	X
Methyl ethyl ketone	X	X
Methyl isobutyl ketone	X	X
Muriatic acid	140	60
Nitric acid, 5%	X	X
Nitric acid, 20%	X	X
Nitric acid, 70%	X	X
Nitric acid, anhydrous	X	X
Nitrous acid, conc.	X	X
Oleum		
Perchloric acid, 10%		
Perchloric acid, 70%		
Phenol	X	X
Phosphoric acid, 50–80%	140	60
Picric acid		
Potassium bromide, 30%	140	60
Salicylic acid		
Silver bromide, 10%		
Sodium carbonate	140	60
Sodium chloride	140	60
Sodium hydroxide, 10%	140	60
Sodium hydroxide, 50%	X	X
Sodium hydroxide, conc.	X	X
Sodium hypochlorite, 20%	X	X
Sodium hypochlorite, conc.	X	X
Sodium sulfide, to 50%	140	60
Stannic chloride	140	60
Stannous chloride	140	60
Sulfuric acid, 10%	140	60
Sulfuric acid, 50%	X	X
Sulfuric acid, 70%	X	X
Sulfuric acid, 90%	X	X
Sulfuric acid, 98%	X	X

(continued)

TABLE 5.6 *Continued*

Chemical	Maximum Temperature	
	°F	°C
Sulfuric acid, 100%	X	X
Sulfuric acid, fuming	X	X
Sulfurous acid	X	X
Thionyl chloride		
Toluene		
Trichloroacetic acid		
White liquor		
Zinc chloride	140	60

The chemicals listed are in the pure state or in a saturated solution unless otherwise indicated. Compatibility is shown to the maximum allowable temperature for which data is available. Incompatibility is shown by an X. A blank space indicates that data is unavailable.

Source: From P.A. Schweitzer. 2004. *Corrosion Resistance Tables*, Vols. 1–4, 5th ed., New York: Marcel Dekker.

5.2.3 Applications

Natural rubber finds its major use in the manufacture of pneumatic tires and tubes, power transmission belts, conveyer belts, gaskets, mountings, hose, chemical tank linings, printing press platens, sound and/or shock absorbers, and seals against air, moisture, sound, and dirt.

Rubber has been used for many years as a lining material for steel tanks, particularly for protection against corrosion by inorganic salt solutions, especially brine, alkalies, and nonoxidizing acids. These linings have the advantage of being readily repaired in place. Natural rubber is also used for lining pipelines used to convey these types of materials. Some of these applications have been replaced by synthetic rubbers that have been developed over the years.

5.3 Isoprene Rubber (IR)

Chemically, natural rubber is natural *cis*-polyisoprene. The synthetic form of natural rubber, synthetic *cis*-polyisoprene, is called *isoprene rubber*. The physical and mechanical properties of IR are similar to the physical and mechanical properties of natural rubber, the one major difference being that isoprene does not have an odor. This feature permits the use of IR in certain food-handling applications.

IR can be compounded, processed, and used in the same manner as natural rubber. Other than the lack of odor, IR has no advantages over natural rubber.

5.4 Neoprene (CR)

Neoprene is one of the oldest and most versatile of the synthetic rubbers. Chemically, it is polychloroprene. Its basic unit is a chlorinated butadiene whose formula is

$$CH_2-\underset{\underset{Cl}{|}}{C}-CH=CH_2$$

The raw material is acetylene which makes this product more expensive than some of the other elastomeric materials.

Neoprene was introduced commercially by DuPont in 1932 as an oil-resistant substitute for natural rubber. Its dynamic properties are very similar to those of natural rubber, but its range of chemical resistance overcomes many of the shortcomings of natural rubber.

As with other elastomeric materials, neoprene is available in a variety of formulations. Depending on the compounding procedure, material can be produced to impart specific properties to meet specific application needs.

Neoprene is also available in a variety of forms. In addition to a neoprene latex that is similar to natural rubber latex, neoprene is produced in a "fluid" form as either a compounded latex dispersion or a solvent solution. Once these materials have solidified or cured, they have the same physical and chemical properties as the solid or cellular forms.

5.4.1 Resistance to Sun, Weather, and Ozone

Neoprene displays excellent resistance to sun, weather, and ozone. Because of its low rate of oxidation, products made of neoprene have high resistance to both outdoor and indoor aging. Over prolonged periods of time in an outdoor environment, the physical properties of neoprene display insignificant change. If neoprene is properly compounded, ozone in atmospheric concentrations has little effect on the product. When severe ozone exposure is expected, as for example around electrical equipment, compositions of neoprene can be provided to resist thousands of parts per million of ozone for hours without surface cracking. Natural rubber will crack within minutes when subjected to ozone concentrations of only 50 ppm.

5.4.2 Chemical Resistance

Neoprene's resistance to attack from solvents, waxes, fats, oils, greases, and many other petroleum-based products is one of its outstanding properties. Excellent service is also experienced when it is in contact with aliphatic compounds (methyl and ethyl alcohols, ethylene glycols, etc.), aliphatic hydrocarbons, and most Freon refrigerants. A minimum amount of swelling and relatively little loss of strength occur when neoprene is in contact with these fluids.

When exposed to dilute mineral acids, inorganic salt solutions, or alkalies, neoprene products show little, if any, change in appearance or change in properties.

Chlorinated and aromatic hydrocarbons, organic esters, aromatic hydroxy compounds, and certain ketones have an adverse effect on neoprene, and consequently, only limited serviceability can be expected with them. Highly oxidizing acid and salt solutions also cause surface deterioration and loss of strength. Included in this category are nitric acid and concentrated sulfuric acid.

Neoprene formulations can be produced that provide products with outstanding resistance to water absorption. These products can be used in continuous or periodic immersion in either freshwater or saltwater without any loss of properties.

Properly compounded neoprene can be buried underground successfully, since moisture, bacteria, and soil chemicals usually found in the earth have little effect on its properties. It is unaffected by soils saturated with seawater, chemicals, oils, gasolines, wastes, and other industrial byproducts. Refer to Table 5.7 for the compatibilities of neoprene with selected corrodents.

TABLE 5.7

Compatibility of Neoprene with Selected Corrodents

Chemical	Maximum Temperature	
	°F	°C
Acetaldehyde	200	93
Acetamide	200	93
Acetic acid, 10%	160	71
Acetic acid, 50%	160	71
Acetic acid, 80%	160	71
Acetic acid, glacial	X	X
Acetic anhydride	X	X
Acetone	X	X
Acetyl chloride	X	X
Acrylic acid	X	X
Acrylonitrile	140	60
Adipic acid	160	71
Allyl alcohol	120	49
Allyl chloride	X	X
Alum	200	93
Aluminum acetate		
Aluminum chloride, aqueous	150	66
Aluminum chloride, dry		
Aluminum fluoride	200	93

(continued)

TABLE 5.7 *Continued*

Chemical	Maximum Temperature	
	°F	°C
Aluminum hydroxide	180	82
Aluminum nitrate	200	93
Aluminum oxychloride		
Aluminum sulfate	200	93
Ammonia gas	140	60
Ammonium bifluoride	X	X
Ammonium carbonate	200	93
Ammonium chloride, 10%	150	66
Ammonium chloride, 50%	150	66
Ammonium chloride, sat.	150	66
Ammonium fluoride, 10%	200	93
Ammonium fluoride, 25%	200	93
Ammonium hydroxide, 25%	200	93
Ammonium hydroxide, sat.	200	93
Ammonium nitrate	200	93
Ammonium persulfate	200	93
Ammonium phosphate	150	66
Ammonium sulfate, 10–40%	150	66
Ammonium sulfide	160	71
Ammonium sulfite		
Amyl acetate	X	X
Amyl alcohol	200	93
Amyl chloride	X	X
Aniline	X	X
Antimony trichloride	140	60
Aqua regia, 3:1	X	X
Barium carbonate	150	66
Barium chloride	150	66
Barium hydroxide	230	110
Barium sulfate	200	93
Barium sulfide	200	93
Benzaldehyde	X	X
Benzene	X	X
Benzenesulfonic acid, 10%	100	38
Benzoic acid	150	66
Benzyl alcohol	X	X
Benzyl chloride	X	X
Borax	200	93
Boric acid	150	66
Bromine gas, dry	X	X
Bromine gas, moist	X	X
Bromine, liquid	X	X
Butadiene	140	66
Butyl acetate	60	16
Butyl alcohol *n*-Butylamine	200	93

(continued)

TABLE 5.7 *Continued*

Chemical	Maximum Temperature	
	°F	°C
Butyric acid	X	X
Calcium bisulfide		
Calcium bisulfite	X	X
Calcium carbonate	200	93
Calcium chlorate	200	93
Calcium chloride	150	66
Calcium hydroxide, 10%	230	110
Calcium hydroxide, sat.	230	110
Calcium hypochlorite	X	X
Calcium nitrate	150	66
Calcium oxide	200	93
Calcium sulfate	150	66
Caprylic acid		
Carbon bisulfide	X	X
Carbon dioxide, dry	200	93
Carbon dioxide, wet	200	93
Carbon disulfide	X	X
Carbon monoxide	X	X
Carbon tetrachloride	X	X
Carbonic acid	150	66
Cellosolve	X	X
Chloroacetic acid, 50% water	X	X
Chloroacetic acid	X	X
Chlorine gas, dry	X	X
Chlorine gas, wet	X	X
Chlorine, liquid	X	X
Chlorobenzene	X	X
Chloroform	X	X
Chlorosulfonic acid	X	X
Chromic acid, 10%	140	60
Chromic acid, 50%	100	38
Chromyl chloride		
Citric acid, 15%	150	66
Citric acid, conc.	150	66
Copper acetate	160	71
Copper carbonate		
Copper chloride	200	93
Copper cyanide	160	71
Copper sulfate	200	93
Cresol	X	X
Cupric chloride, 5%	200	93
Cupric chloride, 50%	160	71
Cyclohexane	X	X
Cyclohexanol	X	X
Dibutyl phthalate		

(continued)

TABLE 5.7   *Continued*

Chemical	Maximum Temperature	
	°F	°C
Dichloroacetic acid	X	X
Dichloroethane (ethylene dichloride)	X	X
Ethylene glycol	100	38
Ferric chloride	160	71
Ferric chloride, 50% in water	160	71
Ferric nitrate, 10–50%	200	93
Ferrous chloride	90	32
Ferrous nitrate	200	93
Fluorine gas, dry	X	X
Fluorine gas, moist	X	X
Hydrobromic acid, dil.	X	X
Hydrobromic acid, 20%	X	X
Hydrobromic acid, 50%	X	X
Hydrochloric acid, 20%	X	X
Hydrochloric acid, 38%	X	X
Hydrocyanic acid, 10%	X	X
Hydrofluoric acid, 30%	X	X
Hydrofluoric acid, 70%	X	X
Hydrofluoric acid, 100%	X	X
Hypochlorous acid	X	X
Iodine solution, 10%	80	27
Ketones, general	X	X
Lactic acid, 25%	140	60
Lactic acid, conc.	90	32
Magnesium chloride	200	93
Malic acid		
Manganese chloride	200	93
Methyl chloride	X	X
Methyl ethyl ketone	X	X
Methyl isobutyl ketone	X	X
Muriatic acid	X	X
Nitric acid, 5%	X	X
Nitric acid, 20%	X	X
Nitric acid, 70%	X	X
Nitric acid, anhydrous	X	X
Nitrous acid, conc.	X	X
Oleum	X	X
Perchloric acid, 10%		
Perchloric acid, 70%	X	X
Phenol	X	X
Phosphoric acid, 50–80%	150	66
Picric acid	200	93
Potassium bromide, 30%	160	71
Salicylic acid		

(continued)

TABLE 5.7 *Continued*

Chemical	Maximum Temperature	
	°F	°C
Silver bromide, 10%		
Sodium carbonate	200	93
Sodium chloride	200	93
Sodium hydroxide, 10%	230	110
Sodium hydroxide, 50%	230	110
Sodium hydroxide, conc.	230	110
Sodium hypochlorite, 20%	X	X
Sodium hypochlorite, conc.	X	X
Sodium sulfide, to 50%	200	93
Stanic chloride	200	93
Stannous chloride	X	X
Sulfuric acid, 10%	150	66
Sulfuric acid, 50%	100	38
Sulfiric acid, 70%	X	X
Sulfuric acid, 90%	X	X
Sulfuric acid, 98%	X	X
Sulfuric acid, 100%	X	X
Sulfiric acid, fuming	X	X
Sulfurous acid	100	38
Thionyl chloride	X	X
Toluene	X	X
Trichloroacetic acid	X	X
White liquor	140	60
Zinc chloride	160	71

The chemicals listed are in the pure state or in a saturated solution unless otherwise indicated. Compatibility is shown to the maximum allowable temperature for which data is available. Incompatibility is shown by an X. A blank space indicates that data is unavailable.

Source: From P.A. Schweitzer. 2004. *Corrosion Resistance Tables*, Vols. 1–4, 5th ed., New York: Marcel Dekker.

5.4.3 Applications

Neoprene products are available in three basic forms:

1. Conventional solid rubber parts
2. Highly compressed cellular materials
3. Free-flowing liquids

Each form has certain specific properties that can be advantageous to final products.

Solid products can be produced by molding, extruding, or calendering. Molding can be accomplished by compression, transfer, injection, blow, vacuum, or wrapped-mandrel methods. Typical products produced by these methods are instrument seals, shoe soles and heels, auto sparkplug boots, radiator hose, boating accessories, appliance parts, O-rings, and other miscellaneous components.

Extrusion processes provide means to economically and uniformly mass produce products quickly. Neoprene products manufactured by these processes include tubing, sealing strips, wire jacketing, filaments, rods, and many types of hose.

Calendered products include sheet stock, belting, and friction and coated fabrics. A large proportion of sheet stock is later die-cut into finished products such as gaskets, pads, and diaphragms.

Cellular forms of neoprene are used primarily for gasketing, insulation, cushioning, and sound and vibration damping. This material provides compressibility not found in solid rubber but still retains the advantageous properties of neoprene. It is available as an open-cell sponge, a closed-cell neoprene, and a foam neoprene.

Open-cell neoprene is a compressible, absorbent material whose cells are uniform and interconnected. This material is particularly useful for gasketing and dustproofing applications where exposure to fluids is not expected.

Closed-cell neoprene is a resilient complex of individual, nonconnecting cells that impart the added advantage of nonabsorbency. This property makes closed-cell neoprene especially suitable for sealing applications where fluid contact is expected such as in wetsuits for divers, shoe soles, automotive deck lid seals, and for other appplications where a compressible nonabsorbent weather-resistant material is required.

Foam neoprene is similar to open-cell neoprene in that it is a compressible material with interconnecting cells. Its main area of application is for cushioning, for example, in mattresses, seating, and carpet underlay. Because of the good heat and oil resistance of neoprene, it has also found application as a railroad car lubricator. This absorbent open-cell structure provides a wicking action to deliver oil to the journal bearings.

Fluid forms of neoprene are important in the manufacture of many products because of their versatility. "Fluid" neoprene is the primary component in such products as adhesives, coatings and paints, sealants, caulks, and fiber binders. It is available in two forms—as a neoprene latex or as a solvent solution. Neoprene latex is an elastomer–water dispersion. It is used primarily in the manufacture of dipped products such as tool grips, household and industrial gloves, meteorological balloons, sealed fractional-horsepower motors, and a variety of rubber-coated metal parts. Other applications include use as a binder for curled animal hair in resilient furniture cushioning, transportation seating, acoustical filtering, and packaging. It is also used extensively in latex-based adhesives, foams, protective coatings, and knife-coated fabrics, as a binder for cellulose and asbestos, and as an elasticizing additive for concrete, mortar, and asphalt. These products produced from neoprene latex possess the same properties as those associated with solid neoprene, including resistance to oil, chemicals, ozone, weather, and flame.

Neoprene solvent solutions are prepared by dissolving neoprene in standard rubber solvents. These solutions can be formulated in a range of viscosities suitable for application by brush, spray, or roller. Major areas of application include coatings for storage tanks, industrial equipment, and chemical processing equipment. These coatings protect the vessels from corrosion by acids, oils, alkalies, and most hydrocarbons.

Neoprene roofing applied in liquid form is used to protect concrete, plywood, and metal decks. The solvent solution can be readily applied and will cure into a roofing membrane that is tough, elastic, and weather resistant.

Solvent-based neoprene adhesives develop quick initial bonds and remain strong and flexible most indefinitely. They can be used to join a wide variety of rigid and flexible materials.

Collapsible nylon containers coated with neoprene are used for transporting and/or storing liquids, pastes, and flowable dry solds. Containers have been designed to hold oils, fuels, molasses, and various bulk-shipped products.

Neoprene, in its many forms, has proved to be a reliable, indispensable substitute for natural rubber, possessing many of the advantageous properties of natural rubber while also overcoming many of its shortcomings.

5.5 Styrene–Butadiene Rubber (SBR, Buna-S, GR-S)

During World War II, a shortage of natural rubber was created when Japan occupied the Far Eastern nations where natural rubber was obtained. Because of the great need for rubber, the U.S. government developed what was originally known as Government Rubber Styrene-Type (GR-S) because it was the most practical to put into rapid production on a wartime scale. It was later designated GR-S.

The rubber is produced by copolymerizing butadiene and styrene. As with natural rubber and the other synthetic elastomers, compounding with other ingredients will improve certain properties. Continued development since World War II has improved its properties considerably over what was initially produced by either Germany or the United States.

5.5.1 Resistance to Sun, Weather, and Ozone

Butadiene–styrene rubber has poor weathering and aging properties. Sunlight will cause it to deteriorate. However, it does have better water resistance than natural rubber.

5.5.2 Chemical Resistance

The chemical resistance of Buna-S is similar to that of natural rubber. It is resistant to water and exhibits fair to good resistance to dilute acids, alkalies, and alcohols. It is not resistant to oils, gasoline, hydrocarbons, or oxidizing agents. Refer to Table 5.8 for the compatibility of SBR with selected corrodents.

TABLE 5.8

Compatibility of SBR with Selected Corrodents

Chemical	Maximum Temperature	
	°F	°C
Acetic acid, 10%	X	X
Acetic acid, 50%	X	X
Acetic acid, 80%	X	X
Acetic acid, glacial	X	X
Acetone	200	93
Ammonium chloride, 10%	200	93
Ammonium chloride, 28%	200	93
Ammonium chloride, 50%	200	93
Ammonium chloride, sat.	200	93
Ammonium sulfate, 10–40%		
Aniline	X	X
Benzene	X	X
Benzoic acid	X	X
Butane	X	X
Butyl acetate	X	X
Butyl alcohol	200	93
Calcium chloride, sat.	200	93
Calcium hydroxide, 10%	200	93
Calcium hydroxide, 20%	200	93
Calcium hydroxide, 30%	200	93
Calcium hydroxide, sat.	200	93
Carbon bisulfide	X	X
Carbon tetrachloride	X	X
Chloroform	X	X
Chromic acid, 10%	X	X
Chromic acid, 30%	X	X
Chromic acid, 40%	X	X
Chromic acid, 50%	X	X
Copper sulfate	200	93
Corn oil	X	X
Ethyl acetate	X	X
Ethyl alcohol	200	93
Ethylene glycol	200	93
Formaldehyde, dil.	200	93
Formaldehyde, 37%	200	93

(continued)

TABLE 5.8 *Continued*

Chemical	Maximum Temperature	
	°F	°C
Formaldehyde, 50%	200	93
Glycerine	200	93
Hydrochloric acid, dil.	X	X
Hydrochloric acid, 20%	X	X
Hydrochloric acid, 35%	X	X
Hydrochloric acid, 38%	X	X
Hydrochloric acid, 50%	X	X
Hydrofluoric acid, dil.	X	X
Hydrofluoric acid, 30%	X	X
Hydrofluoric acid, 40%	X	X
Hydrofluoric acid, 50%	X	X
Hydrofluoric acid, 70%	X	X
Hydrofluoric acid, 100%	X	X
Hydrogen peroxide, dil., all conc.	200	93
Lactic acid, all conc.	200	93
Methyl ethyl ketone	X	X
Methyl isobutyl ketone	X	X
Muriatic acid	X	X
Nitric acid	X	X
Phenol	X	X
Phosphoric acid, 10%	200	93
Potassium hydroxide, to 50%	X	X
Propane	X	X
Sodium chloride	200	93
Sodium hydroxide, all conc.	X	X
Sulfuric acid, all conc.	X	X
Toluene	X	X
Trichloroethylene	X	X
Water, demineralized	210	99
Water, distilled	200	93
Water, salt	200	93
Water, sea	200	93

The chemicals listed are in the pure state or in a saturated solution unless otherwise indicated. Compatibility is shown to the maximum allowable temperature for which data is available. Incompatibility is shown by an X. A blank space indicates that data is unavailable.

Source: From P.A. Schweitzer. 2004. *Corrosion Resistance Tables*, Vols. 1–4, 5th ed., New York: Marcel Dekker.

5.5.3 Applications

The major use of Buna-S is in the manufacture of automobile tires, although Buna-S materials are also used to manufacture conveyor belts, hoses, gaskets, and seals against air, moisture, sound, and dirt.

5.6   Nitrile Rubber (NBR, Buna-N)

Nitrile rubbers are an outgrowth of German Buna-N or Perbunan. They are copolymers of butadiene and acrylonitrile ($CH_2=CH–C=N$) and are one of the four most wildely used elastomers. XNBR is a carboxylic-acrylomtrile-butadiene-nitrile rubber with somewhat improved abrasion resistance over that of the standard NBR nitrile rubbers. The main advantages of nitrile rubbers are their low cost, good oil and abrasion resistance, and good low-temperature and swell characteristics. Their greater resistance to oils, fuel, and solvents compared to that of neoprene is their primary advantage. As with other elastomers, appropriate compounding will improve certain properties.

   Carbonated nitrile rubber (XNBR) incorporates up to 10% of a third comonomer with organic acid functionality. When compared to NBR, XNBR has improved abrasion resistance and strength. XNBR can be difficult to process, and it requires special formulation to prevent sticking to mixer surfaces and premature vulcanization.

5.6.1   Resistance to Sun, Weather, and Ozone

Nitrile rubbers offer poor resistance to sunlight and ozone, and their weathering qualities are not good.

5.6.2   Chemical Resistance

Nitrile rubbers exhibit good resistance to solvents, oil, water, and hydraulic fluids. Very slight swelling occurs in the presence of aliphatic hydrocarbons, fatty acids, alcohols, and glycols. The deterioration of physical properties as a result of this swelling is small, making NBR suitable for gasoline- and oil-resistant applications. NBR has excellent resistance to water. The use of highly polar solvents such as acetone and methyl ethyl ketone, chlorinated hydrocarbons, ozone, nitro hydrocarbons, ether, or esters should be avoided, since these materials will attack the nitrile rubbers.

   XNBR rubbers are used primarily in nonalkaline service. Refer to Table 5.9 for the compatibility of NBR with selected corrodents.

5.6.3   Applications

Because of its exceptional resistance to fuels and hydraulic fluids, Buna-N's major area of application is in the manufacture of aircraft hose, gasoline and oil hose, and self-sealing fuel tanks. Other applications include carburetor diaphragms, gaskets, cables, printing rolls, and machinery mountings.

TABLE 5.9

Compatibility of Nitrile Rubber with Selected Corrodents

Chemical	Maximum Temperature	
	°F	°C
Acetaldehyde	X	X
Acetamide	180	82
Acetic acid, 10%	X	X
Acetic acid, 50%	X	X
Acetic acid, 80%	X	X
Acetic acid, glacial	X	X
Acetic anhydride	X	X
Acetone	X	X
Acetyl chloride	X	X
Acrylic acid	X	X
Acrylonitrile	X	X
Adipic acid	180	82
Allyl alcohol	180	82
Allyl chloride	X	X
Alum	150	66
Aluminum chloride, aqueous	150	66
Aluminum hydroxide	180	82
Aluminum nitrate	190	88
Aluminum sulfate	200	93
Ammonia gas	190	88
Ammonium carbonate	X	X
Ammonium nitrate	150	66
Ammonium phosphate	150	66
Amyl alcohol	150	66
Aniline	X	X
Barium chloride	125	52
Benzene	150	66
Benzoic acid	150	66
Boric acid	150	66
Calcium hypochlorite	X	X
Carbonic acid	100	38
Ethylene glycol	100	38
Ferric chloride	150	66
Ferric nitrate, 10–50%	150	66
Hydrofluoric acid, 70%	X	X
Hydrofluoric acid, 100%	X	X
Hypochlorous acid		
Nitric acid, 20%	X	X
Nitric acid, 70%	X	X
Nitric acid, anhydrous	X	X
Phenol	X	X
Phosphoric acid, 50–80%	150	66
Sodium carbonate	125	52
Sodium chloride	200	93

(continued)

TABLE 5.9 *Continued*

Chemical	Maximum Temperature	
	°F	°C
Sodium hydroxide, 10%	150	66
Sodium hypochlorite, 20%	X	X
Sodium hypochlorite, conc.	X	X
Stannic chloride	150	66
Sulfuric acid, 10%	150	66
Sulfuric acid, 50%	150	66
Sulfuric acid, 70%	X	X
Sulfuric acid, 90%	X	X
Sulfuric acid, 98%	X	X
Sulfuric acid, 100%	X	X
Sulfuric acid, fuming	X	X
Zinc chloride	150	66

The chemicals listed are in the pure state or in a saturated solution unless otherwise indicated. Compatibility is shown to the maximum allowable temperature for which data is available. Incompatibility is shown by an X. A blank space indicates that data is unavailable.

Source: From P.A. Schweitzer. 2004. *Corrosion Resistance Tables*, Vols. 1–4, 5th ed., New York: Marcel Dekker.

5.7 Butyl Rubber (IIR) and Chlorobutyl Rubber (CIIR)

Butyl rubber contains isobutylene

$$
\begin{array}{c}
CH_3 \\
| \\
C-CH_2 \\
| \\
CH_3
\end{array}
$$

as its parent material with small proportions of butadiene or isoprene added. Commercial butyl rubber may contain 5% butadiene as a copolymer. It is a general-purpose synthetic rubber whose outstanding physical properties are low permeability to air (approximately one-fifth that of natural rubber) and high energy absorption.

Chlorobutyl rubber is chlorinated isobutylene–isoprene. It has the same general properties as butyl rubber but with slightly higher allowable operating temperatures.

5.7.1 Resistance to Sun, Weather, and Ozone

Butyl rubber has excellent resistance to sun, weather, and ozone. Its weathering qualities are outstanding as its resistance to water absorption.

5.7.2 Chemical Resistance

Butyl rubber is very nonpolar. It has exceptional resistance to dilute mineral acids, alkalies, phosphate ester oils, acetane, ethylene, ethylene glycol, and water. Resistance to concentrated acids, except nitric and sulfuric, is good. Unlike natural rubber, it is very resistant to swelling by vegetable and animal oils. It has poor resistance to petroleum oils, gasoline, and most solvents (except oxygenated solvents).

CIIR has the same general resistance as natural rubber but can be used at higher temperatures. Unlike butyl rubber, CIIR cannot be used with hydrochloric acid.

Refer to Table 5.10 for the compatibility of butyl rubber with selected corrodents and to Table 5.11 for the compatibility of chlorobutyl rubber with selected corrodents.

TABLE 5.10

Compatibility of Butyl Rubber with Selected Corrodents

Chemical	Maximum Temperature	
	°F	°C
Acetaldehyde	80	27
Acetic acid, 10%	150	66
Acetic acid, 50%	110	43
Acetic acid, 80%	110	43
Acetic acid, glacial	X	X
Acetic anhydride	X	X
Acetone	100	38
Acrylonitrile	X	X
Adipic acid	X	X
Allyl alcohol	190	88
Allyl chloride	X	X
Alum	200	93
Aluminum acetate	200	93
Aluminum chloride, aqueous	200	93
Aluminum chloride, dry	200	93
Aluminum fluoride	180	82
Aluminum hydroxide	100	38
Aluminum nitrate	100	38
Aluminum sulfate	200	93
Ammonium bifluoride	X	X
Ammonium carbonate	190	88
Ammonium chloride, 10%	200	93
Ammonium chloride, 50%	200	93
Ammonium chloride, sat.	200	93
Ammonium fluoride, 10%	150	66
Ammonium fluoride, 25%	150	66
Ammonium hydroxide, 25%	190	88
Ammonium hydroxide, sat.	190	88

(continued)

TABLE 5.10 *Continued*

Chemical	Maximum Temperature	
	°F	°C
Ammonium nitrate	200	93
Ammonium persulfate	190	88
Ammonium phosphate	150	66
Ammonium sulfate, 10–40%	150	66
Amyl acetate	X	X
Amyl alcohol	150	66
Aniline	150	66
Antimony trichloride	150	66
Barium chloride	150	66
Barium hydroxide	190	88
Barium sulfide	190	88
Benzaldehyde	90	32
Benzene	X	X
Benzenesulfonic acid, 10%	90	32
Benzoic acid	150	66
Benzyl alcohol	190	88
Benzyl chloride	X	X
Borax	190	88
Boric acid	150	66
Butyl acetate	X	X
Butyl alcohol	140	60
Butyric acid	X	X
Calcium bisulfite	120	49
Calcium carbonate	150	66
Calcium chlorate	190	88
Calcium chloride	190	88
Calcium hydroxide, 10%	190	88
Calcium hydroxide, sat.	190	88
Calcium hypochlorite	X	X
Calcium nitrate	190	88
Calcium sulfate	100	38
Carbon dioxide, dry	190	88
Carbon dioxide, wet	190	88
Carbon disulfide	190	88
Carbon monoxide	X	X
Carbon tetrachloride	90	32
Carbonic acid	150	66
Cellosolve	150	66
Chloroacetic acid, 50% water	150	66
Chloroacetic acid	100	38
Chlorine gas, dry	X	X
Chlorine, liquid	X	X
Chlorobenzene	X	X
Chloroform	X	X
Chlorosulfonic acid	X	X
Chromic acid, 10%	X	X

(continued)

TABLE 5.10 *Continued*

Chemical	Maximum Temperature	
	°F	°C
Chromic acid, 50%	X	X
Citric acid, 15%	190	88
Citric acid, conc.	190	88
Copper chloride	150	66
Copper sulfate	190	88
Cresol	X	X
Cupric chloride, 5%	150	66
Cupric chloride, 50%	150	66
Cyclohexane	X	X
Dichloroethane (ethylene dichloride)	X	X
Ethylene glycol	200	93
Ferric chloride	175	79
Ferric chloride, 50% in water	160	71
Ferric nitrate, 10–50%	190	88
Ferrous chloride	175	79
Ferrous nitrate	190	88
Fluorine gas, dry	X	X
Hydrobromic acid, dil.	125	52
Hydrobromic acid, 20%	125	52
Hydrobromic acid, 50%	125	52
Hydrochloric acid, 20%	125	52
Hydrochloric acid, 38%	125	52
Hydrocyanic acid, 10%	140	60
Hydrofluoric acid, 30%	150	66
Hydrofluoric acid, 70%	150	66
Hydrofluoric acid, 100%	150	66
Hypochlorous acid	X	X
Lactic acid, 25%	125	52
Lactic acid, conc.	125	52
Magnesium chloride	200	93
Malic acid	X	X
Methyl chloride	90	32
Methyl ethyl ketone	100	38
Methyl isobutyl ketone	80	27
Muriatic acid	X	X
Nitric acid, 5%	200	93
Nitric acid, 20%	150	66
Nitric acid, 70%	X	X
Nitric acid, anhydrous	X	X
Nitrous acid, conc.	125	52
Oleum	X	X
Perchloric acid, 10%	150	66
Phenol	150	66
Phosphoric acid, 50–80%	150	66
Salicylic acid	80	27
Sodium chloride	200	93

(continued)

TABLE 5.10 *Continued*

Chemical	Maximum Temperature	
	°F	°C
Sodium hydroxide, 10%	150	66
Sodium hydroxide, 50%	150	66
Sodium hydroxide, conc.	150	66
Sodium hypochlorite, 20%	X	X
Sodium hypochlorite, conc.	X	X
Sodium sulfide, to 50%	150	66
Stannic chloride	150	66
Stannous chloride	150	66
Sulfuric acid, 10%	200	93
Sulfuric acid, 50%	150	66
Sulfuric acid, 70%	X	X
Sulfuric acid, 90%	X	X
Sulfuric acid, 98%	X	X
Sulfuric acid, 100%	X	X
Sulfuric acid, fuming	X	X
Sulfurous acid	200	93
Thionyl chloride	X	X
Toluene	X	X
Trichloroacetic acid	X	X
Zinc chloride	200	93

The chemicals listed are in the pure state or in a saturated solution unless otherwise indicated. Compatibility is shown to the maximum allowable temperature for which data is available. Incompatibility is shown by an X. A blank space indicates that data is unavailable.

Source: From P.A. Schweitzer. 2004. *Corrosion Resistance Tables*, Vols. 1–4, 5th ed., New York: Marcel Dekker.

TABLE 5.11

Compatibility of Chlorobutyl Rubber with Selected Corrodents

Chemical	Maximum Temperature	
	°F	°C
Acetic acid, 10%	150	60
Acetic acid, 50%	150	60
Acetic acid, 80%	150	60
Acetic acid, glacial	X	X
Acetic anhydride	X	X
Acetone	100	38
Alum	200	93
Aluminum chloride, aqueous	200	93
Aluminum nitrate	190	88
Aluminum sulfate	200	93
Ammonium carbonate	200	93
Ammonium chloride, 10%	200	93

(continued)

TABLE 5.11 *Continued*

Chemical	Maximum Temperature	
	°F	°C
Ammonium chloride, 50%	200	93
Ammonium chloride, sat.	200	93
Ammonium nitrate	200	93
Ammonium phosphate	150	66
Ammonium sulfate, 10–40%	150	66
Amyl alcohol	150	66
Aniline	150	66
Antimony trichloride	150	66
Barium chloride	150	66
Benzoic acid	150	66
Boric acid	150	66
Calcium chloride	160	71
Calcium nitrate	160	71
Calcium sulfate	160	71
Carbon monoxide	100	38
Carbonic acid	150	66
Chloroacetic acid	100	38
Chromic acid, 10%	X	X
Chromic acid, 50%	X	X
Citric acid, 15%	90	32
Copper chloride	150	66
Copper cyanide	160	71
Copper sulfate	160	71
Cupric chloride, 5%	150	66
Cupric chloride, 50%	150	66
Ethylene glycol	200	93
Ferric chloride	175	79
Ferric chloride, 50% in water	100	38
Ferric nitrate, 10–50%	160	71
Ferrous chloride	175	79
Hydrobromic acid, dil.	125	52
Hydrobromic acid, 20%	125	52
Hydrobromic acid, 50%	125	52
Hydrochloric acid, 20%	X	X
Hydrochloric acid, 38%	X	X
Hydrofluoric acid, 70%	X	X
Hydrofluoric acid, 100%	X	X
Lactic acid, 25%	125	52
Lactic acid, conc.	125	52
Magnesium chloride	200	93
Nitric acid, 5%	200	93
Nitric acid, 20%	150	66
Nitric acid, 70%	X	X
Nitric acid, anhydrous	X	X
Nitrous acid, conc.	125	52
Phenol	150	66

(continued)

TABLE 5.11 *Continued*

Chemical	Maximum Temperature	
	°F	°C
Phosphoric acid, 50–80%	150	66
Sodium chloride	200	93
Sodium hydroxide, 10%	150	66
Sodium sulfide, to 50%	150	66
Sulfuric acid, 10%	200	93
Sulfuric acid, 70%	X	X
Sulfuric acid, 90%	X	X
Sulfuric acid, 98%	X	X
Sulfuric acid, 100%	X	X
Sulfuric acid, fuming	X	X
Sulfurous acid	200	93
Zinc chloride	200	93

The chemicals listed are in the pure state or in a saturated solution unless otherwise indicated. Compatibility is shown to the maximum allowable temperature for which data is available. Incompatibility is shown by an X. A blank space indicates that data is unavailable.

Source: From P.A. Schweitzer. 2004. *Corrosion Resistance Tables*, Vols. 1–4, 5th ed., New York: Marcel Dekker.

5.7.3 Applications

Because of its impermeability, butyl rubber finds many uses in the manufacture of inflatable items such as life jackets, life boats, balloons, and inner tubes. The excellent resistance it exhibits in the presence of water and steam makes it suitable for hoses and diaphragms. Applications are also found as flexible electrical insulation, shock and vibration absorbers, curing bags for tire vulcanization, and molding.

5.8 Chlorosulfonated Polyethylene Rubber (Hypalon)

Chlorosulfonated polyethylene synthetic rubber (CSM) is manufactured by DuPont under the trade name Hypalon. In many respects it is similar to neoprene, but it does possess some advantages over neoprene in certain types of service. It has better heat and ozone resistance, better electrical properties, better color stability, and better chemical resistance.

Hypalon, when properly compounded, also exhibits good resistance to wear and abrasion, good flex life, high impact resistance, and good resistance to permanent deformation under heavy loading.

Hypalon has a broad range of service temperatures with excellent thermal properties. General-purpose compounds can operate continuously at temperatures of 248–275°F (120–135°C). Special compounds can be formulated that can be used intermittently up to 302°F (150°C).

On the low-temperature side, conventional compounds can be used continuously down to 0 to −20°F (−18 to −28°C). Special compounds can be produced that will retain their flexibility down to −40°F (−40°C), but to produce such a compound, it is necessary to sacrifice performance of some of the other properties.

5.8.1 Resistance to Sun, Weather, and Ozone

Hypalon is one of the most weather-resistance elastomers available. Oxidation takes place at a very slow rate. Sunlight and ultraviolet light have little, if any, adverse effects on its physical properties. It is also inherently resistant to ozone attack without the need for the addition of special antioxidants or antiozonants to the formulation.

Many elastomers are degraded by ozone concentrations of less than 1 part per million parts of air. Hypalon, however, is unaffected by concentrations as high as 1 part per 100 parts of air.

5.8.2 Chemical Resistance

When properly compounded, Hypalon is highly resistant to attack by hydrocarbon oils and fuels, even at elevated temperatures. It is also resistant to such oxidizing chemicals as sodium hypochlorite, sodium peroxide, ferric chloride, and sulfuric, chromic, and hydrofluoric acids. Concentrated hydrochloric acid (37%) at elevated temperatures above 158°F (70°C) will attack Hypalon but can be handled without adverse effect at all concentrations below this temperature. Nitric acid at room temperature and up to 60% concentration can also be handled without adverse effects.

Hypalon is also resistant to salt solutions, alcohols, and both weak and concentrated alkalies and is generally unaffected by soil chemicals, moisture, and other deteriorating factors associated with burial in the earth. Long-term contact with water has little or no effect on Hypalon. It is also resistant to radiation.

Hypalon has poor resistance to aliphatic, aromatic, and chlorinated hydrocarbons, aldehydes, and ketones. Refer to Table 5.12 for the compatibility of Hypalon with selected corrodents.

5.8.3 Applications

Hypalon finds useful applications in many industries and many fields. Because of its outstanding resistance to oxidizing acids, it has found widespread use as an acid transfer hose. For the same reason, it is used to line railroad tanks cars and other tanks containing acids and other oxidizing chemicals. Its physical and mechanical properties make it suitable for use in hose undergoing continuous flexing and/or those carrying hot water or steam.

TABLE 5.12

Compatibility of Hypalon with Selected Corrodents

| | Maximum Temperature | |
Chemical	°F	°C
Acetaldehyde	60	16
Acetamide	X	X
Acetic acid, 10%	200	93
Acetic acid, 50%	200	93
Acetic acid, 80%	200	93
Acetic acid, glacial	X	X
Acetic anhydride	200	93
Acetone	X	X
Acetyl chloride	X	X
Acrylonitrile	140	60
Adipic acid	140	60
Allyl alcohol	200	93
Aluminum fluoride	200	93
Aluminum hydroxide	200	93
Aluminum nitrate	200	93
Aluminum sulfate	180	82
Ammonia gas	90	32
Ammonium carbonate	140	60
Ammonium chloride, 10%	190	88
Ammonium chloride, 50%	190	88
Ammonium chloride, sat.	190	88
Ammonium fluoride, 10%	200	93
Ammonium hydroxide, 25%	200	93
Ammonium hydroxide, sat.	200	93
Ammonium nitrate	200	93
Ammonium persulfate	80	27
Ammonium phosphate	140	60
Ammonium sulfate, 10–40%	200	93
Ammonium sulfide	200	93
Amyl acetate	60	16
Amyl alcohol	200	93
Amyl chloride	X	X
Aniline	140	60
Antimony trichloride	140	60
Barium carbonate	200	93
Barium chloride	200	93
Barium hydroxide	200	93
Barium sulfate	200	93
Barium sulfide	200	93
Benzaldehyde	X	X
Benzene	X	X
Benzenesulfonic acid, 10%	X	X
Benzoic acid	200	93
Benzyl alcohol	140	60
Benzyl chloride	X	X

(continued)

TABLE 5.12 *Continued*

Chemical	Maximum Temperature	
	°F	°C
Borax	200	93
Boric acid	200	93
Bromine gas, dry	60	16
Bromine gas, moist	60	16
Bromine, liquid	60	16
Butadiene	X	X
Butyl acetate	60	16
Butyl alcohol	200	93
Butyric acid	X	X
Calcium bisulfide	200	93
Calcium carbonate	90	32
Calcium chlorate	90	32
Calcium chloride	200	93
Calcium hydroxide, 10%	200	93
Calcium hydroxide, sat.	200	93
Calcium hypochlorite	200	93
Calcium nitrate	100	38
Calcium oxide	200	93
Calcium sulfate	200	93
Caprylic acid	X	X
Carbon dioxide, dry	200	93
Carbon dioxide, wet	200	93
Carbon disulfide	200	93
Carbon monoxide	X	X
Carbon tetrachloride	200	93
Carbonic acid	X	X
Chloroacetic acid	X	X
Chlorine gas, dry	X	X
Chlorine gas, wet	90	32
Chlorobenzene	X	X
Chloroform	X	X
Chlorosulfonic acid	X	X
Chromic acid, 10%	150	66
Chromic acid, 50%	150	66
Chromyl chloride		
Citric acid, 15%	200	93
Citric acid, conc.	200	93
Copper acetate	X	X
Copper chloride	200	93
Copper cyanide	200	93
Copper sulfate	200	93
Cresol	X	X
Cupric chloride, 5%	200	93
Cupric chloride, 50%	200	93
Cyclohexane	X	X
Cyclohexanol	X	X

(continued)

TABLE 5.12 *Continued*

Chemical	Maximum Temperature	
	°F	°C
Dichloroethane (ethylene dichloride)	X	X
Ethylene glycol	200	93
Ferric chloride	200	93
Ferric chloride, 50% in water	200	93
Ferric nitrate, 10–50%	200	93
Ferrous chloride	200	93
Fluorine gas, dry	140	60
Hydrobromic acid, dil.	90	32
Hydrobromic acid, 20%	100	38
Hydrobromic acid, 50%	100	38
Hydrochloric acid, 20%	160	71
Hydrochloric acid, 38%	140	60
Hydrocyanic acid, 10%	90	32
Hydrofluoric acid, 30%	90	32
Hydrofluoric acid, 70%	90	32
Hydrofluoric acid, 100%	90	32
Hypochlorous acid	X	X
Ketones, general	X	X
Lactic acid, 25%	140	60
Lactic acid, conc.	80	27
Magnesium chloride	200	93
Manganese chloride	180	82
Methyl chloride	X	X
Methyl ethyl ketone	X	X
Methyl lsobutyl ketone	X	X
Muriatic acid	140	60
Nitric acid, 5%	100	38
Nitric acid, 20%	100	38
Nitric acid, 70%	X	X
Nitric acid, anhydrous	X	X
Oleum	X	X
Perchloric acid, 10%	100	38
Perchloric acid, 70%	90	32
Phenol	X	X
Phosphoric acid, 50–80%	200	93
Picric acid	80	27
Potassium bromide, 30%	200	93
Sodium carbonate	200	93
Sodium chloride	200	93
Sodium hydroxide, 10%	200	93
Sodium hydroxide, 50%	200	93
Sodium hydroxide, conc.	200	93
Sodium hypochlorite, 20%	200	93
Sodium hypochlorite, conc.		
Sodium sulfide, to 50%	200	93

(continued)

TABLE 5.12 *Continued*

Chemical	Maximum Temperature	
	°F	°C
Stannic chloride	90	32
Stannous chloride	200	93
Sulfuric acid, 10%	200	93
Sulfuric acid, 50%	200	93
Sulfuric acid, 70%	160	71
Sulfuric acid, 90%	X	X
Sulfuric acid, 98%	X	X
Sulfuric acid, 100%	X	X
Sulfurous acid	160	71
Toluene	X	X
Zinc chloride	200	93

The chemicals listed are in the pure state or in a saturated solution unless otherwise indicated. Compatibility is shown to the maximum allowable temperature for which data is available. Incompatibility is shown by an X. A blank space indicates that data is unavailable.

Source: From P.A. Schweitzer. 2004. *Corrosion Resistance Tables*, Vols. 1–4, 5th ed., New York: Marcel Dekker.

The electrical industry makes use of Hypalon to cover automotive ignition and primary wire, nuclear power station cable, control cable, and welding cable. As an added protection from storms at sea, power and lighting cable on offshore oil platforms are sheathed with Hypalon. Because of its heat and radiation resistance, it is also used as a jacketing material on heating cable embedded in roadways to melt ice and on x-ray machine cable leads. It is also used in applicance cords, insulating hoods and blankets, and many other electrical accessories.

In the automotive industry, advantage is taken of Hypalon's color stability and good weathering properties by using the elastomer for exterior parts on cars, trucks, and other commercial vehicles. Its resistance to heat, ozone, oil, and grease makes it useful for application under the hood for such components as emission control hose, tubing, ignition wire jacketing, spark plug boots, and air-conditioning and power-steering hoses. The ability to remain soil-free and easily cleanable makes it suitable for tire whitewalls.

When combined with cork, Hypalon provides a compressible yet set-resistant gasket suitable for automobile crankcase and rocker pans. Hypalon protects the cork from oxidation at elevated temperatures and also provides excellent resistance to oil, grease, and fuels.

The construction industry has made use of Hypalon for sheet roofing, pond liners, reservoir covers, curtain-wall gaskets, floor tiles, escalator rails, and decorative and maintenance coatings. In these applications, the properties of color stability, excellent weatherability, abrasion resistance, useful temperature range, light weight, flexibility, and good aging characteristics are of importance.

Application is also found in the coating of fabrics that are used for inflatable structures, flexible fuel tanks, tarpaulins, and hatch and boat covers. These products offer the advantages of being lightweight and colorful. Consumer items such as awnings, boating garb, convertible tops, and other products also make use of fabrics coated with this elastomer.

5.9 Polybutadiene Rubber (BR)

Butadiene (CH_2=CH–CH=CH_2) has two unsaturated linkages and can be readily polymerized. When butadiene or its derivatives becomes polymerized, the units link together to form long chains that each contains over 1000 units. Simple butadiene does not yield a good grade of rubber, apparently because the chains are too smooth. Better results are obtained by introducing side groups into the chain either by modifying butadiene or by making a copolymer of butadiene and some other compound.

5.9.1 Resistance to Sun, Weather, and Ozone

Although polybutadiene has good weather resistance, it will deteriorate when exposed to sunlight for prolonged periods of time. It also exhibits poor resistance to ozone.

5.9.2 Chemical Resistance

The chemical resistance of polybutadiene is similar to that of natural rubber. It shows poor resistance to aliphatic and aromatic hydrocarbons, oil, and gasoline, but it displays fair to good resistance in the presence of mineral acids and oxygenated compounds. Refer to Table 5.13 for the compatibility of polybutadiene rubber with selected corrodents.

TABLE 5.13

Compatibility of Polybutadiene (BR) with Selected Corrodents

	Maximum Temperature	
Chemical	°F	°C
Alum	90	32
Alum ammonium	90	32
Alum ammonium sulphate	90	32
Alum chrome	90	32
Alum potassium	90	32
Alum chloride, aqueous	90	32

(continued)

TABLE 5.13 *Continued*

Chemical	Maximum Temperature	
	°F	°C
Aluminum sulfate	90	32
Ammonia gas	90	32
Ammonium chloride, 10%	90	32
Ammonium chloride, 28%	90	32
Ammonium chloride, 50%	90	32
Ammonium chloride, sat.	90	32
Ammonium nitrate	90	32
Ammonium sulfate, 10–40%	90	32
Calcium chloride, sat.	80	27
Calcium hypochlorite, sat.	90	32
Carbon dioxide, wet	90	32
Chlorine gas, wet	X	X
Chrome alum	90	32
Chromic acid, 10%	X	X
Chromic acid, 30%	X	X
Chromic acid, 40%	X	X
Chromic acid, 50%	X	X
Copper chloride	90	32
Copper sulfate	90	32
Fatty acids	90	32
Ferrous chloride	90	32
Ferrous sulfate	90	32
Hydrochloric acid, dil.	80	27
Hydrochloric acid, 20%	90	32
Hydrochloric acid, 35%	90	32
Hydrochloric acid, 38%	90	32
Hydrochloric acid, 50%	90	32
Hydrochloric acid fumes	90	32
Hydrogen peroxide, 90%	90	32
Hydrogen sulfide, dry	90	32
Nitric acid, 5%	80	27
Nitric acid, 10%	80	27
Nitric acid, 20%	80	27
Nitric acid, 30%	80	27
Nitric acid, 40%	X	X
Nitric acid, 50%	X	X
Nitric acid, 70%	X	X
Nitric acid, anhydrous	X	X
Nitrous acid, conc.	80	27
Ozone	X	X
Phenol	80	27
Sodium bicarbonate, to 20%	90	32
Sodium bisulfate	80	27
Sodium bisulfite	90	32
Sodium carbonate	90	32
Sodium chlorate	80	27

(continued)

TABLE 5.13 *Continued*

Chemical	Maximum Temperature	
	°F	°C
Sodium hydroxide, 10%	90	32
Sodium hydroxide, 15%	90	32
Sodium hydroxide, 30%	90	32
Sodium hydroxide, 50%	90	32
Sodium hydroxide, 70%	90	32
Sodium hydroxide, conc.	90	32
Sodium hypochlorite, to 20%	90	32
Sodium nitrate	90	32
Sodium phosphate, acid	90	32
Sodium phosphate, alkaline	90	32
Sodium phosphate, neutral	90	32
Sodium silicate	90	32
Sodium sulfide, to 50%	90	32
Sodium sulfite, 10%	90	32
Sodium dioxide, dry	X	X
Sulfur trioxide	90	32
Sulfuric acid, 10%	80	27
Sulfuric acid, 30%	80	27
Sulfuric acid, 50%	80	27
Sulfuric acid, 60%	80	27
Sulfuric acid, 70%	90	32
Toluene	X	X

The chemicals listed are in the pure state or in a saturated solution unless otherwise indicated. Compatibility is shown to the maximum allowable temperature for which data is available. Incompatibility is shown by an X. A blank space indicates that data is unavailable.

Source: From P.A. Schweitzer. 2004. *Corrosion Resistance Tables*, Vols. 1–4, 5th ed., New York: Marcel Dekker.

5.9.3 Applications

Very rarely is polybutadiene used by itself. It is generally used as a blend with other elastomers to impart better resiliency, abrasion resistance, and/or low-temperature properties, particularly in the manufacture of automobile tire treads, shoe heels and seals, gaskets, seals, and belting.

5.10 Ethylene–Acrylic (EA) Rubber

Ethylene–acrylic rubber is produced from ethylene and acrylic acid. As with other synthetic elastomers, the properties of the EA rubbers can be altered by compounding. Basically, EA is a cost-effective, hot-oil-resistant rubber with good low-temperature properties.

5.10.1 Resistance to Sun, Weather, and Ozone

EA elastomers have extremely good resistance to sun, weather, and ozone. Long-term exposures have no effect on these rubbers.

5.10.2 Chemical Resistance

Ethylene–acrylic elastomers exhibit very good resistance to hot oils and hydrocarbon- or glycol-based proprietary lubricants and to transmission and power steering fluids. The swelling characteristics of EA will be retained better than those of the silicone rubbers after oil immersion.

Ethylene–acrylic rubber also has outstanding resistance to hot water. Its resistance to water absorption is very good. Good resistance is also displayed to dilute acids, aliphatic hydrocarbons, gasoline, and animal and vegetable oils.

Ethylene–acrylic rubber is not recommended for immersion in esters, ketones, highly aromatic hydrocarbons, or concentrated acids. Neither should it be used in applications calling for long-term exposure to high-pressure steam.

5.10.3 Applications

Ethylene–acrylic rubber is used in such products as gaskets, hoses, seals, boots, damping components, low-smoke floor tiling, and cable jackets for offshore oil platforms, ships, and building plenum installations. Ethylene–acrylic rubber in engine parts provides good resistance to heat, fluids, and wear as well as good low-temperature sealing ability.

5.11 Acrylate–Butadiene Rubber (ABR) and Acrylic Ester–Acrylic Halide (ACM) Rubbers

Acrylate–butadiene and acrylic ester–acrylic halide rubbers are similar to ethylene–acrylic rubbers. Because of its fully saturated backbone, the polymer has excellent resistance to heat, oxidation, and ozone. It is one of the few elastomers with higher heat resistance than EPDM.

5.11.1 Resistance to Sun, Weather, and Ozone

Acrylate–butadiene and acrylic ester–acrylic halide rubbers exhibit good resistance to sun, weather, and ozone.

5.11.2 Chemical Resistance

Acrylate–butadiene and ACM rubbers have excellent resistance to aliphatic hydrocarbons (gasoline, kerosene) and offer good resistance to water, acids, synthetic lubricants, and silicate hydraulic fluids. They are unsatisfactory for use in contact with alkali, aromatic hydrocarbons (benzene, toluene), halogenated hydrocarbons, alcohol, and phosphate hydraulic fluids.

5.11.3 Applications

These rubbers are used where resistance to atmospheric conditions and heat are required.

5.12 Ethylene–Propylene Rubbers (EPDM and EPT)

Ethylene–propylene rubber is a synthetic hydrocarbon-based rubber made either from ethylene–propylene diene monomer or ethylene–propylene terpolymer. These monomers are combined in such a manner as to produce an elastomer with a completely saturated backbone and pendant unsaturation for sulfur vulcanization. As a result of this configuration, vulcanizates of EPDM elastomers are extremely resistant to attack by ozone, oxygen, and weather.

Ethylene–propylene rubber possesses many properties superior to those of natural rubber and conventional general-purpose elastomers. In some applications, it will perform better than other materials, whereas in other applications, it will last longer or require less maintenance and may even cost less.

EPDM has exceptional heat resistance, being able to operate at temperatures of 300–350°F (148–176°C), while also finding application at temperatures as low as −70°F (−56°C). Experience has shown that EPDM has exceptional resistance to steam. Hoses manufactured from EPDM have had lives several times longer than that of hoses manufactured from other elastomers.

The dynamic properties of EPDM remain constant over a wide temperature range, making this elastomer suitable for a variety of applications. It also has very high resistance to sunlight, aging, and weather, excellent electrical properties, good mechanical properties, and good chemical resistance. However, being hydrocarbon based, it is not resistant to petroleum-based oils or flame.

This material may be processed and vulcanized by the same techniques and with the same equipment as those used for processing other general-purpose elastomers. As with other elastomers, compounding plays an important part in tailoring the properties of EPDM to meet the needs of a specific application. Each of the properties of the elastomer can be enhanced or reduced by the addition or deletion of chemicals and fillers. Because of this, the properties discussed must be considered in general terms.

Ethylene–propylene terpolymer (EPT) is a synthetic hydrocarbon-based rubber produced from an ethylene–proplene terpolymer. It is very similar in physical and mechanical properties to EPDM.

Ethylene–propylene rubber has relatively high resistance to heat. Standard formulations can be used continuously at temperatures of 250–300°F (121–148°C) in air. In the absence of air, such as in a steam hose lining or cable insulation covered with an outer jacket, higher temperatures can be tolerated. It is also possible by special compounding to produce material that can be used in services up to 350°F (176°C). Standard compounds can be used in intermittent service at 350°F (176°C).

5.12.1 Resistance to Sun, Weather, and Ozone

Ethylene–propylene rubber is particularly resistant to sun, weather, and ozone attack. Excellent weather resistance is obtained whether the material is formulated in color, white, or black. The elastomer remains free of surface crazing and retains a high percentage of its properties after years of exposure. Ozone resistance is inherent in the polymer, and for all practical purposes, it can be considered immune to ozone attack. It is not necessary to add any special compounding ingredients to produce this resistance.

5.12.2 Chemical Resistance

Ethylene–propylene rubber resists attack from oxygenated solvents such as acetone, methyl ethyl ketone, ethyl acetate, weak acids and alkalies, detergents, phosphate esters, alcohols, and glycols. It exhibits exceptional resistance to hot water and high-pressure steam. The elastomer, being hydrocarbon-based, is not resistant to hydrocarbon solvents and oils, chlorinated hydrocarbons, or turpentine. However, by proper compounding, its resistance to oil can be improved to provide adequate service life in many applications where such resistance is required. Ethylene–propylene terpolymer rubbers, in general, are resistant to most of the same corrodents as EPDM but do not have as broad a resistance to mineral acids and some organics.

Refer to Table 5.14 for the compatibility of EPDM with selected corrodents and to Table 5.15 compatibility of EPT with selected corrodents.

5.12.3 Applications

Extensive use is made of ethylene–propylene rubber in the automotive industry. Because of its paintability, this elastomer is used as the gap-filling panel between the grills and bumper, that provides a durable and elastic

TABLE 5.14

Compatibility of EDPM Rubber with Selected Corrodents

	Maximum Temperature	
Chemical	°F	°C
Acetaldehyde	200	93
Acetamide	200	93
Acetic acid, 10%	140	60
Acetic acid, 50%	140	60
Acetic acid, 80%	140	60
Acetic acid, glacial	140	60
Acetic anhydride	X	X
Acetone	200	93
Acetyl chloride	X	X
Acrylonitrile	140	60
Adipic acid	200	93
Allyl alcohol	200	93
Allyl chloride	X	X
Alum	200	93
Aluminum fluoride	190	88
Aluminum hydroxide	200	93
Aluminum nitrate	200	93
Aluminum sulfate	190	88
Ammonia gas	200	93
Ammonium bifluoride	200	93
Ammonium carbonate	200	93
Ammonium chloride, 10%	200	93
Ammonium chloride, 50%	200	93
Ammonium chloride, sat.	200	93
Ammonium fluoride, 10%	200	93
Ammonium fluoride, 25%	200	93
Ammonium hydroxide, 25%	100	38
Ammonium hydroxide, sat.	100	38
Ammonium nitrate	200	93
Ammonium persulfate	200	93
Ammonium phosphate	200	93
Ammonium sulfate, 10–40%	200	93
Ammonium sulfide	200	93
Amyl acetate	200	93
Amyl alcohol	200	93
Amyl chloride	X	X
Aniline	140	60
Antimony trichloride	200	93
Aqua regia, 3:1	X	X
Barium carbonate	200	93
Barium chloride	200	93
Barium hydroxide	200	93
Barium sulfate	200	93
Barium sulfide	140	60

(continued)

TABLE 5.14 *Continued*

Chemical	Maximum Temperature	
	°F	°C
Benzaldehyde	150	66
Benzene	X	X
Benzenesulfonic acid, 10%	X	X
Benzoic acid	X	X
Benzyl alcohol	X	X
Benzyl chloride	X	X
Borax	200	93
Boric acid	190	88
Bromine gas, dry	X	X
Bromine gas, moist	X	X
Bromine, liquid	X	X
Butadiene	X	X
Butyl acetate	140	60
Butyl alcohol	200	93
Butyric acid	140	60
Calcium bisulfite	X	X
Calcium carbonate	200	93
Calcium chlorate	140	60
Calcium chlorite	200	93
Calcium hydroxide, 10%	200	93
Calcium hydroxide, sat.	200	93
Calcium hypochlorite	200	93
Calcium nitrate	200	93
Calcium oxide	200	93
Calcium sulfate	200	93
Carbon bisulfide	X	X
Carbon dioxide, dry	200	93
Carbon dioxide, wet	200	93
Carbon disulfide	200	93
Carbon monoxide	X	X
Carbon tetrachloride	200	93
Carbonic acid	X	X
Cellosolve	200	93
Chloroacetic acid	160	71
Chlorine gas, dry	X	X
Chlorine gas, wet	X	X
Chlorine, liquid	X	X
Chlorobenzene	X	X
Chloroform	X	X
Chlorosulfonic acid	X	X
Chromic acid, 50%	X	X
Citric acid, 15%	200	93
Citric acid, conc.	200	93
Copper acetate	100	38
Copper carbonate	200	93

(continued)

TABLE 5.14 *Continued*

Chemical	Maximum Temperature	
	°F	°C
Copper chloride	200	93
Copper cyanide	200	93
Copper sulfate	200	93
Cresol	X	X
Cupric chloride, 5%	200	93
Cupric chloride, 50%	200	93
Cyclohexane	X	X
Cyclohexanol	X	X
Dichloroethane (ethylene dichloride)	X	X
Ethylene glycol	200	93
Ferric chloride	200	93
Ferric chloride, 50% in water	200	93
Ferric nitrate, 10–50%	200	93
Ferrous chloride	200	93
Ferrous nitrate	200	93
Fluorine gas, moist	60	16
Hydrobromic acid, dil.	90	32
Hydrobromic acid, 20%	140	60
Hydrobromic acid, 50%	140	60
Hydrochloric acid, 20%	100	38
Hydrochloric acid, 38%	90	32
Hydrocyanic acid, 10%	200	93
Hydrofluoric acid, 30%	60	16
Hydrofluoric acid, 70%	X	X
Hydrofluoric acid, 100%	X	X
Hypochlorous acid	200	93
Iodine solution, 10%	140	60
Ketones, general	X	X
Lactic acid, 25%	140	60
Lactic acid, conc.		
Magnesium chloride	200	93
Malic acid	X	X
Methyl chloride	X	X
Methyl ethyl ketone	80	27
Methyl isobutyl ketone	60	16
Nitric acid, 5%	60	16
Nitric acid, 20%	60	16
Nitric acid, 70%	X	X
Nitric acid, anhydrous	X	X
Oleum	X	X
Perchloric acid, 10%	140	60
Phosphoric acid, 50–80%	140	60
Picric acid	200	93
Potassium bromide, 30%	200	93

(continued)

TABLE 5.14 *Continued*

Chemical	Maximum Temperature °F	°C
Salicylic acid	200	93
Sodium carbonate	200	93
Sodium chloride	140	60
Sodium hydroxide, 10%	200	93
Sodium hydroxide, 50%	180	82
Sodium hydroxide, conc.	180	82
Sodium hypochlorite, 20%	200	93
Sodium hypochlorite, conc.	200	93
Sodium sulfide, to 50%	200	93
Stannic chloride	200	93
Stannous chloride	200	93
Sulfuric acid, 10%	150	66
Sulfuric acid, 50%	150	66
Sulfuric acid, 70%	140	66
Sulfuric acid, 90%	X	X
Sulfuric acid, 98%	X	X
Sulfuric acid, 100%	X	X
Sulfuric acid, fuming	X	X
Toluene	X	X
Trichloroacetic acid	80	27
White liquor	200	93
Zinc chloride	200	93

The chemicals listed are in the pure state or in a saturated solution unless otherwise indicated. Compatibility is shown to the maximum allowable temperature for which data is available. Incompatibility is shown by an X. A blank space indicates that data is unavailable.

Source: From P.A. Schweitzer. 2004. *Corrosion Resistance Tables*, Vols. 1–4, 5th ed., New York: Marcel Dekker.

TABLE 5.15

Compatibility of EPT with Selected Corrodents

Chemical	Maximum Temperature °F	°C
Acetamide	200	93
Acetic acid, 10%	X	X
Acetic acid, 50%	X	X
Acetic acid, 80%	X	X
Acetic acid, glacial	X	X
Acetone	X	X
Ammonia gas	140	60
Ammonium chloride, 10%	180	82
Ammonium chloride, 28%	180	82

(continued)

TABLE 5.15 *Continued*

Chemical	Maximum Temperature	
	°F	°C
Ammonium chloride, 50%	180	82
Ammonium chloride, sat.	180	82
Ammonium hydroxide, 10%	140	60
Ammonium hydroxide, 25%	140	60
Ammonium hydroxide, sat.	140	60
Ammonium sulfate, 10–40%	180	82
Amyl alcohol	180	82
Aniline	X	X
Benzene	X	X
Benzoic acid	140	60
Bromine water, dil.	X	X
Bromine water, sat.	X	X
Butane	X	X
Butyl acetate	X	X
Butyl alcohol	180	82
Calcium chloride dil.	180	82
Calcium chloride sat.	180	82
Calcium hydroxide, 10%	180	82
Calcium hydroxide, 20%	180	82
Calcium hydroxide, 30%	180	82
Calcium hydroxide, sat.	210	99
Cane sugar liquors	180	82
Carbon bisulfide	X	X
Carbon tetrachloride	X	X
Carbonic acid	180	82
Castor oil	160	71
Cellosolve	X	X
Chlorine water, sat.	80	27
Chlorobenzene	X	X
Chloroform	X	X
Chromic acid, 10%	X	X
Chromic acid, 30%	X	X
Chromic acid, 40%	X	X
Chromic acid, 50%	X	X
Citric acid, 5%	180	82
Citric acid, 10%	180	82
Citric acid, 15%	180	82
Citric acid, conc.	180	82
Copper sulfate	180	82
Corn oil	X	X
Cottonseed oil	80	27
Cresol	100	38
Diacetone alcohol	210	99
Dibutyl phthalate	100	38
Ethers, general		

(continued)

TABLE 5.15 *Continued*

Chemical	Maximum Temperature	
	°F	°C
Ethyl acetate	X	X
Ethyl alcohol	180	82
Ethylene chloride	X	X
Ethylene glycol	180	82
Formaldehyde, dil.	180	82
Formaldehyde, 37%	180	82
Formaldehyde, 50%	140	60
Glycerine	180	82
Hydrochloric acid, dil.	210	99
Hydrochloric acid, 20%	X	X
Hydrochloric acid, 35%	X	X
Hydrochloric acid, 38%	X	X
Hydrochloric acid, 50%	X	X
Hydrofluoric acid, dil.	210	99
Hydrofluoric acid, 30%	140	60
Hydrofluoric acid, 40%	X	X
Hydrofluoric acid, 50%	X	X
Hydrofluoric acid, 70%	X	X
Hydrofluoric acid, 100%	X	X
Hydrogen peroxide, all conc.	X	X
Lactic acid, all conc.	210	99
Methyl ethyl ketone	X	X
Methyl isobutyl ketone	X	X
Monochlorobenzene	X	X
Muriatic acid	X	X
Nitric acid	X	X
Oxalic acid	140	60
Phenol	80	27
Phosphoric acid, all conc.	180	82
Potassium hydroxide, to 50%	210	99
Potassium sulfate, 10%	210	99
Propane	X	X
Silicone oil	200	93
Sodium carbonate	180	82
Sodium chloride	180	82
Sodium hydroxide, to 70%	200	93
Sodium hypochlorite, all conc.	X	X
Sulfuric acid, to 70%	200	93
Sulfuric acid, 93%	X	X
Toluene	X	X
Trichloroethylene	X	X
Water, demineralized	210	99
Water, distilled	210	99
Water, salt	210	99
Water, sea	210	99

(continued)

TABLE 5.15 *Continued*

| | Maximum Temperature | |
Chemical	°F	°C
Whiskey	180	82
Zinc chloride	180	82

The chemicals listed are in the pure state or in a saturated solution
unless otherwise indicated. Compatibility is shown to the maximum
allowable temperature for which data is available. Incompatibility is
shown by an X. A blank space indicates that data is unavailable.

Source: From P.A. Schweitzer. 2004. *Corrosion Resistance Tables*,
Vols. 1–4, 5th ed., New York: Marcel Dekker.

element. Under hood components such as radiator hose, ignition wire
insulation, overflow tubing, window washer tubing, exhaust emission
control tubing, and various other items make use of EPDM because of its
resistance to heat, chemicals, and ozone. Other automotive applications
include body mounts, spring mounting pads, miscellaneous body seals,
floor mats, and pedal pads. Each application takes advantage of one or more
specific properties of the elastomer.

Appliance manufacturers, especially washer manufacturers, have also
found wide use for EPDM. Its heat and chemical resistance combined with
its physical properties make it ideal for such appliances as door seals and
cushions, drain and water-circulating hoses, bleach tubing, inlet nozzles,
boots, seals, gaskets, diaphragms, vibration isolators, and a variety of
grommets. This elastomer is also used in dishwashers, refrigerators, ovens,
and a variety of small appliances.

Ethylene–propylene rubber finds application in the electrical industry and
in the manufacture of electrical equipment. One of the primary applications
is as an insulating material. It is used for medium-voltage (up to 35 kV) and
secondary network power cable, coverings for line and multiplex
distribution wire, jacketing and insulation for types S and SJ flexible cords,
and insulation for automotive ignition cable.

Accessory items such as molded terminal covers, plugs, transformer
connectors, line-tap and switching devices, splices, and insulating and
semiconductor tape are also produced from EPDM.

Medium- and high-voltage underground power distribution cable
insulated with EPDM offers many advantages. It provides excellent
resistance to tearing and failure caused by high voltage contaminants and
stress. Its excellent electrical properties make it suitable for high-voltage
cable insulation. It withstands heavy corona discharge without
sustaining damage.

Manufacturers of other industrial products take advantage of the heat and chemical resistance, physical durability, ozone resistance, and dynamic properties of EPDM. Typical applications include high-pressure steam hose, high-temperature conveyor belting, water and chemical hose, hydraulic hose for phosphate-type fluids, vibration mounts, industrial tires, tugboat and dock bumpers, tanks and pump linings, O-rings, gaskets, and a variety of molded products. Standard formulations of EPDM are also used for such consumer items as garden hose, bicycle tires, sporting goods, marine accessories, and tires for garden equipment.

5.13 Styrene–Butadiene–Styrene (SBS) Rubber

Styrene–butadiene–styrene (SBS) rubbers are either pure or oil-modified block copolymers. They are most suitable as performance modifiers in blends with thermoplastics or as a base rubber for adhesive, sealant, or coating formulations. SBS compounds are formulations containing block copolymer rubber and other suitable ingredients. These compounds have a wide range of properties and provide the benefits of rubberiness and easy processing on standard thermoplastic processing equipment.

5.13.1 Resistance to Sun, Weather, and Ozone

SBS rubbers are not resistant to ozone, particularly when they are in a stressed condition. Neither are they resistant to prolonged exposure to sun or weather.

5.13.2 Chemical Resistance

The chemical resistance of SBS rubbers is similar to that of natural rubber. They have excellent resistance to water, acids, and bases. Prolonged exposure to hydrocarbon solvents and oils will cause deterioration; however, short exposures can be tolerated.

5.13.3 Applications

The specific formulation will determine the applicability of various products. Applications include a wide variety of general-purpose rubber items and use in the footwear industry. These rubbers are used primarily in blends with other thermoplastic materials and as performance modifiers.

5.14 Styrene–Ethylene–Butylene–Styrene (SEBS) Rubber

Styrene–ethylene–butylene–styrene (SEBS) rubbers are either pure or oil-modified block copolymer rubbers. These rubbers are used as performance modifiers in blends with thermoplastics or as the base rubber for adhesive, sealant, o coating formulations. Formulations of SEBS compounds provide a wide range of properties with the benefits of rubbermess and easy processing on standard thermoplastic processing equipment.

5.14.1 Resistance to Sun, Weather, and Ozone

SEBS rubbers and compounds exhibit excellent resistance to ozone. For prolonged outdoor exposure, the addition of an ultraviolet absorber or carbon black pigment or both is recommended.

5.14.2 Chemical Resistance

The chemical resistance of SEBS rubbers is similar to that of natural rubber. They have excellent resistance to water, acids, and bases. Soaking in hydrocarbon solvents and oils will deteriorate the rubber, but short exposures can be tolerated.

5.14.3 Applications

SEBS rubbers find applications for a wide variety of general-purpose rubber items as well as in automotive, sporting goods, and other products. Many applications are found in the electrical industry for such items as flexible cords, welding and booster cables, flame-resistant appliance wiring materials, and automotive primary wire insulation.

5.15 Polysulfide Rubbers (ST and FA)

Polysulfide rubbers are manufactured by combining ethylene ($CH_2=CH_2$) with an alkaline polysulfide. The sulfur forms part of the polymerized molecule. They are also known as Thiokol rubbers. In general, these elastomers do not have great elasticity, but they do have good resistance to heat and are resistant to most solvents. Compared to nitrile rubber, they have poor tensile strength, a pungent odor, poor rebound, high creep under strain, and poor abrasion resistance.

Modified organic polysulfides are made by substituting other unsaturated compounds for ethylene that results in compounds that have little objectionable odor.

5.15.1 Resistance to Sun, Weather, and Ozone

FA polysulfide rubber compounds display excellent resistance to ozone, weathering, and exposure to ultraviolet light. Their resistance is superior to that of ST polysulfide ribbers. If high concentrations of ozone are to be present, the use of 0.5 part of nickel dibutyldithiocarbamate (NBC) per 100 parts of FA polysulfide rubber will improve the ozone resistance.

ST polysulfide rubber compounded with carbon black is resistant to ultraviolet light and sunlight. Its resistance to ozone is good but can be improved by the addition of NBC, however, this addition can degrade the material's compression set. ST polysulfide rubber also possesses satisfactory weather resistance.

5.15.2 Chemical Resistance

Polysulfide rubbers posses outstanding resistance to solvents. They exhibit excellent resistance to oils, gasoline, and aliphatic and aromatic hydrocarbon solvents, very good water resistance, good alkali resistance, and fair acid resistance. FA polysulfide rubbers are somewhat more resistant to solvents than ST rubbers. Compounding of FA polymers with NBR will provide high resistance to aromatic solvents and improve the physical properties of the blend. For high resistance to esters and ketones, neoprene W is compounded with FA polysulfide rubber to produce improved physical properties.

ST polysulfide rubbers exhibit better resistance to chlorinated organics than FA polysulfide rubbers. Contact of either rubber with strong, concentrated inorganic acids such as sulfuric, nitric, or hydrochloric acid should be avoided.

Refer to Table 5.16 for the compatibility of polysulfide ST rubber with selected corrodents.

TABLE 5.16

Compatibility of Polysulfide ST Rubber with Selected Corrodents

Chemical	Maximum Temperature	
	°F	°C
Acetamide	X	X
Acetic acid, 10%	80	27
Acetic acid, 50%	80	27
Acetic acid, 80%	80	27
Acetic acid, glacial	80	27
Acetone	80	27
Ammonia gas	X	X
Ammonium chloride, 10%	150	66
Ammonium chloride, 28%	150	66
Ammonium chloride, 50%	150	66
Ammonium chloride, sat.	90	32
Ammonium hydroxide, 10%	X	X
Ammonium hydroxide, 25%	X	X
Ammonium hydroxide, sat.	X	X
Ammonium sulfate, 10–40%	X	X
Amyl alcohol	80	27
Aniline	X	X

(continued)

TABLE 5.16 *Continued*

Chemical	Maximum Temperature	
	°F	°C
Benzene	X	X
Benzoic acid	150	66
Bromine water, dil.	80	27
Bromine water, sat.	80	27
Butane	150	66
Butyl acetate	80	27
Butyl alcohol	80	27
Calcium chloride, dil.	150	66
Calcium chloride, sat.	150	66
Calcium hydroxide, 10%	X	X
Calcium hydroxide, 20%	X	X
Calcium hydroxide, 30%	X	X
Calcium hydroxide, sat.	X	X
Cane sugar liquors	X	X
Carbon bisulfide	X	X
Carbon tetrachloride	X	X
Carbonic acid	150	66
Castor oil	80	27
Cellosolve	80	27
Chlorine water, sat.	X	X
Chlorobenzene	X	X
Chloroform	X	X
Chromic acid, 10%	X	X
Chromic acid, 30%	X	X
Chromic acid, 40%	X	X
Chromic acid, 50%	X	X
Citric acid, 5%	X	X
Citric acid, 10%	X	X
Citric acid, 15%	X	X
Citric acid, conc.	X	X
Copper sulfate	X	X
Corn oil	90	32
Cottonseed oil	90	32
Cresol	X	X
Diacetone alcohol	80	27
Dibutyl phthalate	80	27
Ethers, general	90	32
Ethyl acetate	80	27
Ethyl alcohol	80	27
Ethylene chloride	80	27
Ethylene glycol	150	66
Formaldehyde, dil.	80	27
Formaldehyde, 37%	80	27
Formaldehyde, 50%	80	27
Glycerine	80	27

(continued)

TABLE 5.16 *Continued*

Chemical	Maximum Temperature	
	°F	°C
Hydrochloric acid, dil.	X	X
Hydrochloric acid, 20%	X	X
Hydrochloric acid, 35%	X	X
Hydrochloric acid, 38%	X	X
Hydrochloric acid, 50%	X	X
Hydrofluoric acid, dil.	X	X
Hydrofluoric acid, 30%	X	X
Hydrofluoric acid, 40%	X	X
Hydrofluoric acid, 50%	X	X
Hydrofluoric acid, 70%	X	X
Hydrofluoric acid, 100%	X	X
Hydrogen peroxide, all conc.	X	X
Lactic acid, all conc.	X	X
Methyl ethyl ketone	150	66
Methyl isobutyl ketone	80	27
Monochlorobenzene	X	X
Muriatic acid	X	X
Nitric acid, all conc.	X	X
Oxalic acid, all conc.	X	X
Phenol, all conc.	X	X
Phosphoric acid, all conc.	X	X
Potassium hydroxide, to 50%	80	27
Potassium sulfate, 10%	90	32
Propane	150	66
Silicone oil	X	X
Sodium carbonate	X	X
Sodium chloride	80	27
Sodium hydroxide, all conc.	X	X
Sodium hypochlorite, all conc.	X	X
Sulfuric acid, all conc.	X	X
Toluene	X	X
Trichloroethylene	X	X
Water, demineralized	80	27
Water, distilled	80	27
Water, salt	80	27
Water, sea	80	27
Whiskey	X	X
Zinc chloride	X	X

The chemicals listed are in the pure state or in a saturated solution unless otherwise indicated. Compatibility is shown to the maximum allowable temperature for which data is available. Incompatibility is shown by an X. A blank space indicates that data is unavailable.

Source: From P.A. Schweitzer. 2004. *Corrosion Resistance Tables*, Vols. 1–4, 5th ed., New York: Marcel Dekker.

5.15.3 Applications

FA polysulfide rubber is one of the elastomeric materials commonly used for the fabrication of rubber rollers for printing and coating equipment. The major reason for this is its high degree of resistance to many types of solvents, including ketones, esters, aromatic hydrocarbons, and plasticizers, that are used as vehicles for various printing inks and coatings.

Applications are also found in the fabrication of hose and hose liners for the handling of aromatic solvents, esters, ketones, oils, fuels, gasoline, paints, lacquers, and thinners. Large amounts of material are also used to produce caulking compounds, cements, paint can gaskets, seals, and flexible mountings.

The impermeability of the polysuldife rubbers to air and gas has promoted the use of these materials for inflatable products such as life jackets, life rafts, balloons, and other inflatable items.

5.16 Urethane Rubbers (AU)

The first commercial thermoplastic elastomers (TPE) were the thermoplastic urethanes (TPU). Their general structure is A–B–A–B, where A represents a hard crystalline block derived by chain extension of a diisocyanate with a glycol. The soft block is represented by B and can be derived from either a polyester or a polyether. Figure 5.1 shows typical TPU structures, both polyester and polyether types.

The urethane linkages in the hard blocks are capable of a high degree of inter- and intramolecular hydrogen bonding. Such bonding increases the

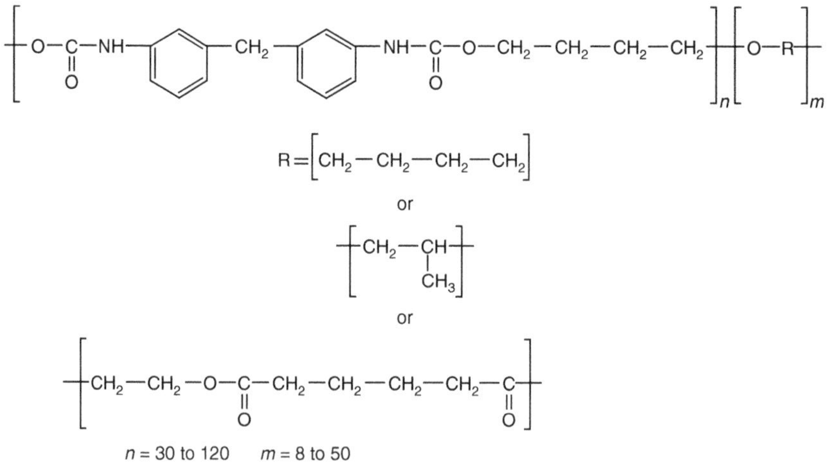

$$R = \left[CH_2 - CH_2 - CH_2 - CH_2 \right]$$

or

$$\left[CH_2 - \underset{\underset{CH_3}{|}}{CH} \right]$$

or

$$\left[CH_2 - CH_2 - O - \underset{\underset{O}{\|}}{C} - CH_2 - CH_2 - CH_2 - CH_2 - \underset{\underset{O}{\|}}{C} \right]$$

n = 30 to 120 m = 8 to 50

FIGURE 5.1
Typical urethane rubber chemical structure.

crystallinity of the hard phase and can influence the mechanical properties—hardness, modulus, tear strength—of the TPU.

As with other block copolymers, the nature of the soft segments determines the elastic behavior and low-temperature performance. TPUs based on polyester soft blocks have excellent resistance to nonpolar fluids and high tear strength and abrasion resistance. Those based on polyether soft blocks have excellent resistance (low heat buildup, or hysteresis), thermal stability, and hydrolytic stability.

TPUs slowly deteriorate but noticeably between 265 and 340°F (130 and 170°C) by both chemical degradation and morphological changes. Melting of the hard phase causes morphological changes and is reversible, whereas, oxidative degradation is slow and irreversible. As the temperature increases, both processes become progressively more rapid. TPUs with polyether soft blocks have greater resistance to thermal and oxidative attack than TPU based on polyester blocks.

TPUs are noted for their outstanding abrasion resistance and low coefficient of friction on other surfaces. They have specific gravities comparable to that of carbon black-filled rubber, and they do not enjoy the density advantage over thermoset rubber compounds.

There are other factors that can have an adverse effect on aging. Such factors as elevated temperature, dynamic forces, contact with chemical agents, or any combination of these factors will lead to a loss in physical and mechanical properties.

Urethane rubbers exhibit excellent recovery from deformation. Parts made of urethane rubbers may build up heat when subjected to high-frequency deformation, but their high strength and load-bearing capacity may permit the use of sections thin enough to dissipate the heat as fast as it is generated.

Urethane rubbers are produced from a number of polyurethane polymers. The properties exhibited are dependent upon the specific polymer and the compounding. Urethane (AU) rubber is a unique material that combines many of the advantages of rigid plastics, metals, and ceramics, yet still has the extensibility and elasticity of rubber. It can be formulated to provide a variety of products with a wide range of physical properties.

Compositions with a Shore A hardness of 95 (harder than a typewriter platen) are elastic enough to withstand stretching to more than four times their own lengths.

At room temperature, a number of raw polyurethane polymers are liquid, simplifying the production of many large and intricately shaped molded products. When cured, these elastomeric parts are hard enough to be machined on standard metalworking equipment. Cured urethane does not require fillers or reinforcing agents.

5.16.1 Resistance to Sun, Weather, and Ozone

Urethane rubbers exhibit excellent resistance to ozone attack and have good resistance to weathering. However, extended exposure to ultraviolet light will reduce their physical properties and will cause the rubber to darken. This can be prevented by the use of pigments or ultraviolet screening agents.

5.16.2 Chemical Resistance

Urethane rubbers are resistant to most mineral and vegetable oils, greases and fuels, and to aliphatic and chlorinated hydrocarbons. This makes these materials particularly suited for service in contact with lubricating oils and automotive fuels.

Tests conducted at the Hanford Laboratories of General Electric Co. indicated that adiprene urethane rubber, as manufactured by DuPont, offered the greatest resistance to the effects of gamma-ray irradiation of the many elastomers and plastics tested. Satisfactory service was given even when the material was exposed to the relatively large gamma-ray dosage of 1×10^9 roentgens. When irradiated with gamma rays, these rubbers were more resistant to stress cracking than other elastomers and retained a large proportion of their original flexibility and toughness.

Aromatic hydrocarbons, polar solvents, esters, ethers, and ketones will attack urethane. Alcohols will soften and swell urethane rubbers.

Urethane rubbers have limited service in weak acid solutions and cannot be used in concentrated acids. They are not resistant to steam or caustic, but they are resistant to the swelling and deteriorating effects of being immersed in water.

Refer to Table 5.17 for the compatibility of urethane rubber with selected corrodents and to Table 5.18 the compatibility of polyether urethane rubber with selected corrodents.

TABLE 5.17

Compatibility of Urethane Rubbers with Selected Corrodents

	Maximum Temperature	
Chemical	°F	°C
Acetamide	X	X
Acetic acid, 10%	X	X
Acetic acid, 50%	X	X
Acetic acid, 80%	X	X
Acetic acid, glacial	X	X
Acetone	X	X
Ammonia gas		
Ammonium chloride, 10%	90	32
Ammonium chloride, 28%	90	32
Ammonium chloride, 50%	90	32

(continued)

TABLE 5.17 *Continued*

Chemical	Maximum Temperature	
	°F	°C
Ammonium chloride, sat.	90	32
Ammonium hydroxide, 10%	90	32
Ammonium hydroxide, 25%	90	32
Ammonium hydroxide, sat.	80	27
Ammonium sulfate, 10–40%		
Amyl alcohol	X	X
Aniline	X	X
Benzene	X	X
Benzoic acid	X	X
Bromine water, dil.	X	X
Bromine water, sat.	X	X
Butane	100	38
Butyl acetate	X	X
Butyl alcohol	X	X
Calcium chloride, dil.	X	X
Calcium chloride, sat.	X	X
Calcium hydroxide, 10%,	X	X
Calcium hydroxide, 20%	90	32
Calcium hydroxide, 30%	90	32
Calcium hydroxide, sat.	90	32
Cane sugar liquors		
Carbon bisulfide		
Carbon tetrachloride	X	X
Carbonic acid	90	32
Castor oil	90	32
Cellosolve	X	X
Chlorine water, sat.	X	X
Chlorobenzene	X	X
Chloroform	X	X
Chromic acid, 10%	X	X
Chromic acid, 30%	X	X
Chromic acid, 40%	X	X
Chromic acid, 50%	X	X
Citric acid, 5%	X	X
Citric acid, 10%	X	X
Citric acid, 15%		
Citric acid, conc.		
Copper sulfate	90	32
Corn oil		
Cottonseed oil	90	32
Cresol	X	X
Dextrose	X	X
Dibutyl phthalate	X	X
Ethers, general	X	X
Ethyl acetate	X	X

(continued)

TABLE 5.17 *Continued*

Chemical	Maximum Temperature	
	°F	°C
Ethyl alcohol	X	X
Ethylene chloride		
Ethylene glycol	90	32
Fomaldehyde, dil.	X	X
Fomaldehyde, 37%	X	X
Formaldehyde, 50%	X	X
Glycerine	90	32
Hydrochloric acid, dil.	X	X
Hydrochloric acid, 20%	X	X
Hydrochloric acid, 35%	X	X
Hydrochloric acid, 38%	X	X
Hydrochloric acid, 50%	X	X
Isoprapyl alcohol	X	X
Jet fuel, JP 4	X	X
Jet fuel, JP 5	X	X
Kerosene	X	X
Lard oil	90	32
Linseed oil	90	32
Magnesium hydroxide	90	32
Mercury	90	32
Methyl alcohol	90	32
Methyl ethyl ketone	X	X
Methyl isobutyl ketone	X	X
Mineral oil	90	32
Muriatic acid	X	X
Nitric acid	X	X
Oils and fats	80	27
Phenol	X	X
Potassium bromide, 30%	90	32
Potassium chloride, 30%	90	32
Potassium hydroxide, to 50%	90	32
Potassium sulfate, 10%	90	32
Silver nitrate	90	32
Soaps	X	X
Sodium carbonate	X	X
Sodium chloride	80	27
Sodium hydroxide, to 50%	90	32
Sodium hypochlorite, all conc.	X	X
Sulfuric acid, all conc.	X	X
Toluene	X	X
Trichloroethylene	X	X
Trisodium phosphate	90	32
Turpentine	X	X
Water, salt	X	X
Water, sea	X	X

(continued)

TABLE 5.17 *Continued*

Chemical	Maximum Temperature	
	°F	°C
Whiskey	X	X
Xylene	X	X

The chemicals listed are in the pure state or in a saturated solution unless otherwise indicated. Compatibility is shown to the maximum allowable temperature for which data is available. Incompatibility is shown by an X. A blank space indicates that data is unavailable.

Source: From P.A. Schweitzer. 2004. *Corrosion Resistance Tables*, Vols. 1–4, 5th ed., New York: Marcel Dekker.

TABLE 5.18

Compatibility of Polyether Urethane Rubbers with Selected Corrodents

Chemical	Maximum Temperature	
	°F	°C
Acetamide	X	X
Acetic acid, 10%	X	X
Acetic acid, 50%	X	X
Acetic acid, 80%	X	X
Acetic acid, glacial	X	X
Acetone	X	X
Acetyl chloride	X	X
Ammonium chloride, 10%	130	54
Ammonium chloride, 28%	130	54
Ammonium chloride, 50%	130	54
Ammonium chloride, sat.	130	54
Ammonium hydroxide, 10%	110	43
Ammonium hydroxide, 25%	110	43
Ammonium hydroxide, sat.	100	38
Ammonium sulfate, 10–40%	120	49
Amyl alcohol	X	X
Aniline	X	X
Benzene	X	X
Benzoic acid	X	X
Benzyl alcohol	X	X
Borax	90	32
Butane	80	27
Butyl acetate	X	X
Butyl alcohol	X	X
Calcium chloride, dil.	130	54
Calcium chloride, sat.	130	54

(continued)

TABLE 5.18 *Continued*

Chemical	Maximum Temperature	
	°F	°C
Calcium hydroxide, 10%	130	54
Calcium hydroxide, 20%	130	54
Calcium hydroxide, 30%	130	54
Calcium hydroxide, sat.	130	54
Calcium hypochlorite	130	54
Carbon bisulfide	X	X
Carbon tetrachloride	X	X
Carbonic acid	80	27
Castor oil	130	54
Caustic potash	130	54
Chloroacetic acid	X	X
Chlorobenzene	X	X
Chloroform	X	X
Chromic acid, 10%	X	X
Chromic acid, 30%	X	X
Chromic acid, 40%	X	X
Chromic acid, 50%	X	X
Citric acid, 5%	90	32
Citric acid, 10%	90	32
Citric acid, 15%	90	32
Cottonseed oil	130	38
Cresylic acid	X	X
Cyclohexane	80	27
Dioxane	X	X
Diphenyl	X	X
Ethyl acetate	X	X
Ethyl benzene	X	X
Ethyl chloride	X	X
Ethylene oxide	X	X
Ferric chloride, 75%	100	38
Fluorosilicic acid	100	38
Formic acid	X	X
Freon, F-11	X	X
Freon, F-12	130	38
Glycerine	130	38
Hydrochloric acid, dil.	X	X
Hydrochloric acid, 20%	X	X
Hydrochloric acid, 35%	X	X
Hydrochloric acid, 38%	X	X
Hydrochloric acid, 50%	X	X
Hydrofluoric acid, dil.	X	X
Hydrofluoric acid, 30%	X	X
Hydrofluoric acid, 40%	X	X
Hydrofluoric acid, 50%	X	X
Hydrofluoric acid, 70%	X	X
Hydrofluoric acid, 100%	X	X

(continued)

TABLE 5.18 *Continued*

Chemical	Maximum Temperature	
	°F	°C
Isobutyl alcohol	X	X
Isopropyl acetate	X	X
Lactic acid, all conc.	X	X
Methyl ethyl ketone	X	X
Methyl isobutyl ketone	X	X
Monochlorobenzene	X	X
Muriatic acid	X	X
Nitrobenzene	X	X
Olive oil	120	49
Phenol	X	X
Potassium acetate	X	X
Potassium chloride, 30%	130	54
Potassium hydroxide	130	54
Potassium sulfate, 10%	130	54
Propane	X	X
Propyl acetate	X	X
Propyl alcohol	X	X
Sodium chloride	130	54
Sodium hydroxide, 30%	140	60
Sodium hypochlorite	X	X
Sodium peroxide	X	X
Sodium phosphate, acid	130	54
Sodium phosphate, alk.	130	54
Sodium phosphate, neut.	130	54
Toluene	X	X
Trichloroethylene	X	X
Water, demineralized	130	54
Water, distilled	130	54
Water, salt	130	54
Water, sea	130	54
Xylene	X	X

The chemicals listed are in the pure state or in a saturated solution unless otherwise indicated. Compatibility is shown to the maximum allowable temperature for which data is available. Incompatibility is shown by an X. A blank space indicates that data is unavailable.

Source: From P. A. Schweitzer. 2004. *Corrosion Resistance Tables*, Vols. 1–4, 5th ed., New York: Marcel Dekker.

5.16.3 Applications

The versatility of urethane has led to a wide variety of applications. Products made from urethane rubber are available in three basic forms: solid, cellular, and films and coatings.

Included under the solid category are those products that are cast or molded, comprising such items as large rolls, impellers, abrasion-resistant parts for textile machines, O-rings, electrical encapsulations, gears, tooling pads, and small intricate parts.

Cellular products are items such as shock mountings, impact mountings, shoe soles, contact wheels for belt grinders, and gaskets.

It is possible to apply uniform coatings or films of urethane rubber to a variety of substrate materials including metal, glass, wood, fabrics, and paper. Examples of products to which a urethane film or coating is often applied are tarpaulins, bowling pins, pipe linings, tank linings, and exterior coatings for protection against atmospheric corrosion. These films also provide abrasion resistance.

Filtration units, clarifiers, holding tanks, and treatment sumps constructed of reinforced concrete are widely used in the treatment of municipal, industrial, and thermal generating station wastewater. In many cases, particularly in anaerobic, industrial, and thermal generating systems, urethane linings are used to protect the concrete from severe chemical attack and prevent seepage into the concrete of chemicals that can attack the reinforcing steel. These linings provide protection from abrasion and erosion and act as a waterproofing system to combat leakage of the equipment resulting from concrete movement and shrinkage.

The use of urethane rubbers in manufactured products has been established as a result of their many unique properties of high tensile and tear strengths, resiliency, impact resistance, and load-bearing capacity. Other products made from urethane rubbers include bearings, gear couplings, mallets and hammers, solid tires, conveyer belts, and many other miscellaneous items.

5.17 Polyamides

Polyamides are produced under a variety of trade names, the most popular of which are Nylons (Nylon 6, etc.) manufactured by DuPont. Although the polyamides find their greatest use as textile fibers, they can also be formulated into thermoplastic molding compounds with many attractive properties. Their relatively high price tends to restrict their use.

There are many varieties of polyamides in production, but the four major types are Nylon 6, Nylon 6/6, Nylon 11, and Nylon 12. Of these, Nylons 11 and 12 find application as elastomeric materials.

The highest performance class of TPE are block copolymeric elastomeric polyamides (PEBAs). Amide linkages connect the hard and soft segments of these TPEs, and the soft segment may have a polyester or a polyether structure. The structures of these commercial PEBAs are shown in Figure 5.2. The amide linkages connecting the hard and soft blocks are more resilient to chemical attack than either an ester or urethane bond. Consequently, the PEBAs have higher temperature resistance than the urethane elastomers, and their cost is greater.

$$\overset{C-(CH_2)_6-C}{\underset{O}{||}}\!\!\!\overset{}{\underset{O}{||}}\!\!\!\Big[NH-(CH_2)_{10}-\overset{C}{\underset{O}{||}}\Big]\!\!\Big[NH(CH_2)_8-\overset{CO}{\underset{}{}}\Big]_x\Big[(CH_2)_y-O\Big]_z$$

Hard   Soft

$$\Big[(CH_2)_5-\overset{C}{\underset{O}{||}}\Big]_x NH-B-NH\overset{C}{\underset{O}{||}}-A-\overset{C}{\underset{O}{||}}-NH-B-NH$$

$$\Big[CH_2\Big]O-\overset{C}{\underset{O}{||}}\Big[CH_2\Big]\overset{C}{\underset{O}{||}}-NH\Big[\text{—}\langle\!\!\bigcirc\!\!\rangle-CH_2-\langle\!\!\bigcirc\!\!\rangle-NH-A-C-CH_2-\overset{C}{\underset{O}{||}}-O\Big]$$

Soft   Hard

Where

A = C_{19} to C_{21} dicarboxylic acid

$$B = \Big[CH_2\Big]_2 O\Big[\Big[CH_2\Big]_4 O\Big]\Big[CH_2\Big]_3$$

FIGURE 5.2
Structures of three elastomeric polyamides

The hard and soft blocks' structure also contributes to their performance characteristics. The soft segment may consist of polyester, polyether, or polyetherester chains. Polyether chains provide better low-temperature properties and resistance to hydrolysis, whereas, polyester chains provide better fluid resistance and resistance to oxidation at elevated temperatures.

5.17.1 Resistance to Sun, Weather, and Ozone

Polyamides are resistant to sun, weather, and ozone. Many metals are coated with polyamide to provide protection from harsh weather.

5.17.2 Chemical Resistance

Polyamides exhibit excellent resistance to a broad range of chemicals and harsh environments. They have good resistance to most inorganic alkalies, particularly ammonium hydroxide and ammonia even at elevated temperatures and sodium potassium hydroxide at ambient temperatures. They also display good resistance to almost all inorganic salts and almost all hydrocarbons and petroleum-based fuels.

They are also resistant to organic acids (citric, lactic, oleic, oxalic, stearic, tartaric, and uric) and most aldehydes and ketones at normal temperatures. They display limited resistance to hydrochloric, sulfonic, and phosphoric acids at ambient temperatures.

Refer to Table 5.19 for the compatibility of Nylon 11 with selected corrodents.

TABLE 5.19

Compatibility of Nylon 11 with Selected Corrodents

Chemical	Maximum Temperature	
	°F	°C
Acetamide	140	60
Acetic acid, 10%	X	X
Acetic acid, 50%	X	X
Acetic acid, 80%	X	X
Acetic acid, glacial	X	X
Acetone		
Ammonia gas		
Ammonium chloride, 10%	200	93
Ammonium chloride, 28%	200	93
Ammonium chloride, 50%	200	93
Ammonium chloride, sat.		
Ammonium hydroxide, 10%	200	93
Ammonium hydroxide, 25%	200	93
Ammonium hydroxide, sat.	200	93
Ammonium sulfate, 10–40%	80	27
Amyl alcohol	100	38
Aniline	X	X
Benzene	200	93
Benzoic acid, 10%	200	93
Bromine, gas	X	X
Bromine, liquid	X	X
Butane	140	60
Butyl acetate		
Butyl alcohol	200	93
Calcium chloride, dil.	200	93
Calcium chloride, sat.	200	93
Calcium hydroxide, 10%	140	60
Calcium hydroxide, 20%	140	60
Calcium hydroxide, 30%	140	60
Calcium hydroxide, sat.	140	60
Calcium sulfate	X	X
Carbon bisulfide	80	27
Carbon tetrachloride	140	60
Carbonic acid	80	27
Caustic potash, 50%	140	60
Chlorine gas	X	X
Chlorine water, sat.	X	X
Chlorobenzene		
Chloroform	X	X
Chromic acid, 10%	X	X
Chromic acid, 30%	X	X
Chromic acid, 40%	X	X
Chromic acid, 50%	X	X
Citric acid, 5%	200	93
Citric acid, 10%	200	93

(continued)

TABLE 5.19 *Continued*

Chemical	Maximum Temperature	
	°F	°C
Citric acid, 15%	200	93
Citric acid, conc.	200	93
Copper nitrate	X	X
Corn oil		
Cottonseed oil	200	93
Cresol	X	X
Cyclohexanone	200	93
Dibutyl phthalate	80	27
Ethers, general		
Ethyl acetate		
Ethyl alcohol	200	93
Ethylene chloride	200	93
Ethylene glycol	200	93
Ethylene oxide	100	38
Ferric chloride	X	X
Formic acid	X	X
Glycerine	200	93
Hydrochloric acid, dil.	80	27
Hydrochloric acid, 20%	X	X
Hydrochloric acid, 35%	X	X
Hydrochloric acid, 38%	X	X
Hydrochloric acid, 50%	X	X
Hydrofluoric acid, dil.	X	X
Hydrofluoric acid, 30%	X	X
Hydrofluoric acid, 40%	X	X
Hydrofluoric acid, 50%	X	X
Hydrofluoric acid, 70%	X	X
Hydrofluoric acid, 100%	X	X
Hydrogen sulfide	X	X
Iodine	X	X
Lactic acid, all conc.	200	93
Methyl ethyl ketone	160	71
Methyl isobutyl ketone	100	38
Monochlorobenzene	X	X
Naphtha	120	49
Nitric acid	X	X
Oxalic acid	140	60
Phenol	X	X
Phosphoric acid, to 10%	80	27
above 10%	X	X
Potassium hydroxide, 50%	200	93
Potassium sulfate, 10%	80	27
Propane	80	27
Silicone oil	80	27
Sodium carbonate	220	104
Sodium chloride	200	93

(continued)

TABLE 5.19 *Continued*

Chemical	Maximum Temperature	
	°F	°C
Sodium hydroxide, all conc.	200	93
Sodium hypochlorite, all conc.	X	X
Sulfunc acid, all conc.	X	X
Toluene	140	60
Trichloroethylene	160	71
Water, acid mine	80	27
Water, distilled	80	27
Water, salt	140	60
Water, sea	140	60
Whiskey	80	27
Zinc chloride	X	X

The chemicals listed are in the pure state or in a saturated solution unless otherwise indicated. Compatibility is shown to the maximum allowable temperature for which data is available. Incompatibility is shown by an X. A blank space indicates that data is unavailable.

Source: From P. A. Schweitzer. 2004. *Corrosion Resistance Tables*, Vols. 1–4, 5th ed., New York: Marcel Dekker.

5.17.3 Applications

Polyamides find many diverse applications resulting from their many advantageous properties. A wide range of flexibility permits material to be produced that is soft enough for high-quality bicycle seats and other materials whose strength and rigidity are comparable to those of many metals. Superflexible grades are also available that are used for shoe soles, gaskets, diaphragms, and seals. Because of the high elastic memory of polyamides, these parts can withstand repeated stretching and flexing over long periods of time.

Since polyamides can meet specification SAE J844 for air-brake hose, coiled tubing is produced for this purpose for use on trucks. Coiled air-brake hose produced from this material has been used on trucks that have traveled over two million miles without a single reported failure. High-pressure hose and fuel lines are also produced from this material.

Corrosion-resistant and wear-resistant coverings for aircraft control cables, automotive cables, and electrical wire are also produced.

5.18 Polyester (PE) Elastomer

This elastomer combines the characteristics of thermoplastics and elastomers. It is structurally strong, resilient, and resistant to impact and flexural fatigue. Its physical and mechanical properties vary depending upon the hardness of the elastomer. Hardnesses range from 40 to 72 on the Shore D scale. The standard hardnesses to which PE elastomers are formulated are 40, 55, 63, and 72 Shore D. Combined are such features as resilience, high resistance to deformation under moderate strain conditions, outstanding flex-fatigue resistance, good abrasion resistance, retention of flexibility at low temperatures, and good retention of properties at elevated temperatures. Polyester elastomers can successfully replace other thermoset rubbers at lower costs in many applications by taking advantage of their higher strength and by using thinner cross sections.

5.18.1 Resistance to Sun, Weather, and Ozone

Polyester rubbers possess excellent resistance to ozone. When formulated with appropriate additives, their resistance to sunlight aging is very good. Resistance to general weathering is good.

5.18.2 Chemical Resistance

In general, the fluid resistance of polyester rubbers increases with increasing hardness. Since these rubbers contain no plasticizers, they are not susceptible to the solvent extraction or heat volatilization of such additives. Many fluids and chemicals will extract plasticizers from elastomers, causing a significant increase in stiffness (modulus) and volume shrinkage.

Overall, PE elastomers are resistant to the same classes of chemicals and fluids as polyurethanes are. However, PE has better high-temperature properties than polyurethanes and can be used satisfactorily at higher temperatures in the same fluids.

Polyester elastomers have excellent resistance to nonpolar materials such as oils and hydraulic fluids, even at elevated temperatures. At room temperature, elastomers are resistant to most polar fluids such as acids, base, amines, and glycols. Resistance is very poor at temperatures of 158°F (70°C) or above. These rubbers should not be used in applications requiring continuous exposure to polar fluids at elevated temperatures.

Polyester elastomers also have good resistance to hot, moist atmospheres. Their hydrolytic stability can be further improved by compounding.

Refer to Table 5.20 for the compatibility of polyester elastomers with selected corrodents.

5.18.3 Applications

Applications for PE elastomers are varied. Large quantities of PE materials are used for liners for tanks, ponds, swimming pools, and drums. Because of their low permeability to air, they are also used for inflatables. Their chemical resistance to oils and hydraulic fluids coupled with their high

TABLE 5.20

Compatibility of Polyester Elastomers with Selected Corrodents

Chemical	Maximum Temperature	
	°F	°C
Acetic acid, 10%	80	27
Acetic acid, 50%	80	27
Acetic acid, 80%	80	27
Acetic acid, glacial	100	38
Acetic acid, vapor	90	32
Ammonium chloride, 10%	90	32
Ammonium chloride, 28%	90	32
Ammonium chloride, 50%	90	32
Ammonium chloride, sat.	90	32
Ammonium chloride, 10–40%	80	27
Amyl acetate	80	27
Aniline	X	X
Beer	80	27
Benzene	80	27
Borax	80	27
Boric acid	80	27
Bromine, liquid	X	X
Butane	80	27
Butyl acetate	80	27
Calcium chloride, dil.	80	27
Calcium chloride, sat.	80	27
Calcium hydroxide, 10%	80	27
Calcium hydroxide, 20%	80	27
Calcium hydroxide, 30%	80	27
Calcium hydroxide, sat.	80	27
Calcium hypochlorite, 5%	80	27
Carbon bisulfide	80	27
Carbon tetrachloride	X	X
Castor oil	80	27
Chlorine gas, dry	X	X
Chlorine gas, wet	X	X
Chlorobenzene	X	X
Chloroform	X	X
Chlorosulfonic acid	X	X
Citric acid, 5%	80	27
Citric acid, 10%	80	27
Citric acid, 15%	80	27
Citric acid, conc.	80	27
Copper sulfate	80	27
Cottonseed oil	80	27
Cupric chloride, to 50%	80	27
Cyclohexane	80	27
Dibutyl phthalate	80	27

(continued)

TABLE 5.20 *Continued*

Chemical	Maximum Temperature	
	°F	°C
Dioctyl phthalate	80	27
Ethyl acetate	80	27
Ethyl alcohol	80	27
Ethylene chloride	X	X
Ethylene glycol	80	27
Formaldehyde, dil.	80	27
Formaldehyde, 37%	80	27
Formic acid, 10–85%	80	27
Glycerine	80	27
Hydrochloric acid, dil.	80	27
Hydrochloric acid, 20%	X	X
Hydrochloric acid, 35%	X	X
Hydrochloric acid, 38%	X	X
Hydrochloric acid, 50%	X	X
Hydrofluoric acid, dil.	X	X
Hydrofluoric acid, 30%	X	X
Hydrofluoric acid, 40%	X	X
Hydrofluoric acid, 50%	X	X
Hydrofluoric acid, 70%	X	X
Hydrofluoric acid, 100%	X	X
Hydrogen	80	27
Hydrogen sulfide, dry	80	27
Lactic acid	80	27
Methyl ethyl ketone	80	27
Methylene chloride	X	X
Mineral oil	80	27
Muriatic acid	X	X
Nitric acid	X	X
Nitrobenzene	X	X
Ozone	80	27
Phenol	X	X
Potassium dichromate, 30%	80	27
Potassium hydroxide	80	27
Pyridine	80	27
Soap solutions	80	27
Sodium chloride	80	27
Sodium hydroxide, all conc.	80	27
Stannous chloride, 15%	80	27
Stearic acid	80	27
Sulfuric acid, all conc.	X	X
Toluene	80	27
Water, demineralized	160	71
Water, distilled	160	71
Water, salt	160	71
Water, sea	160	71

(continued)

TABLE 5.20 *Continued*

	Maximum Temperature	
Chemical	°F	°C
Xylene	80	27
Zinc chloride	80	27

The chemicals listed are in the pure state or in a saturated solution unless otherwise indicated. Compatibility is shown to the maximum allowable temperature for which data is available. Incompatibility is shown by an X. A blank space indicates that data is unavailable.

Source: From P.A. Schweitzer. 2004. *Corrosion Resistance Tables*, Vols. 1–4, 5th ed., New York: Marcel Dekker.

heat resistance make PE elastomers very suitable for automotive hose applications.

Since the PE elastomers do not contain any plasticizers, hose and tubing produced from them do not stiffen with age. Other PE products include seals, gaskets, speciality belting, noise-damping devices, low-pressure tires, industrial solid tires, wire and cable jacketing, pump parts, electrical connectors, flexible shafts, sports equipment, piping clamps and cushions, gears, flexible couplings, and fasteners.

5.19 Thermoplastic Elastomers (TPE) Olefinic Type (TEO)

Thermoplastic elastomers contain sequences of hard and soft repeating units in the polymer chain. Elastic recovery occurs when the hard segments act to pull back the more soft and rubbery segments. Cross-linking is not required. The six generic classes of TPEs are, in order of increasing cost and performance, styrene block copolymers, polyolefin blends, elastomeric alloys, thermoplastic urethanes, thermoplastic copolyesters, and thermoplastic polyamides.

The family of thermoplastic olefins (TEO) are simple blends of a rubbery polymer (such as EPDM or NBR) with a thermoplastic such as PP or PVC. Each polymer will have its own phase, and the rubber phase will have little or no cross-linking (that is, vulcanization). The polymer present in the larger amount will usually be the continuous phase, and the thermoplastic is favored because of its lower viscosity. The discontinuous phase should have a smaller particle size for the best performance of the TEO. EPDM rubber and PP are the constituents of the most common TEOs. Blends of NBR and PVC are also significant but less common in Europe and North America than in Japan.

TEOs are similar to thermoset rubbers because they can be compounded with a variety of the same additives and fillers to meet specific applications.

These additives include carbon black, plasticizers, antioxidants, and fillers, all that tend to concentrate in the soft rubber phase of the TEO.

5.19.1 Resistance to Sun, Weather, and Ozone

TPEs possess good resistance to sun and ozone and have excellent weatherability. Their water resistance is excellent, showing essentially no property changes after prolonged exposure to water at elevated temperatures.

5.19.2 Chemical Resistance

As a result of the low level of cross-linking, a TEO is highly vulnerable to fluids with a similar solubility parameter (or polarity). The EPDM/PP TEOs have very poor resistance to hydrocarbon fluids such as the alkanes, alkenes, or alkyl substituted benzenes, especially at elevated temperatures.

The absence of unsaturation in the polymer backbone of both EPDM and PP makes these TEO blends very resistant to oxidation.

The nonpolar nature of EPDM/PP TEOs make them highly resistant to water, aqueous solutions, and other polar fluids such as alcohols and glycols, but they swell excessively with toss of properties when exposed to halocarbons and hydrocarbons such as oils and fuels. Blends with NBR and PVC are more resistant to aggressive fluids with the exception of the halocarbons.

5.19.3 Applications

These elastomeric compounds are found in a variety of applications, including reinforced hose, seals, gaskets and profile extrusions, flexible and supported tubing, automotive trim, functional parts and under-the-hood components, mechanical goods, and wire and cable jacketing.

5.20 Silicone (SI) and Fluorosilicone (FSI) Rubbers

Silicone rubbers, also known as polysiloxanes, are a series of compounds whose polymer structure consists of silicon and oxygen atoms rather than the carbon structures of most other elastomers. The silicones are derivatives of silica, SiO_2 or $O=Si=O$. When the atoms are combined so that the double linkages are broken and methyl groups enter the linkages, silicone rubber is produced:

$$\begin{bmatrix} CH_3 & CH_3 \\ | & | \\ Si & O & Si & O \\ | & | \\ CH_3 & CH_3 \end{bmatrix}_n$$

5.20.1 Resistance to Sun, Weather, and Ozone

Silicone and fluorosilicone rubbers display excellent resistance to sun, weathering, and ozone. Their properties are virtually unaffected by long-term exposure.

5.20.2 Chemical Resistance

Silicone rubbers can be used in contact with dilute acids and alkalies, alcohols, animal and vegetable oils, and lubricating oils. They are also resistant to aliphatic hydrocarbons, but aromatic solvents such as benzene, toluene, gasoline, and chlorinated solvents will cause excessive swelling. Although they have excellent resistance to water and weathering, they are not resistant to high-pressure and high-temperature steam.

The fluorosilicone rubbers have better chemical resistance than the silicone rubbers. They have excellent resistance to aliphatic hydrocarbons and good resistance to aromatic hydrocarbons, oil and gasoline, animal and vegetable oils, dilute acids and alkalies, and alcohols and fair resistance to concentrated alkalies.

Refer to Table 5.21 for the compatibility of silicone rubbers with selected corrodents.

TABLE 5.21

Compatibility of Silicone Rubbers with Selected Corrodents

	Maximum Temperature	
Chemical	°F	°C
Acetamide	80	27
Acetic acid, 10%	90	32
Acetic acid, 20%	90	32
Acetic acid, 50%	90	32
Acetic acid, 80%	90	32
Acetic acid, glacial	90	32
Acetic acid, vapors	90	32
Acetone	110	43
Acetone, 50% water	110	43
Acetophenone	X	X
Acrylic acid, 75%	80	27
Acrylonitrile	X	X
Aluminum acetate	X	X
Aluminum phosphate	400	204
Aluminum sulfate	410	210
Ammonia gas	X	X
Ammonium chloride, 10%	80	27
Ammonium chloride, 29%	80	27
Ammonium chloride, sat.	80	27

(continued)

TABLE 5.21 *Continued*

Chemical	Maximum Temperature	
	°F	°C
Ammonium hydroxide, 10%	210	99
Ammonium hydroxide, sat.	400	204
Ammonium nitrate	80	27
Amyl acetate	X	X
Amyl alcohol	X	X
Aniline	80	27
Aqua regia, 3:1	X	X
Barium sulfide	400	204
Benzene	X	X
Benzyl chloride	X	X
Boric acid	390	189
Butyl alcohol	80	27
Calcium acetate	X	X
Calcium bisulfite	400	204
Calcium chloride, all conc.	300	149
Calcium hydroxide, to 30%	210	99
Calcium hydroxide, sat.	400	204
Carbon bisulfide	X	X
Carbon monoxide	400	204
Carbonic acid	400	204
Chlorobenzene	X	X
Chlorosulfonic acid	X	X
Dioxane	X	X
Ethane	X	X
Ethers, general	X	X
Ethyl acetate	170	77
Ethyl alcohol	400	204
Ethyl chloride	X	X
Ethylene chloride	X	X
Ethylene diamine	400	204
Ethylene glycol	400	204
Ferric chloride	400	204
Fluosilicic acid	X	X
Formaldehyde, all conc.	200	93
Fuel oil	X	X
Gasoline	X	X
Glucose (corn syrup)	400	204
Glycerine	410	210
Green liquor	400	204
Hexane	X	X
Hydrobromic acid	X	X
Hydrochloric acid, dil.	90	32
Hydrochloric acid, 20%	90	32
Hydrochloric acid, 35%	X	X
Hydrofluoric acid	X	X
Hydrogen peroxide, all conc.	200	93

(continued)

TABLE 5.21 *Continued*

Chemical	Maximum Temperature	
	°F	°C
Lactic acid, all conc.	80	27
Lead acetate	X	X
Lime sulfur	400	204
Linseed oil	X	X
Magnesium chloride	400	204
Magnesium sulfate	400	204
Mercury	80	27
Methyl alcohol	410	210
Methyl cellosolve	X	X
Methyl chloride	X	X
Methyl ethyl ketone	X	X
Methylene chloride	X	X
Mineral oil	300	149
Naptha	X	X
Nickel acetate	X	X
Nickel chloride	400	204
Nickel sulfate	400	204
Nitric acid, 5%	80	27
Nitric acid, 10%	80	27
Nitric acid, 20%	X	X
Nitric acid, anhydrous	X	X
Nitrous acid–sulfuric acid, 50:50	X	X
Nitrobenzene	X	X
Nitrogen	400	204
Nitromethane	X	X
Oils, vegetable	400	204
Oleic acid	X	X
Oleum	X	X
Oxalic acid, to 50%	80	27
Ozone	400	204
Palmitic acid	X	X
Paraffin	X	X
Peanut oil	400	204
Perchloric acid	X	X
Phenol	X	X
Phosphoric acid	X	X
Picric acid	X	X
Potassium chloride, 30%	400	204
Potassium cyanide, 30%	410	210
Potassium dichromate	410	210
Potassium hydroxide, to 50%	210	99
Potassium hydroxide, 90%	80	27
Potassium nitrate, to 80%	400	204
Potassium sulfate, 10%	400	204
Potassium sulfate, pure	400	204
Propane	X	X

(continued)

TABLE 5.21 *Continued*

Chemical	Maximum Temperature	
	°F	°C
Propyl acetate	X	X
Propyl alcohol	400	204
Propyl nitrate	X	X
Pyridine	X	X
Silver nitrate	410	210
Sodium acetate	X	X
Sodium bisulfite	410	210
Sodium borate	400	204
Sodium carbonate	300	149
Sodium chloride, 10%	400	204
Sodium hydroxide, all conc.	90	32
Sodium peroxide	X	X
Sodium sulfate	400	204
Sodium thiosulfate	400	204
Stannic chloride	80	27
Styrene	X	X
Sulfite liquors	X	X
Sulfuric acid	X	X
Sulfurous acid	X	X
Tartaric acid	400	204
Tetrahydroturan	X	X
Toluene	X	X
Tributyl phosphate	X	X
Turpentine	X	X
Vinegar	400	204
Water, acid mine	210	99
Water, demineralized	210	99
Water, distilled	210	99
Water, salt	210	99
Water, sea	210	99
Xylene	X	X
Zinc chloride	400	204

The chemicals listed are in the pure state or in a saturated solution unless otherwise indicated. Compatibility is shown to the maximum allowable temperature for which data is available. Incompatibility is shown by an X. A blank space indicates that data is unavailable.

Source: From P. A. Schweitzer. 2004. *Corrosion Resistance Tables*, Vols. 1–4, 5th ed., New York: Marcel Dekker.

5.20.3 Applications

Because of their unique thermal stability and/or their insulating values, silicone rubbers find many uses in the electrical industries, primarily in appliances, heaters, furnaces, aerospace devices, and automotive parts.

Their excellent weathering qualities and wide temperature range have also resulted in their employment as caulking compounds.

When silicone or fluorosilicone rubbers are infused with a high-density conductive filler, an electric path is created. These conductive elastomers are used as part of an EMI/RFI/EMP shielding process in forms such as O-rings and gaskets to provide

1. Shielding for containment to prevent the escape of EMI internally generated by the device
2. Shielding for exclusion to prevent the intrusion of EMI/RF/EMP created by outside sources into the protected device
3. Exclusion or containment plus pressure or vacuum sealing to provide both EMI/EMP attenuation and pressure containment and/or weatherproofing
4. Grounding and contacting to provide a dependable low-impedance connection to conduct electric energy to ground, often used where mechanical mating is imperfect or impractical

The following equipment is either capable of generating EMI or susceptible to EMI:

Aircraft and aerospace electronics

Analog instrumentation

Automotive electronics

Business machines

Communication systems

Digital instrumentation and process control systems

Home appliances

Radio-frequency instrumentation and radar

Medical electronics

Military and marine electronics

Security systems (military and commercial)

5.21 Vinylidene Fluoride (HFP, PVDF)

Polyvinylidene fluoride (PVDF) is a homopolymer of 1,1-difluoroethene with alternating CH_2 and CF_2 groups along the polymer chain. These groups impart a unique polarity that influences its solubility and electrical properties. The polymer has the characteristic stability of fluoropolymers when exposed to aggressive thermal, chemical, and ultraviolet conditions.

In general, PVDF is one of the easiest fluoropolymers to process and can be easily recycled without affecting its physical and chemical properties. As with other elastomeric materials, compounding can be used to improve certain specific properties. Cross-linking of the polymer chain and control of the molecular weight are also done to improve particular properties.

PVDF possesses mechanical strength and toughness, high abrasion resistance, high thermal stability, high dielectric strength, high purity, resistance, to most chemicals and solvents, resistance to ultraviolet and nuclear radiation, resistance to weathering, and resistance to fungi. It can be used in applications intended for repeated contact with food per Title 21, Code of Federal Regulations, Chapter 1, Part 177.2520. PVDF is also permitted for use in processing or storage areas on contact with meat or poultry food products prepared under federal inspection according to the U.S. Department of Agriculture (USDA). Use is also permitted under "3-A Sanitary Standards for Multiple-Use Plastic Materials Used as Product Contact Surfaces for Dairy Equipment Serial No. 2000." This material has the ASTM designation of MFP.

5.21.1 Resistance to Sun, Weather, and Ozone

PVDF is highly resistant to the effects of sun, weather, and ozone. Its mechanical properties are retained, whereas the percent elongation to break decreases to a lower level and then remains constant.

5.21.2 Chemical Resistance

In general, PVDF is completely resistant to chlorinated solvents, aliphatic solvents, weak bases and salts, strong acids, halogens, strong oxidants, and aromatic solvents. Strong bases will attack the material.

The broader molecular weight of PVDF gives it greater resistance to stress cracking than many other materials, but it is subject to stress cracking in the presence of sodium hydroxide.

PVDF also exhibits excellent resistance to nuclear radiation. The original tensile strength is essentially unchanged after exposure to 1000 Mrads of gamma irradiation from a cobalt-60 source at 122°F (50°C) and in high vacuum (10^{-6} torr). Because of cross-linking, the impact strength and elongation are slightly reduced. This resistance makes PVDF useful in plutonium reclamation operations.

Refer to Table 5.22 for the compatibility of PVDF with selected corrodents.

5.21.3 Applications

PVDF finds many applications where its properties of corrosion resistance, wide allowable operating temperature range, mechanical strength and toughness, high abrasion resistance, high dielectric strength, and resistance to weathering, ultraviolet light, radiation, and fungi are useful.

In the electrical and electronics fields, PVDF is used for multiwire jacketing, plenum cables, heat-shrinkable tubing, anode lead wire, computer

TABLE 5.22

Compatibility of PVDF with Selected Corrodents

	Maximum Temperature	
Chemical	°F	°C
Acetaldehyde	150	66
Acetamide	90	32
Acetic acid, 10%	300	149
Acetic acid, 50%	300	149
Acetic acid, 80%	190	88
Acetic acid, glacial	190	88
Acetone	X	X
Acetyl chloride	120	49
Acrylic acid	150	66
Acrylonitrile	130	54
Adipic acid	280	138
Allyl alcohol	200	93
Allyl chloride	200	93
Alum	180	82
Aluminum acetate	250	121
Aluminum chloride, aqueous	300	149
Aluminum chloride, dry	270	132
Aluminum fluoride	300	149
Aluminum hydroxide	260	127
Aluminum nitrate	300	149
Aluminum oxychloride	290	143
Aluminum sulfate	300	149
Ammonia gas	270	132
Ammonium bifluoride	250	121
Ammonium carbonate	280	138
Ammonium chloride, 10%	280	138
Ammonium chloride, 50%	280	138
Ammonium chloride, sat.	280	138
Ammonium fluoride, 10%	280	138
Ammonium fluoride, 25%	280	138
Ammonium hydroxide, 25%	280	138
Ammonium hydroxide, sat.	280	138
Ammonium nitrate	280	138
Ammonium persulfate	280	138
Ammonium phosphate	280	138
Ammonium sulfate, 10–40%	280	138
Ammonium sulfide	280	138
Ammonium sulfite	280	138
Amyl acetate	190	88
Amyl alcohol	280	138
Amyl chloride	280	138
Aniline	200	93
Antimony trichloride	150	66
Aqua regia, 3:1	130	54
Barium carbonate	280	138

(continued)

TABLE 5.22 *Continued*

Chemical	Maximum Temperature	
	°F	°C
Barium chloride	280	138
Barium hydroxide	280	138
Barium sulfate	280	138
Barium sulfide	280	138
Benzaldehyde	120	49
Benzene	150	66
Benzenesulfonic acid, 10%	100	38
Benzoic acid	250	121
Benzyl alcohol	280	138
Benzyl chloride	280	138
Borax	280	138
Boric acid	280	138
Bromine gas, dry	210	99
Bromine gas, moist	210	99
Bromine, liquid	140	60
Butadiene	280	138
Butyl acetate	140	60
Butyl alcohol	280	138
n-Butylamine	X	X
Butyric acid	230	110
Calcium bisulfide	280	138
Calcium bisulfite	280	138
Calcium carbonate	280	138
Calcium chlorate	280	138
Calcium chloride	280	138
Calcium hydroxide, 10%	270	132
Calcium hydroxide, sat.	280	138
Calcium hypochlorite	280	138
Calcium nitrate	280	138
Calcium oxide	250	121
Calcium sulfate	280	138
Caprylic acid	220	104
Carbon bisulfide	80	27
Carbon dioxide, dry	280	138
Carbon dioxide, wet	280	138
Carbon disulfide	80	27
Carbon monoxide	280	138
Carbon tetrachloride	280	138
Carbonic acid	280	138
Cellosolve	280	138
Chloroacetic acid, 50% water	210	99
Chloroacetic acid	200	93
Chlorine gas, dry	210	99
Chlorine gas, wet, 10%	210	99
Chlorine, liquid	210	99
Chlorobenzene	220	104

(continued)

TABLE 5.22 *Continued*

Chemical	Maximum Temperature	
	°F	°C
Chloroform	250	121
Chlorosulfonic acid	110	43
Chromic acid, 10%	220	104
Chromic acid, 50%	250	121
Chromyl chloride	110	43
Citric acid, 15%	250	121
Citric acid, conc.	250	121
Copper acetate	250	121
Copper carbonate	250	121
Copper chloride	280	138
Copper cyanide	280	138
Copper sulfate	280	138
Cresol	210	99
Cupric chloride, 5%	270	132
Cupric chloride, 50%	270	132
Cyclohexane	250	121
Cyclohexanol	210	99
Dibutyl phthalate	80	27
Dichloroacetic acid	120	49
Dichloroethane (ethylene dichloride)	280	138
Ethylene glycol	280	138
Ferric chloride	280	138
Ferric chloride, 50% in water	280	138
Ferric nitrate, 10–50%	280	138
Ferrous chloride	280	138
Ferrous nitrate	280	138
Fluorine gas, dry	80	27
Fluorine gas, moist	80	27
Hydrobromic acid, dil.	260	127
Hydrobromic acid, 20%	280	138
Hydrobromic acid, 50%	280	138
Hydrochloric acid, 20%	280	138
Hydrochloric acid, 38%	280	138
Hydrocyanic acid, 10%	280	138
Hydrofluoric acid, 30%	260	127
Hydrofluoric acid, 70%	200	93
Hydrofluoric acid, 100%	200	93
Hypochlorous acid	280	138
Iodine solution, 10%	250	121
Ketones, general	110	43
Lactic acid, 25%	130	54
Lactic acid, conc.	110	43
Magnesium chloride	280	138
Malic acid	250	121
Manganese chloride	280	138

(continued)

TABLE 5.22 *Continued*

Chemical	Maximum Temperature	
	°F	°C
Methyl chloride	X	X
Methyl ethyl ketone	X	X
Methyl isobutyl ketone	110	43
Muriatic acid	280	138
Nitric acid, 5%	200	93
Nitric acid, 20%	180	82
Nitric acid, 70%	120	49
Nitric acid, anhydrous	150	66
Nitrous acid, conc.	210	99
Oleum	X	X
Perchloric acid, 10%	210	99
Perchloric acid, 70%	120	49
Phenol	200	93
Phosphoric acid, 50–80%	220	104
Picric acid	80	27
Potassium bromide, 30%	280	138
Salicylic acid	220	104
Silver bromide, 10%	250	121
Sodium carbonate	280	138
Sodium chloride	280	138
Sodium hydroxide, 10%	230	110
Sodium hydroxide, 50%	220	104
Sodium hydroxide, conc.	150	66
Sodium hypochlorite, 20%	280	138
Sodium hypochlorite, conc.	280	138
Sodium sulfide, to 50%	280	138
Stannic chloride	280	138
Stannous chloride	280	138
Sulfuric acid, 10%	250	121
Sulfuric acid, 50%	220	104
Sulfuric acid, 70%	220	104
Sulfuric acid, 90%	210	99
Sulfuric acid, 98%	140	60
Sulfuric acid, 100%	X	X
Sulfuric acid, fuming	X	X
Sulfurous acid	220	104
Thionyl chloride	X	X
Toluene	X	X
Trichloroacetic acid	130	54
White liquor	80	27
Zinc chloride	260	127

The chemicals listed are in the pure state or in a saturated solution unless otherwise indicated. Compatibility is shown to the maximum allowable temperature for which data is available. Incompatibility is shown by an X. A blank space indicates that data is unavailable.

Source: From P.A. Schweitzer. 2004. *Corrosion Resistance Tables*, Vols. 1–4, 5th ed., New York: Marcel Dekker.

wiring, and cable and cable ties. Because of its acceptance in the handling of foods and pharmaceuticals, transfer hose are lined with PVDF. Its corrosion resistance is also a factor in these applications.

In fluid-handling systems, PVDF finds applications as gasketing material, valve diaphragms, and membranes for microporous filters and ultrafiltration.

As a result of its resistance to fungi and its exceptional corrosion resistance, it is also used as the insulation material for underground anode bed installations.

5.22   Fluoroelastomers (FKM)

Fluoroelastomers are fluorine-containing hydrocarbon polymers with a saturated structure obtained by polymerizing fluorinated monomers such as vinylidene fluoride, hexafluoropropene, and tetrafluoroethylene. The result is a high-performance synthetic rubber with exceptional resistance to oils and chemicals at elevated temperatures. Initially, this material was used to produce O-rings for use in severe conditions. Although this remains a major area of application, these compounds have found wide use in other applications because of their chemical resistance at high temperatures and other desirable properties.

As with other rubbers, FKM are capable of being compounded with various additives to enhance specific properties for particular applications. FKM are suitable for all rubber processing applications, including compression molding, injection molding, injection/compression molding, transfer molding, extrusion, calendering, spreading, and dipping.

These compounds possess the rapid recovery from deformation, or resilience, of a true elastomer and exhibit mechanical properties of the same order of magnitude as those of conventional synthetic rubbers.

FKM are manufactured under various trade names by different manufacturers. Three typical materials are listed below.

Trade Name	Manufacturer
Viton	Dupont
Technoflon	Ausimont
Fluorel	3M

These elastomers have the ASTM designation FKM.

Fluoroelastomers have been approved by the U.S. Food and Drug Administration for use in repeated contact with food products. More details are available in the *Federal Register*, Vol. 33, No. 5, Tuesday January 9, 1968, Part 121—Food Additives, Subpart F—Food Additives Resulting from Contact with Containers or Equipment and Food Additives Otherwise Affecting Food—Rubber Articles Intended for Repeated Use.

The biological resistance of FKM is excellent. A typical compound tested against specification MIL-E-5272C showed no fungus growth after 30 days. This specification covers four common fungus groups.

5.22.1 Resistance to Sun, Weather, and Ozone

Because of their chemically saturated structure, FKM exhibit excellent weathering resistance to sunlight and especially to ozone. After 13 years of exposure in Florida in direct sunlight, samples showed little or no change in properties or appearance. Similar results were experienced with samples exposed to various tropical conditions in Panama for a period of 10 years. Products made of this elastomer are unaffected by ozone concentrations as high as 100 ppm. No cracking occurred in a bent loop test after one year of exposure to 100 ppm of ozone in air at 100°F (38°C) or in a sample held at 356°F (180°C) for several hundred hours. This property is particularly important considering that standard tests, for example, in the automotive industry, require resistance only to 0.5 ppm ozone.

5.22.2 Chemical Resistance

Fluoroelastomers provide excellent resistance to oils, fuels, lubricants, most mineral acids, many aliphatic and aromatic hydrocarbons (carbon tetrachloride, benzene, toluene, xylene) that act as solvents for other rubbers, gasoline, naphtha, chlorinated solvents, and pesticides. Special formulation can be produced to obtain resistance to hot mineral acids, steam, and hot water.

These elastomers are not suitable for use with low molecular weight esters and ethers, ketones, certain amines, or hot anhydrous hydrofluoric or chlorosulfonic acids. Their solubility in low molecular weight ketones is an advantage in producing solution coatings of FKM.

Refer to Table 5.23 for the compatibility of FKM with selected corrodents.

5.22.3 Applications

The main applications for fluoroelastomers are in those products requiring resistance to high operating temperatures together with high chemical resistance to aggressive fluids and to those characterized by severe operating conditions that no other elastomer can withstand. By proper formulation, cured items can be produced that will meet the rigid specifications of the industrial, aerospace, and military communities.

Recent changes in the automotive industry that have required reduction in environmental pollution, reduced costs, energy saving, and improved reliability have resulted in higher operating temperatures, which in turn

TABLE 5.23

Compatibility of Fluoroelastomers with Selected Corrodents

Chemical	Maximum Temperature	
	°F	°C
Acetaldehyde	X	X
Acetamide	210	99
Acetic acid, 10%	190	88
Acetic acid, 50%	180	82
Acetic acid, 80%	180	82
Acetic acid, glacial	X	X
Acetic anhydride	X	X
Acetone	X	X
Acetyl chloride	400	204
Acrylic acid	X	X
Acrylonitrile	X	X
Adipic acid	190	88
Allyl alcohol	190	88
Allyl chloride	100	38
Alum	190	88
Aluminum acetate	180	82
Aluminum chloride, aqueous	400	204
Aluminum fluoride	400	204
Aluminum hydroxide	190	88
Aluminum nitrate	400	204
Aluminum oxychloride	X	X
Aluminum sulfate	390	199
Ammonia gas	X	X
Ammonium bifluoride	140	60
Ammonium carbonate	190	88
Ammonium chloride, 10%	400	204
Ammonium chloride, 50%	300	149
Ammonium chloride, sat.	300	149
Ammonium fluoride, 10%.	140	60
Ammonium fluoride, 25%	140	60
Ammonium hydroxide, 25%	190	88
Ammonium hydroxide, sat.	190	88
Ammonium nitrate	X	X
Ammonium persulfate	140	60
Ammonium phosphate	180	82
Ammonium sulfate, 10–40%	180	82
Ammonium sulfide	X	X
Amyl acetate	X	X
Amyl alcohol	200	93
Amyl chloride	190	88
Aniline	230	110
Antimony trichloride	190	88
Aqua regia, 3:1	190	88
Barium carbonate	250	121

(continued)

TABLE 5.23 *Continued*

Chemical	Maximum Temperature	
	°F	°C
Barium chloride	400	204
Barium hydroxide	400	204
Barium sulfate	400	204
Barium sulfide	400	204
Benzaldehyde	X	X
Benzene	400	204
Benzenesulfonic acid, 10%	190	88
Benzoic acid	400	204
Benzyl alcohol	400	204
Benzyl chloride	400	204
Borax	190	88
Boric acid	400	204
Bromine gas, dry, 25%	180	82
Bromine gas, moist, 25%	180	82
Bromine, liquid	350	177
Butadiene	400	204
Butyl acetate	X	X
Butyl alcohol	400	204
n-Butylamine	X	X
Butyric acid	120	49
Calcium bisulfide	400	204
Calcium bisulfite	400	204
Calcium carbonate	190	88
Calcium chlorate	190	88
Calcium chloride	300	149
Calcium hydroxide, 10%	300	149
Calcium hydroxide, sat.	400	204
Calcium hypochlorite	400	204
Calcium nitrate	400	204
Calcium sulfate	200	93
Carbon bisulfide	400	204
Carbon dioxide, dry	80	27
Carbon dioxide, wet	X	X
Carbon disulfide	400	204
Carbon monoxide	400	204
Carbon tetrachloride	350	177
Carbonic acid	400	204
Cellosolve	X	X
Chloroacetic acid, 50% water	X	X
Chloroacetic acid	X	X
Chlorine gas, dry	190	88
Chlorine gas, wet	190	88
Chlorine, liquid	190	88
Chlorobenzene	400	204
Chloroform	400	204

(continued)

TABLE 5.23 *Continued*

Chemical	Maximum Temperature	
	°F	°C
Chlorosulfonic acid	X	X
Chromic acid, 10%	350	177
Chromic acid, 50%	350	177
Citric acid, 15%	300	149
Citric acid, conc.	400	204
Copper acetate	X	X
Copper carbonate	190	88
Copper chloride	400	204
Copper cyanide	400	204
Copper sulfate	400	204
Cresol	X	X
Cupric chloride, 5%	180	82
Cupric chloride, 50%	180	82
Cyclohexane	400	204
Cyclohexanol	400	204
Dibutyl phthalate	80	27
Dichloroethane (ethylene dichloride)	190	88
Ethylene glycol	400	204
Ferric chloride	400	204
Ferric chloride, 50% in water	400	204
Ferric nitrate, 10–50%	400	204
Ferrous chloride	180	82
Ferrous nitrate	210	99
Fluorine gas, dry	X	X
Fluorine gas, moist	X	X
Hydrobromic acid, dil.	400	204
Hydrobromic acid, 20%	400	204
Hydrobromic acid, 50%	400	204
Hydrochloric acid, 20%	350	177
Hydrochloric acid, 38%	350	177
Hydrocyanic acid, 10%	400	204
Hydrofluoric acid, 30%	210	99
Hydrofluoric acid, 70%	350	177
Hydrofluoric acid, 100%	X	X
Hypochlorous acid	400	204
Iodine solution, 10%	190	88
Ketones, general	X	X
Lactic acid, 25%	300	149
Lactic acid, conc.	400	204
Magnesium chloride	390	199
Malic acid	390	199
Manganese chloride	180	82
Methyl chloride	190	88
Methyl ethyl ketone	X	X
Methyl isobutyl ketone	X	X

(continued)

TABLE 5.23 *Continued*

Chemical	Maximum Temperature	
	°F	°C
Muriatic acid	350	177
Nitric acid, 5%	400	204
Nitric acid, 20%	400	204
Nitric acid, 70%	190	88
Nitric acid, anhydrous	190	88
Nitrous acid, conc.	90	32
Oleum	190	88
Perchloric acid, 10%	400	204
Perchloric acid, 70%	400	204
Phenol	210	99
Phosphoric acid, 50–80%	300	149
Picric acid	400	204
Potassium bromide, 30%	190	88
Salicylic acid	300	149
Sodium carbonate	190	88
Sodium chloride	400	204
Sodium hydroxide, 10%	X	X
Sodium hydroxide, 50%	X	X
Sodium hydroxide, conc.	X	X
Sodium hypochlorite, 20%	400	204
Sodium hypochlorite, conc.	400	204
Sodium sulfide, to 50%	190	88
Stannic chloride	400	204
Stannous chloride	400	204
Sulfuric acid, 10%	350	177
Sulfuric acid, 50%	350	177
Sulfuric acid, 70%	350	177
Sulfuric acid, 90%	350	177
Sulfuric acid, 98%	350	177
Sulfuric acid, 100%	180	88
Sulfuric acid, fuming	200	93
Sulfurous acid	400	204
Thionyl chloride	X	X
Toluene	400	204
Trichloroacetic acid	190	88
White liquor	190	88
Zinc chloride	400	204

The chemicals listed are in the pure state or in a saturated solution unless otherwise indicated. Compatibility is shown to the maximum allowable temperature for which data is available. Incompatibility is shown by an X. A blank space indicates that data is unavailable.

Source: From P.A. Schweitzer. 2004. *Corrosion Resistance Tables*, Vols. 1–4, 5th ed., New York: Marcel Dekker.

require a higher performance elastomer. The main innovations resulting from these requirements are

Turbocharging

More compact, more efficient, and faster engines

Catalytic exhausts

Cx reduction

Soundproofing

In addition, the use of lead-free fuels, alternative fuels, sour gasoline lubricants, and antifreeze fluids has caused automotive fluids to be more corrosive to elastomers. At the present time, fluoroelastomers are being applied as shaft seats, valve stem seals, O-ring (water-cooled cylinders and injection pumps), engine head gaskets, filter casing gaskets, diaphragms for fuel pumps, water pump gaskets, turbocharge lubricating circuit bellows, carburetor accelerating pump diaphragms, carburetor needle-valve tips, fuel hose, and seals for exhaust gas pollution control equipment.

In the field of aerospace applications, the reliability of materials under extreme exposure conditions is of prime importance. The high- and low-temperature properties of the fluoroelastomers have permitted them to give reliable performance in a number of aircraft and missile components, specifically manifold gaskets, coated fabrics, firewall seals, heat-shrinkable tubing and fittings for wire and cable, mastic adhesive sealants, protective coatings, and numerous types of O-ring seals.

The ability of fluoroelastomers to seal under extreme vacuum conditions in the range of 10^{-9} mm Hg is an additional feature that makes these materials useful for components used in space.

The exploitation of oilfields in difficult areas, such as deserts or offshore sites, has increased the problems of high temperatures and pressures, high viscosities, and high acidity. These extreme operating conditions require elastomers that have high chemical resistance, thermal stability, and overall reliability to reduce maintenance. The same problems exist in the chemical industry. Fluoroelastomers provide a solution to these problems and are used for O-rings, V-rings, U-rings, gaskets, valve seats, diaphragms for metering pumps, hose, expansion joints, safety clothing and gloves, linings for valves, and maintenance coatings.

Another important application of these elastomers is in the production of coatings and linings. Their chemical stability solves the problems of chemical corrosion by making it possible to use them for such purposes as a protective lining for power station stacks operated with high-sulfur fuels, a coating on rolls for the textile industry to permit scouring of fabrics, and tank linings for the chemical industry.

5.23 Ethylene–Tetrafluoroethylene (ETFE) Elastomer

This elastomer is sold under the trade name Tefzel by DuPont. ETFE is a modified partially fluorinated copolymer of ethylene and polytetrafluoroethylene (PTFE). Since it contains more than 75% TFE by weight, it has better resistance to abrasion and cut-through than TFE while retaining most of the corrosion resistance properties.

5.23.1 Resistance to Sun, Weather, and Ozone

Ethylene–tetrafluoroethylene has outstanding resistance to sunlight, ozone, and weather. This feature, coupled with its wide range of corrosion resistance, makes the material particularly suitable for outdoor applications subject to atmospheric corrosion.

5.23.2 Chemical Resistance

Ethylene–tetrafluoroethylene is inert to strong mineral acids, inorganic bases, halogens, and metal salt solutions. Even carboxylic acids, anhydrides, aromatic and aliphatic hydrocarbons, alcohols, aldehydes, ketones, ethers, esters, chloro-carbons, and classic polymer solvents have little effect on ETFE.

Very strong oxidizing acids near their boiling points, such as nitric acid at high concentration, will affect ETFE in varying degrees, as will organic bases such as amines and sulfonic acids.

Refer to Table 5.24 for the compatibility of ETFE with selected corrodents.

TABLE 5.24

Compatibility of ETFE with Selected Corrodents

	Maximum Temperature	
Chemical	°F	°C
Acetaldehyde	200	93
Acetamide	250	121
Acetic acid, 10%	250	121
Acetic acid, 50%	250	121
Acetic acid, 80%	230	110
Acetic acid, glacial	230	110
Acetic anhydride	300	149
Acetone	150	66
Acetyl chloride	150	66
Acrylonitrile	150	66
Adipic acid	280	138
Allyl alcohol	210	99
Allyl chloride	190	88

(continued)

TABLE 5.24 *Continued*

Chemical	Maximum Temperature	
	°F	°C
Alum	300	149
Aluminum chloride, aqueous	300	149
Aluminum chloride, dry	300	149
Aluminum fluoride	300	149
Aluminum hydroxide	300	149
Aluminum nitrate	300	149
Aluminum oxychloride	300	149
Aluminum sulfate	300	149
Ammonium bifluoride	300	149
Ammonium carbonate	300	149
Ammonium chloride, 10%	300	149
Ammonium chloride, 50%	290	143
Ammonium chloride, sat.	300	149
Ammonium fluoride, 10%.	300	149
Ammonium fluoride, 25%	300	149
Ammonium hydroxide, 25%	300	149
Ammonium hydroxide, sat.	300	149
Ammonium nitrate	230	110
Ammonium persulfate	300	149
Ammonium phosphate	300	149
Ammonium sulfate, 10–40%	300	149
Ammonium sulfide	300	149
Amyl acetate	250	121
Amyl alcohol	300	149
Amyl chloride	300	149
Aniline	230	110
Antimony trichloride	210	99
Aqua regia, 3:1	210	99
Barium carbonate	300	149
Barium chloride	300	149
Barium hydroxide	300	149
Barium sulfate	300	149
Barium sulfide	300	149
Benzaldehyde	210	99
Benzene	210	99
Benzenesulfonic acid, 10%	210	99
Benzoic acid	270	132
Benzyl alcohol	300	149
Benzyl chloride	300	149
Borax	300	149
Boric acid	300	149
Bromine gas, dry	150	66
Bromine water, 10%	230	110
Butadiene	250	121
Butyl acetate	230	110

(continued)

TABLE 5.24 *Continued*

Chemical	Maximum Temperature	
	°F	°C
Butyl alcohol	300	149
n-Butylamine	120	49
Butyric acid	250	121
Calcium bisulfide	300	149
Calcium carbonate	300	149
Calcium chlorate	300	149
Calcium chloride	300	149
Calcium hydroxide, 10%	300	149
Calcium hydroxide, sat.	300	149
Calcium hypochlorite	300	149
Calcium nitrate	300	149
Calcium oxide	260	127
Calcium sulfate	300	149
Caprylic acid	210	99
Carbon bisulfide	150	66
Carbon dioxide, dry	300	149
Carbon dioxide, wet	300	149
Carbon disulfide	150	66
Carbon monoxide	300	149
Carbon tetrachloride	270	132
Carbonic acid	300	149
Cellosolve	300	149
Chloroacetic acid, 50% water	230	110
Chloroacetic acid	230	110
Chlorine gas, dry	210	99
Chlorine gas, wet	250	121
Chlorine, water	100	38
Chlorobenzene	210	99
Chloroform	230	110
Chlorosulfonic acid	80	27
Chromic acid, 10%	150	66
Chromic acid, 50%	150	66
Chromyl chloride	210	99
Citric acid, 15%	120	49
Copper chloride	300	149
Copper cyanide	300	149
Copper sulfate	300	149
Cresol	270	132
Cupric chloride, 5%	300	149
Cyclohexane	300	149
Cyclohexanol	250	121
Dibutyl phthalate	150	66
Dichloroacetic acid	150	66
Ethylene glycol	300	149
Ferric chloride, 50% in water	300	149

(continued)

TABLE 5.24 *Continued*

Chemical	Maximum Temperature	
	°F	°C
Ferric nitrate, 10–50%	300	149
Ferrous chloride	300	149
Ferrous nitrate	300	149
Fluorine gas, dry	100	38
Fluorine gas, moist	100	38
Hydrobromic acid, dil.	300	149
Hydrobromic acid, 20%	300	149
Hydrobromic acid, 50%	300	149
Hydrochloric acid, 20%	300	149
Hydrochloric acid, 38%	300	149
Hydrocyanic acid, 10%	300	149
Hydrofluoric acid, 30%	270	132
Hydrofluoric acid, 70%	250	121
Hydrofluoric acid, 100%	230	110
Hypochlorous acid	300	149
Lactic acid, 5%	250	121
Lactic acid, conc.	250	121
Magnesium chloride	300	149
Malic acid	270	132
Manganese chloride	120	49
Methyl chloride	300	149
Methyl ethyl ketone	230	110
Methyl isobutyl ketone	300	149
Muriatic acid	300	149
Nitric acid, 5%	150	66
Nitric acid, 20%	150	66
Nitric acid, 70%	80	27
Nitric acid, anhydrous	X	X
Nitrous acid, conc.	210	99
Oleum	150	66
Perchloric acid, 10%	230	110
Perchloric acid, 70%	150	66
Phenol	210	99
Phosphoric acid, 50–80%	270	132
Picric acid	130	54
Potassium bromide, 30%	300	149
Salicylic acid	250	121
Sodium carbonate	300	149
Sodium chloride	300	149
Sodium hydroxide, 10%	230	110
Sodium hydroxide, 50%	230	110
Sodium hypochlorite, 20%	300	149
Sodium hypochlorite, conc.	300	149
Sodium sulfide, to 50%	300	149
Stannic chloride	300	149

(continued)

TABLE 5.24 *Continued*

Chemical	Maximum Temperature	
	°F	°C
Stannous chloride	300	149
Sulfuric acid, 10%	300	149
Sulfuric acid, 50%	300	149
Sulfuric acid, 70%	300	149
Sulfuric acid, 90%	300	149
Sulfuric acid, 98%	300	149
Sulfuric acid, 100%	300	149
Sulfuric acid, fuming	120	49
Sulfurous acid	210	99
Thionyl chloride	210	99
Toluene	250	121
Trichloroacetic acid	210	99
Zinc chloride	300	149

The chemicals listed are in the pure state or in a saturated solution unless otherwise indicated. Compatibility is shown to the maximum allowable temperature for which data is available. Incompatibility is shown by an X. A blank space indicates that data is unavailable.

Source: From P.A. Schweitzer. 2004. *Corrosion Resistance Tables*, Vols. 1–4, 5th ed., New York: Marcel Dekker.

5.23.3 Applications

The principal applications for ETFE are found in such products as gaskets, packings, and seals (O-rings, lip and X-rings) in areas where corrosion is a problem. The material is also used for sleeve, split curled, and thrust bearings, and for bearing pads for pipe and equipment support where expansion and contraction or movement may occur.

5.24 Ethylene–Chlorotrifluoroethylene (ECTFE) Elastomer

Ethylene–chlorotrifluoroethylene (ECTFE) elastomer is a 1:1 alternating copolymer of ethylene and chlorotrifluoroethylene. This chemical structure gives the polymer a unique combination of properties. It possesses excellent chemical resistance, good electrical properties, and broad-use temperature range [from cryogenic to 340°F (171°C)] and meets the requirements of the UL-94V-0 vertical flame test in thicknesses as low as 7 mils. ECTFE is a tough material with excellent impact strength over its entire operating temperature range. Of all the fluoropolymers, ECTFE ranks among the best for abrasion resistance.

5.24.1 Resistance to Sun, Weather, and Ozone

Ethylene–chlorotrifluoroethylene is extremely resistant to sun, weather, and ozone attack. Its physical properties undergo very little change after long exposures.

5.24.2 Chemical Resistance

The chemical resistance of ECTFE is outstanding. It is resistant to most of the common corrosive chemicals encountered in industry. Included in this list of chemicals are strong mineral and oxidizing acids, alkalies, metal etchants, liquid oxygen, and practically all organic solvents except hot amines (aniline, dimethylamine, etc.). No known solvent dissolves or stress cracks ECTFE at temperatures up to 250°F (120°C).

Some halogenated solvents can cause ECTFE to become slightly plasticized when it comes into contact with them. Under normal circumstances, this does not impair the usefulness of the polymer. When the part is removed from contact with the solvent and allowed to dry, its mechanical properties return to their original values, indicating that no chemical attack has taken place.

As with other fluoropolymers, ECTFE will be attacked by metallic sodium and potassium.

The useful properties of ECTFE are maintained on exposure to cobalt-60 radiation of 200 Mrads.

Refer to Table 5.25 for the compatibility of ECTFE with selected corrodents.

TABLE 5.25

Compatibility of ECTFE with Selected Corrodents

Chemical	Maximum Temperature	
	°F	°C
Acetic acid, 10%	250	121
Acetic acid, 50%	250	121
Acetic acid, 80%	150	66
Acetic acid, glacial	200	93
Acetic anhydride	100	38
Acetone	150	66
Acetyl chloride	150	66
Acrylonitrile	150	66
Adipic acid	150	66
Allyl chloride	300	149
Alum	300	149
Aluminum chloride, aqueous	300	149
Aluminum chloride, dry		
Aluminum fluoride	300	149
Aluminum hydroxide	300	149

(continued)

TABLE 5.25 *Continued*

Chemical	Maximum Temperature	
	°F	°C
Aluminum nitrate	300	149
Aluminum oxychloride	150	66
Aluminum sulfate	300	149
Ammonia gas	300	149
Ammonium bifluoride	300	149
Ammonium carbonate	300	149
Ammonium chloride, 10%	290	143
Ammonium chloride, 50%	300	149
Ammonium chloride, sat.	300	149
Ammonium fluoride, 10%	300	149
Ammonium fluoride, 25%	300	149
Ammonium hydroxide, 25%	300	149
Ammonium hydroxide, sat.	300	149
Ammonium nitrate	300	149
Ammonium persulfate	150	66
Ammonium phosphate	300	149
Ammonium sulfate, 10–40%	300	149
Ammonium sulfide	300	149
Amyl acetate	160	71
Amyl alcohol	300	149
Amyl chloride	300	149
Aniline	90	32
Antimony trichloride	100	38
Aqua regia, 3:1	250	121
Barium carbonate	300	149
Barium chloride	300	149
Barium hydroxide	300	149
Barium sulfate	300	149
Barium sulfide	300	149
Benzaldehyde	150	66
Benzene	150	66
Benzenesulfonic acid, 10%	150	66
Benzoic acid	250	121
Benzyl atcohol	300	149
Benzyl chloride	300	149
Borax	300	149
Boric acid	300	149
Bromine gas, dry	X	X
Bromine, liquid	150	66
Butadiene	250	121
Butyl acetate	150	66
Butyl alcohol	300	149
Butyric acid	250	121
Calcium bisulfide	300	149
Calcium bisulfite	300	149
Calcium carbonate	300	149

(continued)

TABLE 5.25 *Continued*

| | Maximum Temperature | |
Chemical	°F	°C
Calcium chlorate	300	149
Calcium chloride	300	149
Calcium hydroxide, 10%	300	149
Calcium hydroxide, sat.	300	149
Calcium hypochlorite	300	149
Calcium nitrate	300	149
Calcium oxide	300	149
Calcium sulfate	300	149
Caprylic acid	220	104
Carbon bisulfide	80	27
Carbon dioxide, dry	300	149
Carbon dioxide, wet	300	149
Carbon disulfide	80	27
Carbon monoxide	150	69
Carbon tetrachloride	300	149
Carbonic acid	300	149
Cellosolve	300	149
Chloroacetic acid, 50% water	250	121
Chloroacetic acid	250	121
Chlorine gas, dry	150	66
Chlorine gas, wet	250	121
Chlorine, liquid	250	121
Chlorobenzene	150	66
Chloroform	250	121
Chlorosulfonic acid	80	27
Chromic acid, 10%	250	121
Chromic acid, 50%	250	121
Citric acid, 15%	300	149
Citric acid, conc.	300	149
Copper carbonate	150	66
Copper chloride	300	149
Copper cyanide	300	149
Copper sulfate	300	149
Cresol	300	149
Cupric chloride, 5%	300	149
Cupric chloride, 50%	300	149
Cyclohexane	300	149
Cyclohexanol	300	149
Ethylene glycol	300	149
Ferric chloride	300	149
Ferric chloride, 50% in water	300	149
Ferric nitrate 10–50%	300	149
Ferrous chloride	300	149
Ferrous nitrate	300	149
Fluorine gas, dry	X	X
Fluorine gas, moist	80	27

(continued)

TABLE 5.25 *Continued*

Chemical	Maximum Temperature	
	°F	°C
Hydrobromic acid, dil.	300	149
Hydrobromic acid, 20%	300	149
Hydrobromic acid, 50%	300	149
Hydrochloric acid, 20%	300	149
Hydrochloric acid, 38%	300	149
Hydrocyanic acid, 10%	300	149
Hydrofluoric acid, 30%	250	121
Hydrofluoric acid, 70%	240	116
Hydrofluoric acid, 100%	240	116
Hypochlorous acid	300	149
Iodine solution, 10%	250	121
Lactic acid, 25%	150	66
Lactic acid, conc.	150	66
Magnesium chloride	300	149
Malic acid	250	121
Methyl chloride	300	149
Methyl ethyl ketone	150	66
Methyl isobutyl ketone	150	66
Muriatic acid	300	149
Nitric acid, 5%	300	149
Nitric acid, 20%	250	121
Nitric acid, 70%	150	66
Nitric acid, anhydrous	150	66
Nitrous acid, conc.	250	121
Oleum	X	X
Perchloric acid, 10%	150	66
Perchloric acid, 70%	150	66
Phenol	150	66
Phosphoric acid, 50–80%	250	121
Picric acid	80	27
Potassium bromide, 30%	300	149
Salicylic acid	250	121
Sodium carbonate	300	149
Sodium chloride	300	149
Sodium hydroxide, 10%	300	149
Sodium hydroxide, 50%	250	121
Sodium hydroxide, conc.	150	66
Sodium hypochlorite, 20%	300	149
Sodium hypochlorite, conc.	300	149
Sodium sulfide, to 50%	300	149
Stannic chloride	300	149
Stannous chloride	300	149
Sulfuric acid, 10%	250	121
Sulfuric acid, 50%	250	121
Sulfuric acid, 70%	250	121
Sulfuric acid, 90%	150	66

(continued)

TABLE 5.25 *Continued*

| Chemical | Maximum Temperature | |
	°F	°C
Sulfuric acid, 98%	150	66
Sulfuric acid, 100%	80	27
Sulfuric acid, fuming	300	149
Sulfurous acid	250	121
Thionyl chloride	150	66
Toluene	150	66
Trichloroacetic acid	150	66
White liquor	250	121
Zinc chloride	300	149

The chemicals listed are in the pure state or in a saturated solution unless otherwise indicated. Compatibility is shown to the maximum allowable temperature for which data is available. Incompatibility is shown by an X. A blank space indicates that data is unavailable.

Source: From P.A. Schweitzer. 2004. *Corrosion Resistance Tables*, Vols. 1–4, 5th ed., New York: Marcel Dekker.

5.24.3 Applications

This elastomer finds many applications in the electrical industry as wire and cable insulation and jacketing; plenum cable insulation, oil well wire and cable insulation; logging wire jacketing and jacketing for cathodic protection; aircraft, mass transit and automotive wire; connectors; coil forms; resistor sleeves; wire tie wraps; tapes, tubing; and flexible printed circuitry and flat cable.

Applications are also found in other industries as diaphragms, flexible tubing, closures, seals, gaskets, convoluted tubing, and hose, particularly in the chemical, cryogenic, and aerospace industries.

Materials of ECTFE are also used for lining vessels, pumps, and other equipment.

5.25 Perfluoroelastomers (FPM)

Perfluoroelastomers provide the elastomeric properties of fluoroelastomers and the chemical resistance of PTFE. These compounds are true rubbers. Compared with other elastomeric compounds, they are more resistant to swelling and embrittlement and retain their elastomeric properties over the long term. In difficult environments, there are no other elastomers that can outperform the FPMs. These synthetic rubbers provide the sealing force of a true elastomer and the chemical inertness and thermal stability of polytetra–fluoroethylene.

As with other elastomers, perfluoroelastomers are compounded to modify certain of their properties. Such materials as carbon black, perfluormated oil, and various fillers are used for this purpose.

The ASTM designations for these elastomers is FPM.

One such perfluoroelastomer is sold by DuPont under the trade name Kalrez. As with other perfluoroelastomers, Kalrez is available in different compounds to meet specific needs.

Compound 4079

This is a carbon black filled compound having low compression set with excellent chemical resistance, good mechanical properties, and outstanding hot air aging properties. It exhibits low swell in organic and inorganic acids and aldehydes and has good response to temperature cycling effects. This compound is not recommended for use in hot water/steam applications or in contact with certain hot aliphatic amines, ethylene oxide, and propylene oxide.

It has a maximum continuous operating temperature of 600°F/316°C with short-term exposures at higher temperatures.

Applications include O-rings, diaphragms, seals, and other parts used in the process and aircraft industries.

Compound 1050LF

This compound has good water/steam resistance, excellent amine resistance, and compression set properties. It is the suggested compound for use with ethylene oxide and propylene oxide. It is not recommended for use with organic or inorganic acids at high temperatures. The maximum recommended continuous operating temperature is 550°F/288°C.

This compound finds applications as O-rings, seals, and other parts used in the chemical process industry.

Compound 1050

This compound is carbon black filled. It exhibits good mechanical properties, excellent all around chemical resistance, and is slightly harder than compound 4079. This compound should not be used in pure water/steam applications at elevated temperatures.

The maximum allowable continuous operating temperature is 500°F/260°C with short-term exposures at 550°F/288°C in nonoxidizing environments.

Applications induce O-rings, seals, and other parts in the chemical process industry.

Specialty Compounds

In addition to the three standard compounds, there are six specialty compounds available. These have been modified to meet specific needs, but it is recommended that the manufacturer be consulted before ordering. If the standard compounds do not meet the specified needs, contact the manufacturer for a recommendation on the use of a specialty compound.

5.25.1 Resistance to Sun, Weather, and Ozone

Perfluoroelastomers provide excellent resistance to sun, weather, and ozone. Long-term exposure under these conditions has no effect on them.

5.25.2 Chemical Resistance

Perfluoroelastomers have outstanding chemical resistance. They are virtually immune to chemical attack, at ambient and elevated temperatures. Typical corrodents that perfluoroeiastomers are resistant to include the following:

Chlorine, wet and dry

Fuels (ASTM Reference Fuel C, JP-5 jet fuel, aviation gas, and kerosene)

Heat transfer fluids

Hot mercury

Hydraulic fluids

Inorganic and organic acids (hydrochloric, nitric, sulfuric, and trichloroacetic) and bases (hot caustic soda)

Inorganic salt solutions

Metal halides (titanium tetrachloride, diethylaluminum chloride)

Oil well sour gas (methane, hydrogen sulfide, carbon dioxide, and steam)

Polar solvents (ketones, esters, and ethers)

Steam

Strong organic solvents (benzene, dimethyl formamide, perchloroethylene, and tetrahydrofuran)

Strong oxidizing agents (dinitrogen tetroxide, and fuming nitric acid)

These perfluoroelastomers should not be exposed to molten or gaseous alkali metals such as sodium because a highly exothermic reaction may occur. Service life can be greatly reduced in fluids containing high concentrations of some diamines, nitric acid, and basic phenol when the temperature exceeds 212°F (100°C). Uranium hexafluoride and fully halogenated Freons (F-11 and F-12) cause considerable swelling.

The corrosion resistance given above is for the base polymer. Since the polymer is quite often compounded with fillers and curatives, these additives may interact with the environment, even though the polymer is resistant. Therefore, a knowledge of the additives present is essential in determining the material's suitability for a particular application. A corrosion testing program is the best method whereby this evaluation can be undertaken.

Kalrez compounds have virtually universal chemical resistance. They withstand attack by more than 1600 chemicals, solvents, and plasmas. Table 5.26 lists the compatibilities of Kalrez compounds with selected

TABLE 5.26

Compatibility of Kalrez Compounds
with Selected Corrodents

Abietic Acid
Acetic acid, glacial (4079)
Acetic acid, 30% (4079)
Acetone
Acetonitrile
Acetophenetidine
Acetophenone
Acetyl bromide
Acetyl chloride
Acrylic acid
Acrylonitrile (1050LF)
Alkyl acetone
Alkyl alcohol
Alkyl benzene
Alkyl chloride
Alkyl sulfide
Aluminum acetate
Aluminum bromide
Aluminum chlorate
Aluminum chloride
Aluminum ethylate
Aluminum fluoride
Aluminum fluorosilicate
Aluminum formate
Aluminum hydroxide
Aluminum nitrate
Aluminum oxalate
Aluminum phosphate
Aluminum sulfate
Alum
Ammonia, anhydrous (1050LF)
Ammonium acetate
Ammonium arsenate
Ammonium benzoate
Ammonium bicarbonate
Ammonium bisulfite
Ammonium bromide
Ammonium carbonate
Ammonium chloride
Ammonium citrate
Ammonium dichromate
Ammonium fluoride (1050LF)
Ammonium fluorosilicate
Ammonium formate
Ammonium hydride, conc. (1050LF)
Ammonium iodide
Ammonium lactate

(continued)

TABLE 5.26 *Continued*

Ammonium nitrate
Ammonium nitrite
Ammonium oxalate
Ammonium salicylate
Ammonium sulfate
Ammonium sulfide
Ammonium sulfite
Ammonium thiosulfate
Amyl acetate
Amyl alcohol
Amyl chloride
Amyl mercaptan
Amyl naphthalene
Amyl nitrate
Amyl nitrite
Amyl phenol
Aniline hydrdochloride
Aniline sulfate
Aniline sulfite
Animal fats
Animal oils
Anthraquinone
Antimony sulfate
Antimony tribromide
Antimony trichloride
Antimony trioxide
Aqua regia
Arsenic oxide
Arsenic trichloride
Arsenic trioxide
Arsenic trisulfide
Ascorbic acid
Barium carbonate
Barium chlorate
Barium chloride, aqueous
Barium cyanide
Barium hydroxide
Barium iodide
Barium nitrate
Barium oxide
Barium peroxide
Barium salts
Beet sugar liquors
Benzaldehyde
Benzene
Benzenesulfonic acid
Benzocatechol
Benzoyl chloride
Benzyl chloride

(continued)

TABLE 5.26 *Continued*

Bismuth carbonate
Bismuth nitrate
Bismuth oxychloride
Boric acid
Brine
Bromic acid
Bromine, anhydrous
Butadiene
Butane
Butyl acetate
Butyl chloride
Butyl ether
Butyl lactate
Butyl mercaptan
Butyl chloride
Cadmium chloride
Cadmium nitrate
Cadmium sulfate
Cadmium sulfide
Calcium acetate
Calcium arsenate
Calcium bicarbonate
Calcium bisulfide
Calcium hydrosulfide
Calcium carbide
Calcium carbonate
Calcium chlorate
Calcium chloride
Calcium chromate
Calcium fluoride
Calcium hydroxide
Calcium hypochlorite
Calcium nitrate
Calcium oxide
Calcium peroxide
Calcium phosphate
Calcium stearate
Calcium sulfate
Calcium sulfide
Calcium sulfite
Cane sugar liquors
Capric acid
Carbamate
Carbon bisulfide
Carbon dioxide
Carbon disulfide
Carbon fluorides
Carbon monoxide
Caustic lime

(continued)

TABLE 5.26 *Continued*

Caustic potash
Chloric acid
Chlorine, dry
Chloroacetic acid
Chloroacetone
Chloroform
Chlorosilanes
Chlorosulfonic acid
Chromic acid
Chromic chloride
Chromic fluorides
Chromic hydroxide
Chromic nitrates
Chromic oxides
Chromic phosphate
Chromic sulfate
Coconut oil
Cod liver oil
Coke oven gas
Copper carbonate
Copper chloride
Copper cyanide
Copper nitrate
Copper oxide
Copper sulfate
Corn oil
Coconut oil
Cresylic acid
Crude oil
Cutting oils
Cyclohexane
Cyclohexanol
Cyclohexanone
Cyclohexene
Denatured alcohol
Detergent solution
Dextrain
Dextrose
Diacetone
Diallyl ether
Diallyl phthalate
Dichloroacetic acid
Dichloroaniline
o-Dichlorobenzene
Dichloroethane
Dichloroethylene
Dichloromethane
Dichlorophenol
Diesel oil

(continued)

TABLE 5.26 *Continued*

Diethanolamine (1050LF)
Diethylamine (1050LF)
Diethyl sulfate
Diethylbenzene
Diethylene glycol
Difluoroethane
Difluoromonochloroethane
Diisobutyl ketone
Diisobutylcarbinol
Diisobutylene
Diisopropyl ether
Diisopropyl ketone
Epichlorohydrin
Ethane
Ethanol
Ethanolamine (1050LF)
Ethers
Ethyl acetate
Ethyl acetoacetate
Ethyl acrylate
Ethyl alcohol
Ethylamine (1050LF)
Ethyl benzene
Ethyl benzoate
Ethyl bromide
Ethyl butylate (4079)
Ethyl cellulose
Ethyl chloride
Ethyl ether
Ethyl formate (4079)
Ethyl nitrite
Ethyl oxalate
Ethylene
Ethylene chloride
Ethylene dibromide
Ethylene dichloride
Ethylene glycol
Ethylene oxide (2035)
Fatty acids
Ferric acetate
Ferric hydroxide
Ferric sulfate
Ferrous carbonate
Ferrous chloride
Fuel oils
Fuming sulfuric acid
Gallic acid
Gasoline
Gelatin

(continued)

TABLE 5.26   *Continued*

Glauber's salt
Gluconic acid
Glucose
Glue
Glycerine (glycerol)
Helium
Heptane
Hexane
Hydrolodic acid (4079)
Hydrobromic acid
Hydrobromic acid, 40%
Hydrochloric acid, 37%, cold
Hydrochloric acid, conc.
Hydrochloric acid, 37%, hot
Hydrocyanic acid
Hydrofluoric acid, anhydrous
Hydrofluoric acid, conc., cold conc., hot (4079)
Hydrogen peroxide, 90%
Iodic acid
Iodine
Iodoform
Isopropyl chloride
Isopropyl ether
Isopropylamine (1050LF)
Kerosene
Lacquer solvents
Lacquers
Lactic acid, cold
Lactic acid, hot
Lauric acid
Lead, molten
Lead acetate
Lead arsenate
Lead bromide
Lead carbonate
Lead chloride
Lead chromate
Lead dioxide
Lead nitrate
Lead oxide
Lime bleach
Lime sulfur
Linoleic acid
Linseed oil
Lithium carbonate
Lithium chloride
Lithium hydroxide
Lithium nitrate
Lithium nitrite

(continued)

TABLE 5.26 *Continued*

Lithium salicylate
Lithopone
Lye
Magnesium chloride
Magnesium hydroxide
Magnesium sulfate
Magnesium sulfite
Maleic acid
Manganese acetate
Maleic anhydride
Manganese carbonate
Manganese dioxide
Manganese chloride
Manganese phosphate
Manganese sulfate, aqueous
Mercuric cyanide
Mercuric iodide
Mercuric nitrate
Mercuric sulfate
Mercuric sulfite
Mercurous nitrate
Mercury
Mercury chloride
Methane
Methyl acetate
Methyl acrylate
Methyl alcohol (methanol)
Methyl benzoate
Methyl butyl ketone
Methyl chloride
Methyl ether
Methyl ethyl ketone
Methyl isobutyl ketone
Methyl salicylate
Methylene bromide
Methylene chloride
Methylene iodide
Mineral oil
Mixed acids
Morpholine
Motor oils
Mustard gas
Myristic acid
Naphtha
Naphthalene
Natural gas
Neon
Nickel acetate aqueous
Nickel chloride aqueous

(continued)

TABLE 5.26   *Continued*

Nickel cyanide
Nickel nitrate
Nickel salts
Nicotine
Nicotine sulfate
Niter cake
Nitric acid, 0–50% (4079)
Nitric acid, 50–100% (4079)
Nitroaniline
Nitrobenzene
Nitrocellulose
Nitrochlorobenzene
Nitrogen
Nitrogen oxides
Nitrogen peroxide
Nitroglycerine
Nitroglycerol
Nitromethane
Nitrophenol
Nitrotoluene
Nitrous acid
Nonane
n-Octane
Octyl acetate
Octyl alcohol
Octyl chloride
Olefins
Oleic acid
Oleum
Olive oil
Oxalic acid
Oxygen, cold (4079)
Paint thinner
Palmitic acid
Paraffins
Peanut oil
Pectin (liquor)
Penicillin
Peracetic acid
Perchloric acid (4079)
Perchloroethylene
Petroleum
Petroleum ether
Petroleum crude (1050LF)
Phenyl acetate
Phosgene
Phosphoric acid, 20%
Phosphoric acid, 45%
Phosphorous, molten

(continued)

TABLE 5.26 *Continued*

Phosphorous oxychloride
Phosphorous trichloride
Phthalic acid
Phthalic anhydride
Pickling solution
Picric acid
Pine oil
Pine tar
Pinene
Piperazine (1050LF)
Piperidine
Plating solution, chrome
Potassium acid sulfate
Potassum alum
Potassum aluminum sulfate
Potassium bicarbonate
Potassium bichromate
Potassium bifluoride
Potassium bisulfate
Potassium bisulfite
Potassium bitartrate
Potassium bromide
Potassium carbonate
Potassium chlorate
Potassium chloride
Potassium chromates
Potassium citrate
Potassium cyanate
Silicone tetrachloride, dry
Silicone tetrachloride, wet
Silver bromide
Silver chloride
Silver cyanide
Silver nitrate
Silver sulfate
Soap solutions
Soda ash
Sodium acetate
Sodium benzoate
Sodium bicarbonate
Sodium bichromate
Sodium bifluoride
Sodium bisulfate
Sodium bisulfide
Sodium bisulfite
Sodium borate
Sodium bromate
Sodium bromide
Sodium carbonate

(continued)

TABLE 5.26 *Continued*

Sodium chlorate
Sodium chloride
Sodium chlorite
Sodium chloroacetate
Sodium chromate
Sodium citrate
Sodium cyanate
Sodium cyanide
Sodium terricyanide
Sodium terrocyanide
Sodium fluoride
Sodium fluorosilicate
Sodium hydride
Sodium hydrosulfide
Sodium hydroxide
Sodium hypochlorite
Sodium lactate
Sodium nitrate
Sodium oleate
Sodium oxalate
Sodium perborate
Sodium percarbonate
Sodium perchlorate
Sodium peroxide
Sodium persulfate
Sodium phenolate
Sodium phosphate
Sodium salicylate
Sodium salts
Sodium silicate
Sodium stannate
Sodium sulfate
Sodium sulfide
Sodium sulfite
Sour crude oil (1050LF)
Sour natural gas (1050LF)
Soybean oil
Stannic ammonium chloride
Stannic chloride, aqueous
Stannic tetrachloride
Stannous bromide
Stannous chloride, aqueous
Stannous fluoride
Stannous sulfate
Steam above 300°F/149°C (1050LF)
Steam below 300°F/149°C (2035)
Stearic acid
Stoddard solvent
Strontium acetate

(continued)

TABLE 5.26 *Continued*

Strontium carbonate
Strontium chloride
Strontium hydroxide
Strontium nitrate, aqueous
Styrene (3018)
Succinic acid
Sucrose solution
Sulfamic acid
Sulfite liquors
Sulfur chloride
Sulfur dioxide, dry
Sulfur dioxide, liquefied
Sulfur dioxide, wet
Sulfur trioxide
Sulfuric acid, conc.
Sulfuric acid, dil.
Sulfurous acid
Sulfuryl chloride
Sulfonated oil
Tallow
Tannic acid
Tar, bituminous
Tartaric acid
Tetraethyl lead
Tetrahydrofuran
Thionyl chloride
Thiourea
Toluene
Transformer oil
Transmission fluid type A
Triacetin
Trichloroacetic acid (4079)
Trichlorobenzene
Trichloroethane
Trichloroethylene
Triethanolamine (1050LF)
Triethyl phosphate
Triethylamine (1050LF)
Trimethylamine (1050LF)
Trimethylbenzene
Trinitrotoluene
Trisodium phosphate
Tung oil
Turbine oil
Turpentine
Uranium sulfate
Uric acid
Valeric acid
Vanilla extract (2035)

(continued)

TABLE 5.26 *Continued*

Vegetable oils
Vinyl benzene
Vinyl benzoate
Vinyl chloride
Vinyl fluoride
Vinylidene chloride
Water, cold
Water, hot (1050LF)
White oil
White pine oil
Wood oil
Xenon
Xylene
Zinc acetate
Zinc chloride
Zinc chromate
Zinc nitrate
Zinc oxide
Zinc sulfate
Zinc sulfide
Zirconium nitrate

Kairez is resistant to the above chemicals up to a maximum temperature of 212°F/100°C. In some chemicals, a higher operating temperature may be allowable. The listing is based on standard compounds 4079 and 1050LF for the majority of the fluids. When no compound is specified, any Kalrex compound may be used.

corrodents. The listing is based on standard compounds 4079 and 1050LF for the majority of the fluids. When no compound is specified, any Kalrez compound may be used. Kalrez will be resistent to the corrodents listed up to a maximum temperature of 212°F/100°C. In contact with some chemicals, a higher operating temperature may be allowable.

5.25.3 Applications

Perfluoraelastomer parts are a practical solution wherever the sealing performance of rubber is desirable but not feasible because of severe chemical or thermal conditions. In the petrochemical industry, FPM is widely used for O-ring seals on equipment. O-rings of FPM are employed in mechanical seals, pump housings, compressor casings, valves, rotameters, and other instruments. Custom-molded parts are also used as valve seats, packings, diaphragms, gaskets, and miscellaneous sealing elements including U-cups and V-rings.

Other industries where FPM contributes importantly are aerospace (versus jet fuels, hydrazine, N_2O_4 and other oxidizers, Freon-21 fluorocarbon, etc.); nuclear power (versus radiation, high temperature); oil, gas, and geothermal drilling (versus sour gas, acidic fluids, amine-containing hydraulic fluids, extreme temperatures and pressures); and analytical and process instruments (versus high vacuum, liquid and gas chromotography exposures, high-purity reagents, high-temperature conditions).

The semiconductor industry makes use of FPM O-rings to seal the aggressive chemical reagents and specialty gases required for producing silicon chips. Also the combination of thermal stability and low outgassing characteristics are desirable in furnaces for growing crystals and in high-vacuum applications.

The chemical transportation industry is also a heavy user of FPM components in safety relief and unloading valves to prevent leakage from tank trucks and trailers, rail cars, ships, and barges carrying hazardous and corrosive chemicals.

Other industries that also extensively use FPM include Pharmaceuticals, agricultural chemicals, oil and gas recovery, and analytical and process control instrumentation.

Because of their cost, perfluoroelastomers are used primarily as seals where their corrosion- and/or heat-resistance properties can be utilized and other elastomeric materials will not do the job or where high maintenance costs results if other elastomeric materials are used.

5.26 Epichlorohydrin Rubber

The epichlorhydrin polymer group includes the homopolymer (CO), the copolymer with ethylene oxide (ECO), and terpolymers (GECO).

5.26.1 Resistance to Sun, Weather, and Ozone

These rubbers have excellent resistance to sun, weathering, and ozone.

5.26.2 Chemical Resistance

Epichlorhydrin rubbers are resistant to hydrocarbon fuels, to swelling in oils, and resistant to acids, bases, water, and aromatic hydrocarbons. However, they are susceptible to devulcanization in the presence of oxidized fuels such as sour gasoline.

5.26.3 Applications

Epichlorhydrin rubber applications include seals and tubes in air-conditioning and fuel systems.

5.27   Ethylene–Vinylacetate Copolymer (EVM)

EVM can only be cured by peroxide. The elastic properties of EVM are due to the absence of crystallinity in the copolymer.

5.27.1   Resistance to Sun, Weather, and Ozone

EVM has excellent resistance to sun, weathering, and ozone.

5.27.2   Chemical Resistance

The resistance of EVM to hydrocarbon fluids is dependent upon the vinyl acetate content. Higher levels of vinyl acetate increase the polarity of the polymer and, therefore, increase the hydrocarbon resistance.

5.27.3   Applications

EVM finds application as wire insulation in the automotive industry. Ethylene vinyl acetate is also used in the production of thermoplastic elastomers as the soft phase.

5.28   Chlorinated Polyethylene (CM)

Unmodified polyethylene cannot be converted into an elastomer because of its molecular crystallinity. Chlorination of the polyethylene polymer disrupts the crystallinity and allows the polymer to become elastic upon vulcanization. CM can be vulcanized by peroxides or certain nitrogen-containing organic compounds that cross-link the chlorine atoms.

5.28.1   Resistance to Sun, Weather, and Ozone

CM has excellent resistance to sun, weathering, and ozone.

5.28.2   Chemical Resistance

Chlorinated polyethylene has reasonably good resistance to fuels and oils and excellent resistance to atmospheric pollutants.

5.28.3   Applications

CM finds application as vinyl siding, window profiles, hoses, protective boots, flexible dust shields, and various wire and cable applications.

6

Comparative Corrosion Resistance of Selected Elastomers

On the following pages, the compatibility of selected elastomers in contact with selected corrodents is found.

The chemicals listed are in the pure state or in a saturated solution unless otherwise indicated. Compatibility is shown to the maximum allowable temperature for which data is available. Incompatibility is shown by X. A blank space indicates that data is unavailable. The information is taken from Reference [1].

It must be remembered that many of these elastomers may be compounded; therefore, it is important that the composition be verified as to its suitability for the application. The tables are based on the pure elastomer (Table 6.1).

TABLE 6.1

Chemical	Butyl °F	Butyl °C	Hypalon °F	Hypalon °C	EPDM °F	EPDM °C	EPT °F	EPT °C	Viton A °F	Viton A °C	Kalrez °F	Kalrez °C	Natural Rubber °F	Natural Rubber °C	Neoprene °F	Neoprene °C	Buna-N °F	Buna-N °C
Acetaldehyde	80	27	60	16	200	93	210	99	X	X	X	X	X	X	200	93	X	X
Acetamide			X	X	200	93	200	93	210	99	X	X	X	X	200	93	180	82
Acetic acid, 10%	150	66	200	93	140	60	X	X	190	88	200	93	150	66	160	71	200	93
Acetic acid, 50%	110	43	200	93	140	60	X	X	180	82	200	93	X	X	160	71	200	93
Acetic acid, 80%	110	43	200	93	140	60	X	X	180	82	90	32	X	X	160	71	210	99
Acetic acid, glacial	90	32	X	X	140	60	X	X	X	X	80	27	X	X	X	X	100	38
Acetic anhydride	150	66	220	104	X	X	X	X	X	X	210	99	X	X	90	32	200	93
Acetone	160	71	X	X	300	149	X	X	X	X	210	99	X	X	X	X	X	X
Acetyle chloride			X	X	X	X	X	X	190	88	210	99	X	X	X	X	X	X
Acrylic acid					X				X	X	210	99	X	X	X	X	X	X
Acrylonitrile	X	X	140	60	140	60	100	38	X	X	110	43	90	32	160	71	X	X
Adipic acid	X	X	140	60	200	93	140	60	180	82	210	99	80	27	160	71	180	82
Allyl alcohol	190	88	200	93	300	149	80	27	190	88			80	27	120	49	180	82
Allyl chloride	X	X			X	X	X	X	100	38			X	X	X	X	X	X
Alum	190	88	200	93	200	93	140	60	190	88	210	99	150	66	200	93	200	93
Aluminum acetate			X	X	200	93	180	82	180	82	210	99			X	X	200	93
Aluminum chloride, aqueous	150	66	250	121	210	99	180	82	190	88	210	99	140	60	200	93	200	93
Aluminum chloride, dry											190	88						
Aluminum fluoride	180	82	200	93	210	99	180	82	180	82	210	99	150	66	200	93	190	88

Chemical	°F	°C	°F	°C	°F	°C	°F	°C	°F	°C	°F	°C	°F	°C	°F	°C	°F	°C
Aluminum hydroxide	100	38	250	121	210	99	140	60	190	88	210	99			180	82	180	82
Aluminum nitrate	190	88	250	121	210	99	180	82	190	88	210	99	150	66	200	93	200	93
Aluminum oxychloride	X								X	X								
Aluminum sulfate	190	88	200	93	210	99	210	99	190	88	210	99	160	71	200	93	210	99
Ammonia gas	X		140	60	140	60	140	60	X	X	210	99	X	X	140	60	190	88
Ammonium bifluoride	X	X			300	149	140	60	140	60	210	99	X	X	X	X	180	82
Ammonium carbonate	190	88	140	60	210	99	180	82	190	88	210	99	150	66	200	93	200	93
Ammonium chloride, 10%	190	88	200	93	210	99	180	82	190	88	210	99	150	66	200	93	200	93
Ammonium chloride, 50%	190	88	200	93	210	99	180	82	190	88	210	99	150	66	190	88	200	93
Ammonium chloride, sat.	190	88	200	93	300	149	180	82	190	88	210	99	150	66	200	93	200	93
Ammonium fluoride, 10%	150	66	200	93	210	99	210	99	140	60	210	99	160	71	100	38	200	93
Ammonium fluoride, 25%	150	66			300	149	140	60	140	60	210	99	80	27	200	93	120	49
Ammonium hydroxide, 25%	190	88	250	121	100	38	140	60	190	88	210	99	X	X	200	93	200	93
Ammonium hydroxide, sat.	190	88	250	121	100	38	140	60	190	88	210	99	90	32	210	99	200	93

(continued)

TABLE 6.1 *Continued*

Chemical	Butyl °F	Butyl °C	Hypalon °F	Hypalon °C	EPDM °F	EPDM °C	EPT °F	EPT °C	Viton A °F	Viton A °C	Kalrez °F	Kalrez °C	Natural Rubber °F	Natural Rubber °C	Neoprene °F	Neoprene °C	Buna-N °F	Buna-N °C
Ammonium nitrate	180	82	200	93	250	121	180	82	X	X	300	149	170	77	200	93	180	82
Ammonium persulfate	190	88	80	27	300	149	210	99	140	60	210	99	150	66	200	93	200	93
Ammonium phosphate	180	82	140	60	300	149	180	82	180	82	210	99	150	66	200	93	200	93
Ammonium sulfate, 10–40%	190	88	200	93	300	149	180	82	180	82	210	99	150	66	200	93	200	93
Ammonium sulfide			200	93	300	149	210	99	X	X	210	99			160	71	180	82
Ammonium sulfite											210	99					160	71
Amyl acetate	X	X	60	16	210	99	X	X	X	X	210	99	X	X	X	X	X	X
Amyl alcohol	180	82	200	93	210	99	180	82	200	93	210	99	150	66	200	93	180	82
Amyl chloride			X	X	X	X	X	X	190	88	210	99	X	X	X	X	X	X
Aniline	150	66	140	60	140	60	X	X	230	110	250	121	X	X	X	X	X	X
Antimony trichloride	150	66	140	60	300	149	X	X	190	88	210	99	X	X	140	60		
Aqua regia, 3:1					X	X	X	X	190	88	210	99	X	X	X	X	X	X
Barium carbonate			200	93	300	149	180	82	250	121	210	99	180	82	160	71	180	82
Barium chloride	190	88	250	121	250	121	180	82	190	88			150	66	200	93	200	93
Barium hydroxide	190	88	250	121	250	121	180	82	190	88	210	99	150	66	200	93	200	93
Barium sulfate			200	93	300	149	180	82	190	88	210	99	180	82	160	71	180	82

Chemical																		
Barium sulfide	190	88	200	93	140	60	140	60	190	88	210	99	150	66	200	93	200	93
Benzaldehyde	90	32	X	X	150	66	X	X	X	X	210	99	X	X	X	X	X	X
Benzene	X	X	X	X	X	X	X	X	190	88	210	99	X	X	X	X	X	X
Benzenesulfonic acid, 10%	90	32	X	X	X	X	X	X	170	77	210	99	X	X	100	38	X	X
Benzoic acid	150	66	200	93	X	X	X	X	190	88	310	154	150	66	200	93	X	X
Benzyl alcohol	190	88	140	60	X	X	X	X	350	177	210	99	X	X	X	X	140	60
Benzyl chloride	X	X	X	X	X	X	X	X	110	43	210	99	X	X	X	X	X	X
Borax	190	88	200	93	300	149	210	99	190	88	210	99	150	66	200	93	180	82
Boric acid	190	88	290	143	190	88	140	60	190	88	210	99	150	66	200	93	180	82
Bromine gas, dry			60	16	X	X	X	X	350	177	210	99			X	X	X	X
Bromine gas, moist			60	16	X	X	X	X	190	88	210	99			X	X	X	X
Bromine, liquid	X	X	60	16	X	X	X	X	X	X	140	60	X	X	X	X	X	X
Butadiene	140	60	X	X	140	60	X	X	250	121	210	99			140	60	200	93
Butyl acetate	X	X	60	16	200	93	180	82	X	X	240	116	X	X	60	16	X	X
Butyl alcohol	140	60	250	121	140	60	X	X	120	49	210	99	150	66	200	93	X	X
n-Butylamine	X	X	X	X	X	X	X	X	X	X	X	X			X	X	80	27
Butyric acid	X	X	X	X	140	60	X	X	120	49	210	99			X	X	X	X
Caicium bisulfide	120	49	250	121	X	X	X	X	190	88	210	99	X	X	180	82	180	82
Calcium bisulfite	150	66	90	32	X	X	X	X	190	88	210	99	120	49	60	16	200	93
Calcium carbonate	190	88	90	32	210	99	180	82	200	93	200	93	180	82	200	93	180	82
Calcium chlorate	190	88	90	32	140	60	140	60	190	88	210	99	150	66	200	93	200	93
Calcium chloride	190	88	200	93	210	99	180	82	190	88	210	99	150	66	200	93	180	82
Calcium hydroxide, 10%	190	88	200	93	210	99	180	82	190	88	210	99	200	93	220	104	180	82

(continued)

TABLE 6.1 Continued

Chemical	Butyl °F	Butyl °C	Hypalon °F	Hypalon °C	EPDM °F	EPDM °C	EPT °F	EPT °C	Viton A °F	Viton A °C	Kalrez °F	Kalrez °C	Natural Rubber °F	Natural Rubber °C	Neoprene °F	Neoprene °C	Buna-N °F	Buna-N °C
Calcium hydroxide, sat.	190	88	250	121	220	104	180	82	190	88	210	99	200	93	220	104	180	82
Calcium hypochlorite	190	88	250	121	210	99	180	82	190	88	210	99	200	93	220	104	80	27
Calcium nitrate	190	88	100	38	300	149	180	82	190	88	210	99	150	66	200	93	200	93
Calcium oxide	100	38	200	93	210	99	180	82	200	93	210	99			160	71	180	82
Calcium sulfate	100	38	250	121	300	149	180	82	200	93	210	99	180	82	160	71	180	82
Caprylic acid											210	99						
Carbon bisulfide			X	X	X	X	X	X	190	88	210	99	X	X	X	X	X	X
Carbon dioxide, dry	190	88	200	93	250	121	180	82	X	X	210	99	150	66	200	93	200	93
Carbon dioxide, wet	190	88	200	93	250	121	180	82	X	X	210	99	150	66	200	93	200	93
Carbon disulfide	190	88	230	110	250	121	180	82	X	X	210	99	X	X	X	X	200	93
Carbon monoxide	X	X	X	X	X	X	X	X	190	88	210	99	X	X	X	X	X	X
Carbon tetrachloride	90	32	200	93	250	121	180	82	190	88	210	99	X	X	200	93	180	82
Carbonic acid	X	X	X	X	X	X	X	X	350	177	210	99	X	X	X	X	X	X
Cellosolve	150	66			300	149	X	X	X	X	210	99	X	X	X	X	X	X
Chloroacetic acid, 50% water	150	66					X	X	X	X	210	99	X	X	X	X	X	X
Chloroacetic acid	160	71	X	X	160	71	X	X	X	X	210	99	X	X	X	X	X	X

Chlorine gas, dry	X	X	X	X	X	X	X	X	190	88	210	99	X	X	X	X	X	X
Chlorine gas, wet		X	90	32	X	X	X	X	190	88					X	38	X	X
Chlorine, liquid	X	X			X	X	X	X	190	88	210	99	X	X	X	X	X	X
Chlorobenzene	X	X	X	X	X	X	X	X	190	88	210	99	X	X	X	X	X	X
Chloroform	X	X	X	X	X	X	X	X	190	88	210	99	X	X	X	X	X	X
Chlorosulfonic acid	X	X	X	X	X	X	X	X	X	X	210	99	X	X	X	X	X	X
Chromic acid, 10%	100	38	150	66	X		X	X	350	177	210	99	X	X	140	60	190	88
Chromic acid, 50%	X	X	160	71	X	X	X	X	350	177	210	99	X	X	100	38	190	88
Chromic chloride											210	99						
Citric acid, 15%	190	88	250	121	210	99	180	82			210	99	110	43	200	93	180	82
Citric acid, conc.	190	88	250	121	210	99	180	82	190	88	210	99	150	66	200	93	180	82
Copper acetate			X	X	100	38	100	38	X	X	210	99			160	71	180	82
Copper carbonate					210	99	210	99	190	88	210	99					X	X
Copper chloride	190	88	200	93	210	99	180	82	190	88	210	99	150	66	200	93	200	93
Copper cyanide			250	121	210	99	210	99	190	88	210	99	160	71	160	71	180	82
Copper sulfate	190	88	250	121	210	99	180	82	190	88	210	99	150	66	200	93	200	93
Cresol	X	X	X	X	X	X	100	38	X	X	210	99	X	X	X	X	X	X
Cupric chloride, 5%			200	93	210	99	210	99	180	82					210	99	210	99
Cupric chloride, 50%			200	93	210	99	210	99	180	82					160	71	180	82
Cyclohexane	X	X	X	X	X	X	X	X	190	88	210	99	X	X	X	X	180	82
Cyclohexanol			X	X	X	X	X	X	190	88	210	99			X	X	X	X

(continued)

TABLE 6.1 *Continued*

Chemical	Butyl °F	Butyl °C	Hypalon °F	Hypalon °C	EPDM °F	EPDM °C	EPT °F	EPT °C	Viton A °F	Viton A °C	Kalrez °F	Kalrez °C	Natural Rubber °F	Natural Rubber °C	Neoprene °F	Neoprene °C	Buna-N °F	Buna-N °C
Dibutyl phthalate	X						X	X	80	27	210	99	X	X	X	X	X	X
Dichloroacetic acid											210	99			X	X		
Dichloroethane	X	X	X	X	X	X	X	X	190	88	210	99	X	X	X	X	X	X
Ethylene glycol	190	88	200	93	200	93	180	82	350	177	210	99	150	66	160	71	200	93
Ferric chloride	190	88	250	121	220	104	180	82	190	88	210	99	150	66	160	71	200	93
Ferric chloride, 50% in water	160	71	250	121	210	99	180	82	180	82	210	99	150	66	160	71	180	82
Ferric nitrate, 10–50%	190	88	250	121	210	99	180	82	190	88	210	99	150	66	200	93	200	93
Ferrous chloride	190	88	250	121	200	93	180	82	180	82	210	99	150	66	90	32	200	93
Ferrous nitrate	190	88			210	99	180	82	210	99			150	66	200	93	200	93
Fluorine gas, dry	X	X	140	60			X	X	X	X	X	X	X	X	X	X	X	X
Fluorine gas, moist					60	16	100	38	X	X	X	X	X	X	X	X	X	X
Hydrobromic acid, dil.	150	66	90	32	90	32	140	60	190	88	210	99	100	38	X	X	X	X
Hydrobromic acid, 20%	160	71	100	38	140	60	140	60	190	88	210	99	110	43	X	X	X	X
Hydrobromic acid, 50%	110	43	100	38	140	60	140	60	190	88	210	99	150	66	X	X	X	X
Hydrochloric acid, 20%	X	X	160	71	100	38	X	X	350	177	210	99	150	66	90	32	130	54
Hydrochloric acid, 38%	X	X	140	60	90	32	X	X	350	177	210	99	160	71	90	32	X	X

	°F	°C	°F	°C	°F	°C	°F	°C	°F	°C	°F	°C	°F	°C	°F	°C	°F	°C
Hydrocyanic acid, 10%	140	60	90	32	200	93	X	X	190	88	210	99	90	32	X	X	200	93
Hydrofluoric acid, 30%	350	177	90	32	60	16	140	60	210	99	210	99	100	38	200	93	X	X
Hydrofluoric acid, 70%	150	66	90	32	X	X	X	X	350	177	210	99	X	X	200	93	X	X
Hydrofluoric acid, 100%	X	X	90	32	X	X	X	X	60	16	210	99	X	X	X	X	X	X
Hypochlorous acid	X	X	X	X	300	149	140	60	190	88	190	88	150	66	X	X	X	X
Iodine solution, 10%					140	60	140	60	190	88					80	27	80	27
Ketones, general			X	X	X	X			X	X	99	99	X	X	X	X	X	X
Lactic acid, 25%	120	49	140	60	140	60	210	99	190	88	210	99	X	X	140	60	X	X
Lactic acid, conc.	120	49	80	27			210	99	150	66	210	99	80	27	90	32	X	X
Magnesium chloride	200	93	250	121	250	121	180	82	180	82	150	66	150	66	210	99	180	82
Malic acid	X	X			X	X	80	26	190	88	210	99	80	27			180	82
Manganese chloride			180	82			210	99	180	82	210	99			200	93	100	38
Methyl chloride	90	32	X	X	X	X	X	X	190	88			X	X	X	X	X	X
Methyl ethyl ketone	100	38	X	X	80	27	X	X	X	X	210	99	X	X	X	X	X	X
Methyl isobutyl ketone	80	27	X	X	60	16	X	X	X	X	210	99	X	X	X	X	X	X
Muriatic acid	X	X	140	60			X	X	350	177	210	99			X	X	X	X
Nitric acid, 5%	160	71	100	38	60	16	X	X	190	88	210	99	X	X	X	X	X	X
Nitric acid, 20%	160	71	100	38	60	16	X	X	190	88	210	99	X	X	X	X	X	X
Nitric acid, 70%	90	32	X	X	X	X	X	X	190	88	160	71	X	X	X	X	X	X

(continued)

TABLE 6.1 *Continued*

Chemical	Butyl °F	Butyl °C	Hypalon °F	Hypalon °C	EPDM °F	EPDM °C	EPT °F	EPT °C	Viton A °F	Viton A °C	Kalrez °F	Kalrez °C	Natural Rubber °F	Natural Rubber °C	Neoprene °F	Neoprene °C	Buna-N °F	Buna-N °C
Nitric acid, anhydrous	X	X	X	X	X	X	X	X	190	88	X	X	X	X	X	X	X	X
Nitrous acid, conc.	120	49							100	38	210	99	X	X	X	X	X	X
Oleum	X	X	X	X	X	X	X	X	190	88	210	99	X	X	X	X		
Perchloric acid, 10%	150	66	100	38	140	60	190	88	190	88	210	99	150	66	X		X	X
Perchloric acid, 70%			90	32			140	60	190	88					X	X	X	X
Phenol	150	66	X	X			80	27	210	99	210	99	X	X	X	X	X	X
Phosphoric acid, 50–80%	150	66	200	93	140	60	180	82	190	88	210	99	110	43			X	X
Picric acid			80	27	300	148	140	60	190	88	210	99	X	X	200	93	130	54
Potassium bromide, 30%			250	121	210	99	180	82	190	88	210	99	160	71	160	71	180	82
Salicylic acid	80	27			210	99	180	82	190	88	210	99						
Silver bromide, 10%											210	99						
Sodium carbonate	180	82	250	121	300	149	180	82	190	88	210	99	180	82	200	93	200	93
Sodium chloride	180	82	240	116	140	60	180	82	190	88	210	99	130	54	200	93	180	82
Sodium hydroxide, 10%	180	82	250	121	210	99	210	99	X	X	210	99	150	66	200	93	160	71

Chemical	E1 °C	E1 °F	E2 °C	E2 °F	E3 °C	E3 °F	E4 °C	E4 °F	E5 °C	E5 °F	E6 °C	E6 °F	E7 °C	E7 °F	E8 °C	E8 °F	E9 °C	E9 °F
Sodium hydroxide, 50%	66	150	93	200	66	150	99	210	X	X	93	200	82	180	121	250	88	190
Sodium hydroxide, conc.	66	150	93	200	66	150	99	210	X	X	27	80	82	180	121	250	82	180
Sodium hypochlorite, 20%	X	X	X	X	32	90			88	190	X	X	149	300	121	250	54	130
Sodium hypochlorite	X	X	X	X	32	90			88	190	X	X	149	300			32	90
Sodium sulfide, to 50%	82	180	93	200	66	150	99	210	88	190	99	210	149	300	121	250	66	150
Stannic chloride	82	180	99	210	66	150	99	210	82	180	99	210	149	300	32	90	66	150
Stannous chloride	82	180	71	160	66	150	99	210	88	190	99	210	138	280	93	200	66	150
Sulfuric acid, 10%	66	150	93	200	66	150	116	240	177	350	99	210	66	150	121	250	66	150
Sulfuric acid, 50%	93	200	93	200	38	100	99	210	177	350	99	210	66	150	121	250	66	150
Sulfuric acid, 70%	X	X	93	200	X	X	66	150	177	350	99	210	60	140	71	160	38	100
Sulfuric acid, 90%	X	X	X	X	X	X	66	150	177	350	27	80	X	X	X	X	X	X
Sulfuric acid, 98%	X	X	X	X	X	X			177	350	X	X	X	X	43	110	X	X
Sulfuric acid, 100%	X	X	X	X	X	X			88	190	X	X	X	X	X	X	X	X
Sulfuric acid, fuming	X	X	X	X			99	210			X	X	X	X	X	X		

(continued)

TABLE 6.1 Continued

Chemical	Butyl		Hypalon		EPDM		EPT		Viton A		Kalrez		Natural Rubber		Neoprene		Buna-N	
	°F	°C	°F	°C	°F	°C	°F	°C	°F	°C	°F	°C	°F	°C	°F	°C	°F	°C
Sulfurous acid	150	66	160	171	X	X	180	82	190	88	210	99	X	X	X	X	X	X
Thionyl chloride	X	X	X	X					X	X	210	99	X	X	X	X	X	X
Toluene	X	X	X	X	X	X	X	X	190	88	80	27	X	X	X	X	150	66
Trichloroacetic acid	X	X			80	27	X	X	190	88	210	99	X	X	X	X	X	X
White liquor					300	149	180	82	190	88	210	99	X	X	140	60	140	60
Zinc chloride	190	88	250	121	300	149	180	82	210	99	210	99	150	66	160	71	190	88

The chemicals listed are in the pure state or in a saturated solution unless otherwise indicated. Compatibility is shown to the maximum allowable temperature for which data is available. Incompatibility is shown by X. A blank space indicates that data is unavailable.

Source: From P.A. Schweitzer. 2004. *Corrosion Resistance Tables*, Vols. 1–4, 5th ed., New York: Marcel Dekker.

Reference

1. P.A. Schweitzer. 2004. *Corrosion Resistance Tables*, Vols. 1–4, 5th ed., New York: Marcel Dekker.

Index